Ulrich Schumann
Rainer Friedrich (Eds.)

**Direct and Large
Eddy Simulation
of Turbulence**

Notes on Numerical Fluid Mechanics
Volume 15

Series Editors: Ernst Heinrich Hirschel, München
Maurizio Pandolfi, Torino
Arthur Rizzi, Stockholm
Bernard Roux, Marseille

Volume 1   Boundary Algorithms for Multidimensional Inviscid Hyperbolic Flows (Karl Förster, Ed.)

Volume 2   Proceedings of the Third GAMM-Conference on Numerical Methods in Fluid Mechanics (Ernst Heinrich Hirschel, Ed.) (out of print)

Volume 3   Numerical Methods for the Computation of Inviscid Transonic Flows with Shock Waves (Arthur Rizzi / Henri Viviand, Eds.)

Volume 4   Shear Flow in Surface-Oriented Coordinates (Ernst Heinrich Hirschel / Wilhelm Kordulla)

Volume 5   Proceedings of the Fourth GAMM-Conference on Numerical Methods in Fluid Mechanics (Henri Viviand, Ed.) (out of print)

Volume 6   Numerical Methods in Laminar Flame Propagation (Norbert Peters / Jürgen Warnatz, Eds.)

Volume 7   Proceedings of the Fifth GAMM-Conference on Numerical Methods in Fluid Mechanics (Maurizio Pandolfi / Renzo Piva, Eds.)

Volume 8   Vectorization of Computer Programs with Applications to Computational Fluid Dynamics (Wolfgang Gentzsch)

Volume 9   Analysis of Laminar Flow over a Backward Facing Step (Ken Morgan / Jaques Periaux / François Thomasset, Eds.)

Volume 10 Efficient Solutions of Elliptic Systems (Wolfgang Hackbusch, Ed.)

Volume 11 Advances in Multi-Grid Methods (Dietrich Braess / Wolfgang Hackbusch / Ulrich Trottenberg, Eds.)

Volume 12 The Efficient Use of Vector Computers with Emphasis on Computational Fluid Dynamics (Willi Schönauer / Wolfgang Gentzsch, Eds.)

Volume 13 Proceedings of the Sixth GAMM-Conference on Numerical Methods in Fluid Mechanics (Dietrich Rues / Wilhelm Kordulla, Eds.)

Volume 14 Finite Approximations in Fluid Mechanics (Ernst Heinrich Hirschel, Ed.)

Volume 15 Direct and Large Eddy Simulation of Turbulence (Ulrich Schumann / Rainer Friedrich, Eds.)

Manuscripts should have well over 100 pages. As they will be reproduced photomechanically they should be typed with utmost care on special stationary which will be supplied on request. In print, the size will be reduced linearly to approximately 75 %. Figures and diagrams should be lettered accordingly so as to produce letters not smaller than 2 mm in print. The same is valid for handwritten formulae. Manuscripts (in English) or proposals should be sent to the general editor Prof. Dr. E. H. Hirschel, Herzog-Heinrich-Weg 6, D-8011 Zorneding.

Ulrich Schumann
Rainer Friedrich (Eds.)

Direct and Large Eddy Simulation of Turbulence

Proceedings of the EUROMECH Colloquium No. 199,
München, FRG, September 30 to October 2, 1985

Friedr. Vieweg & Sohn   Braunschweig/Wiesbaden

CIP-Kurztitelaufnahme der Deutschen Bibliothek

Direct and large eddy simulation of turbulence:
proceedings of the Euromech Colloquium No. 199,
München, FRG, September 30—October 2, 1985 /
Ulrich Schumann; Rainer Friedrich (ed.). —
Braunschweig; Wiesbaden: Vieweg, 1986.
  (Notes on numerical fluid mechanics;
  Vol. 15)
  ISBN 3-528-08089-2

NE: Schumann, Ulrich [Hrsg.]; Euromech
Colloquium ⟨199, 1985, München⟩; GT

All rights reserved
© Friedr. Vieweg & Sohn Verlagsgesellschaft mbH, Braunschweig 1986

No part of this publication may be reproduced, stored in a retrieval
system or transmitted, mechanical, photocopying or otherwise,
without prior permission of the copyright holder.

Produced by Lengericher Handelsdruckerei, Lengerich
Printed in Germany

ISSN   0179-9614

ISBN   3-528-08089-2

PREFACE

This volume contains papers presented to a EUROMECH-Colloquium held in Munich, September 30 to October 2, 1985. The Colloquium is number 199 in a series of colloquia inaugurated by the European Mechanics Committee. The meeting was jointly organized by the 'Lehrstuhl für Strömungsmechanik' at the 'Technische Universität München' and the 'Institut für Physik der Atmosphäre' of the 'Deutsche Forschungs- und Versuchsanstalt für Luft- und Raumfahrt' (DFVLR) in Oberpfaffenhofen.

'Direct' and 'large eddy simulation' are terms which denote two closely connected methods of turbulence research. In a 'direct simulation' (DS), turbulent motion is simulated by numerically integrating the Navier-Stokes equations in three-dimensional space and as a function of time. Besides initial and boundary conditions no physical simplifications are involved. Computer resources limit the resolution in time and space, though simulations with an order of one million discrete points in space are feasible. The simulated flow fields can be considered as true realizations of turbulent flow fields and analysed to answer questions on the basic behaviour of turbulence. Direct simulations are valid as long as all the excited scales remain within the band of resolved scales. This means that viscosity must be strong enough to damp out the not resolved scales or the simulation is restricted to a limited integration-time interval only. In summary, DS provides a tool to investigate turbulent motions from first principles at least for a finite band of scales.

Large eddy simulation (LES) is a combination of direct simulation with statistical modelling. The large 'eddies' are the motion structures at scales which remain after a certain filter has been applied to the turbulent field. This filter is strongly related to the discrete numerical representation of the flow field employed in the simulation algorithm. For a finite difference simulation the large scales must be resolvable by the underlying grid. In contrast, the 'subgrid-scales' are those which are too small for the grid. In LES, these subgrid-scales are modelled by a subgrid-scale (SGS) model. The SGS-model mainly provides the correct damping of those resolved scales which lie at the border of resolution and which are excited by the large scale motions. Also, the SGS-model has to describe turbulent fluxes not represented by the large scales. SGS-models are formally very similar to the models commonly used to describe time- or ensemble-averaged turbulent fields. Thus they have basically the same advantages and disadantages. Approximation errors in SGS-models are of less importance, however, because the SGS-motion carries only a fraction of total turbulent kinetic energy and fluxes. If the grid (and the related filter) is fine enough, this fraction becomes very small. In summary, LES serves as a tool to construct the large scale portion of turbulent fields at high Reynolds numbers.

Direct turbulence simulations were pioneered by Orszag and co-workers since 1969. The first convincing large eddy simulations were reported by Deardorff in 1970 for a turbulent channel flow and applied to atmospheric boundary layers thereafter. His SGS-model employed a proposal by Smagorinsky (1963) for parametrization of subgrid-scale motions in meteorology. Lilly (1966) showed that this SGS-model is consistent with Kolmogorov's inertial-subrange theory. The state of the art as of 1983 was reported by Rogallo & Moin (1984).

The Colloquium was concerned with the discussion of new developments in the field of DS and LES and its applications. The presentations included new aspects on the numerical algorithms to integrate the basic equations, treatments of filtering, boundary and initial conditions, and subgrid-scale models. Also new concepts based on fractals and helicity were outlined which possibly are capable of influencing SGS-modelling in the future. The applications included transition and established turbulence, homogeneous and inhomogeneous turbulence. They were related to basic research as well as to engineering problems and to atmospheric boundary layers.

We were glad to receive much response to our invitation. About fifty scientists attended the colloquium, among them we could welcome most of the active researchers in the field, in particular from the USA and Japan, and about thirty papers have been presented. This was the first meeting of experts from all over the world on this special subject. As no collection of papers exists, which summarizes the present state of the art in DS and LES, it was decided to collect the contributions for the present volume. Each paper has been revised after being reviewed by at least one reviewer. A few contributions have the form of extended abstracts; they summarize results, which were presented at the Colloquium and which are described in more detail in papers under publication. Some contributions are not included because their content has been published just recently. However, we include the contribution of Dr. McComb who was not able to participate in the colloquium. The papers are grouped into sections of related subjects: 1. Transition to turbulence, 2. Subgrid-scale models and basic concepts, 3. Large eddy simulations of wall-bounded shear flows, 4. Direct and large eddy simulations of mixed shear and buoyant flows, 5. Convective or stable atmospheric boundary layers. The volume ends up with summarizing statements on results, trends and recommendations.

We are very grateful to all who supported the Colloquium and the resultant proceedings. We thank the Volkswagen Foundation, the companies Cray Research, Munich and Messerschmitt-Bölkow-Blohm, Munich for financial support of the meeting. Last, but not least we are indebted to all colleagues who contributed to the proceedings or took part in the reviewing process.

April 1986	Ulrich Schumann Oberpfaffenhofen	Rainer Friedrich München

CONTENTS

Page

1. TRANSITION TO TURBULENCE

N. GILBERT, L. KLEISER: Subcritical Transition to Turbulence in Channel Flow 1

T. HERBERT: Vortical Mechanisms in Shear Flow Transition 19

2. SUBGRID-SCALE MODELS AND BASIC CONCEPTS

B. AUPOIX: Subgrid Scale Models for Homogeneous Anisotropic Turbulence . 37

W. D. McCOMB: Application of Renormalization Group (RG) Methods to the Subgrid Modelling Problem. 67

E. LEVICH: Helical Fluctuations, Fractal Dimensions and Path Integral in the Theory of Turbulence 82

3. LARGE EDDY SIMULATIONS OF WALL-BOUNDED SHEAR FLOWS

S. GAVRILAKIS, H. M. TSAI, P. R. VOKE, D. C. LESLIE: Large-Eddy Simulation of Low Reynolds Number Channel Flow by Spectral and Finite Difference Methods 105

K. HORIUTI, A. YOSHIZAWA: Large Eddy Simulation of Turbulent Channel Flow by 1-Equation Model 119

T. KOBAYASHI, M. KANO: Numerical Prediction of Turbulent Plane Couette Flow by Large Eddy Simulation 135

D. LAURENCE: Advective Formulation of Large Eddy Simulation for Engineering Type Flows 147

L. SCHMITT, K. RICHTER, R. FRIEDRICH: Large-Eddy Simulation of Turbulent Boundary Layer and Channel Flow at High Reynolds Number 161

Short Contributions:

J. KIM: Numerical Investigation of a Vortical Structure in a Wall-Bounded Shear Flow 177

P. MOIN: Recent Results on the Structure of Turbulent Shear Flows using Simulation Databases 181

4. DIRECT AND LARGE EDDY SIMULATIONS OF MIXED SHEAR AND BUOYANT FLOWS

T. M. EIDSON, M. Y. HUSSAINI, T. A. ZANG: Simulation of the Turbulent Rayleigh-Benard Problem using a Spectral/Finite Difference Technique . . 188

G. GRÖTZBACH: Application of the TURBIT-3 Subgrid Scale Model to Scales Between Large Eddy and Direct Simulations 210

K. KUWAHARA, S. SHIRAYAMA: Direct Simulation of High-Reynolds-Number Flows by Finite-Difference Methods 227

U. SCHUMANN, S. E. ELGHOBASHI, T. GERZ: Direct Simulation of Stably Stratified Turbulent Homogeneous Shear Flows 245

Short Contribution:

R. W. METCALFE, S. MENON, J.J. RILEY: The Effect of Coherent Modes on the Evolution of a Turbulent Mixing Layer 265

5. CONVECTIVE OR STABLE ATMOSPHERIC BOUNDARY LAYERS

D. J. CARRUTHERS, J. C. R. HUNT, C. J. TURFUS: Turbulent Flow near Density Inversion Layers 271

C.-H. MOENG: A Large Eddy Simulation Model for the Stratus-Topped Boundary Layer 291

F. T. M. NIEUWSTADT, R. A. BROST, T. L. van STIJN: Decay of Convective Turbulence, a Large Eddy Simulation 304

6. SUMMARIZING STATEMENTS ON RESULTS, TRENDS AND RECOMMENDATIONS 318

List of Participants 336
List of Authors 340

SUBCRITICAL TRANSITION TO TURBULENCE IN CHANNEL FLOW

N. Gilbert, L. Kleiser
DFVLR, Institute for Theoretical Fluid Mechanics
Bunsenstr. 10, D-3400 Göttingen, Germany

SUMMARY

The transition process from laminar to turbulent channel flow is investigated by numerical integration of the three-dimensional time-dependent Navier-Stokes equations using a spectral method. The classical peak-valley splitting mode of secondary instability is considered. First results of an investigation of the breakdown stages of the transition process are given. Visualizations of the development of three-dimensional flow structures are presented. In addition long-time integrations of the equations with few Fourier modes in streamwise and spanwise direction are performed. Comparison of statistical quantities of the resulting chaotic solutions with turbulent channel flow data shows unexpected, partly even quantitative agreement.

1. INTRODUCTION

Transition to turbulence in shear flows proceeds through a sequence of increasingly complex stages initiated by flow instabilities. In two-dimensional (2-D) channel and boundary layer flows of low disturbance level the first step of the transition process consists of the amplification of 2-D Tollmien- Schlichting (TS) waves. This primary instability is well understood and is described by the classical linear stability theory [2]. During the past ten years much progress has been made in clarifying the non-linear and three-dimensional stages of the transition process [1],[4],[13]. These stages arise from a secondary instability of 2-D finite amplitude waves against small 3-D disturbances (Orszag & Patera [23], Herbert [6],[7]). Experimental and theoretical work has demonstrated the existence of two different types of secondary instability corresponding to different paths to turbulence: the classical "peak-valley splitting" type (or K-type), and the subharmonic type (or H-type). In the K-type the

Λ-vortices appearing periodically in streamwise and spanwise direction form a regular pattern where one Λ-vortex follows the other at a distance of one TS wavelength. The other type has also been detected both in boundary layers (Kachanov & Levchenko [8], Saric & Thomas [30]) and in plane Poiseuille flow (Ramazanov [26]). It is characterized by a staggered Λ-vortex pattern and, accordingly, the appearance of the subharmonic frequency $\omega_{TS}/2$.

In the present work transition to turbulence is considered in the prototype case of plane Poiseuille flow. A substantial amount of experimental information is available on the transition process and on fully developed turbulence in this flow. Fig. 1 shows the experimentally observed regions where either laminar or turbulent channel flow occurs, dependent upon Reynolds number and disturbance level at the channel entrance (Feliss et al. [5]). With large inlet disturbances, turbulence is observed down to a Reynolds number Re (based on channel half-width and undisturbed centerline velocity) of order 1000 (Patel & Head [25], Nishioka et al. [22]). At very low disturbance levels, laminar flow can be maintained up to the critical Reynolds number Re_{cr} = 5772 of linear stability theory [2]. The subcritical Reynolds number range is of particular interest. Many experimental and theoretical studies have been devoted to subcritical transition. However, at present still no theory exists which is able to describe the subcritical laminar-turbulent boundary.

Transition to turbulence is certainly possible in many ways. A particularly well-studied case, which is more amenable to theoretical treatment than others, is the "ribbon-induced" transition starting with small-amplitude 2-D TS waves. The breakthrough in the experimental investigation of this transition process has been achieved by Nishioka et al. [18],[19]. They observed a characteristic sequence of events quite similar to boundary layer transition [9]. Typical phenomena are the 3-D distortion of the wave front, the development of a peak-valley structure of the rms fluctuations, the generation of mean longitudinal vortices, inflectional velocity profiles and spike signals of the instantaneous velocity, and finally a breakdown of the regular wave motion into irregular fluctuations. The breakdown stages have been investigated by Nishioka et al. in [20],[21]. They found that the final multi-spike stages have properties similar to wall turbulence.

In recent years numerical simulations of the nonlinear, 3-D stages of the transition process have been performed by several groups [23],[29],[31],[32], [10]-[12],[15],[16]. Detailed comparisons of simulation results with measurements demonstrated good agreement up to highly developed 3-D stages, both for Blasius flow [31],[15] and for plane Poiseuille flow [10]-[12]. In the work of this group in particular the development of 3-D flow structures, as obtained from velocity and vorticity fields and from flow visualizations by fluid markers, has been investigated. Both the classical K-type and the subharmonic type [16] have been considered. As an application the theoretical potential of transition control by superposition of periodic disturbances has been explored [11],[15].

In the present contribution first results of our investigation of the spike stage with increased spatial resolution are given. In addition results of long-time integrations with few horizontal modes are presented and compared with experimental results for turbulent channel flow.

2. MATHEMATICAL MODEL AND NUMERICAL DISCRETIZATION

The three-dimensional time-dependent incompressible Navier-Stokes equations are solved in the spatial domain $0 \leq x_j \leq L_j$ ($j = 1,2$), $|x_3| \leq 1$ (Fig. 2). No-slip conditions are applied at the walls $x_3 = \pm 1$ and periodic boundary conditions in the horizontal directions x_1, x_2 (all variables are non-dimensionalized with channel half-width and undisturbed laminar centerline velocity). Either the mean pressure gradient or the mean mass flux is imposed. In the latter case the mean pressure gradient is adjusted at each time step in order to obtain the specified flow rate. We use the model of a streamwise and spanwise periodic flow developing in time. In the transition simulations we apply a frame of reference moving downstream with the phase velocity of the TS wave. Accordingly the roles of x_1 and t are exchanged to compare experimental and numerical results. In analogy with vibrating ribbon experiments the evolution of the velocity field $\underline{v}(\underline{x},t)$ from an initial disturbance superimposed on the basic Poiseuille flow $U(x_3) = 1-x_3^2$ is calculated.

The numerical method is essentially the same as described in [12]. The spatial discretization is based on a spectral method with Fourier expansions in the horizontal coordinates,

$$\underline{v}(\underline{x},t) = \Sigma \Sigma \hat{\underline{v}}(k_1,k_2,x_3,t)\exp(ik_1\alpha_1 x_1 + ik_2\alpha_2 x_2) \quad (2.1)$$

($\alpha_j = 2\pi/L_j$), and Chebyshev polynomial expansions in the normal coordinate. The sum in (2.1) extends over the wavenumber range

$$K = \{ (k_1,k_2): |k_j|<N_j/2, \ |k_1|/N_1 + |k_2|/N_2 < 2/3 \}. \quad (2.2)$$

For time integration, viscous terms are treated implicitly by the Crank-Nicolson method, and nonlinear terms explicitly by the Adams-Bashforth method. As usual, the nonlinear terms are calculated by the pseudospectral approximation. Aliasing errors in the Fourier coefficients are reduced by truncating the expansion (2.1) according to (2.2). The pressure is calculated from a Poisson equation with correct boundary conditions obtained from the condition of vanishing divergence at the boundary, using an influence matrix technique. The continuity equation and boundary conditions are satisfied exactly by the discretized equations. The solution is obtained by solving sequentially a set of 1-D Helmholtz equations. More details on the numerical method are given in [12].

An improved version of the algorithm has been implemented in the computer code TRANSIT on a Cray-1S. The code also covers the flat plate boundary layer case with a Chebyshev collocation method adapted to a semi-infinite domain [15]. The implementation of the influence matrix method has been improved by exploiting properties of Chebyshev expansions. In the spanwise direction the option of symmetric flow fields

$$v_j(-x_2) = v_j(x_2) \ (j=1,3), \quad v_2(-x_2) = -v_2(x_2) \quad (2.3)$$

is implemented by using the appropriate cosine/sine expansions instead of the general form (2.1). This reduces computing time and storage for symmetric problems by a factor of two. It may be easily shown that the symmetry (2.3) of an initial velocity field will be preserved by the Navier-Stokes equations for all t > 0. The transitional flows under consideration show the symmetry (2.3) both in experiments and theoretical analysis. All results presented herein have been obtained with the spanwise-symmetric version (the grid point number N_2 always refers to the number of points on the the full spanwise periodicity length L_2). A high efficiency of the code is obtained by vectorizing the innermost loop which runs over the (k_1,k_2) plane. The typical CPU time per grid point and time step is 10 μs for the symmetric version. With use of an optimized I/O scheme for some of the arrays up to 54^3 grid points can be accomodated with the available memory of 700 000 words.

3. INVESTIGATION OF THE SPIKE STAGE

We first present results of a three-dimensional transition simulation. The physical parameters are the same as in [11] and are adjusted to the experimental conditions of Nishioka et al. [19],[20], Re = 5000, $α_1$ = 1.12 and $α_2$ = 2.1. The oscillation period is T_{TS} ≃ 20 and the phase velocity c_{TS} ≃ 0.28. The initial disturbance consists of a 2-D wave with maximum amplitude A_{TS} = 0.03 and a pair of oblique waves with spanwise wavenumbers $±α_2$ and amplitude 0.001. The amplitude distribution $v_1(x_3)$ of both disturbance parts is antisymmetric. In this simulation the mean mass flux is kept fixed. The numerical resolution is $N_1×N_2×N_3$ = 16×16×90, Δt = 0.05 during 0 ≤ t ≤ 100 and is increased to 30×60×90, Δt = 0.025 for t > 100.

To obtain an overview of the timewise development of the wave amplitude we consider in fig. 3 the evolution of the maximum (across x_3) rms fluctuation $u_1' = (\overline{(v_1-\bar{v}_1)^2})^{1/2}$. Here $\bar{v}_1(x_2,x_3)$ denotes the average of v_1 over x_1, corresponding to a time average in an experiment. We see an early splitting of the curves at different x_2 positions with a peak at $x_2 = L_2/2$ and a valley (minimum) at $x_2 = 0$ in this simulation. The absolute spanwise maxima and minima occur at different positions in later stages, in accordance with the experiment [19]. Up to about t = 120 the results of the present simulation are in good agreement with the earlier results [11] obtained with maximum resolution of $N_1×N_2×N_3$ = 32×32×40. At later times differences appear. The present simulation was continued until t = 140. In the following we mainly discuss the development of the instantaneous velocity field at the peak position.

Fig. 4 shows the energy $E(k_1,k_2)$ (sum of the three velocity components) of the Fourier modes in (2.1) at t = 128 and t = 138 (E(0,0) contains only the disturbance energy). We see that at t = 128 all spanwise modes up to k_2 = 29 have energies > 10^{-9}, while streamwise modes with this energy are confined to k_1 ≤ 7. This indicates the excitation of small spanwise scales during transition. This anisotropy has been accounted for by choosing for N_2 twice the value of N_1. At t = 138 a pile-up of energy near k_{2max} = 30 is seen with an absolute value of the modal energies of order 10^{-7}. Obviously a higher spanwise resolution would be needed at this stage.

In fig. 5 the instantaneous velocity profiles $v_1(x_3)$ over the downstream period are shown at three stages. The profiles become highly distorted and indicate regions of high shear. The same data are represented in fig. 6 by contour lines of v_1, extended to two periods in downstream direction. We see that initially one, then two and later several localized regions of low velocity appear within one period. These regions accelerate downstream and move away from the wall, the outer one rising towards the channel center. At the outer edge of the low-velocity regions we observe high gradients in streamwise and in normal direction. Low-velocity regions which are concentrated in streamwise direction give rise to the typical "spike" signals. An example is given in fig. 7 for the low-velocity region near the centerline at t = 138.

The development of the high-shear layer which separates the low-velocity regions from the high-speed fluid above is illustrated in fig. 8 (two periods in streamwise direction). Fig. 8a,b shows the contour lines of the shear $\partial v_1/\partial x_3$ at t = 128 and t = 134. The width of the high-shear layer is of order 0.1. At t = 128 already two local maxima appear. The high-shear layer starts to wrinkle and later has several local maxima, with the front part decaying and lifting towards the channel center. These patches of high shear are usually identified with "hairpin eddies" [9],[20]. Fig. 8c,d illustrates the location of the high-shear regions in 3-D space (lines near the wall are suppressed, only one half of the spanwise period is shown, and the intersection of the constant-shear surface with the symmetry plane is indicated). The front part of the high-shear layer is confined to a narrow region around the peak. It becomes quite elongated in downstream direction and is overtaking the central part of the high-shear layer from the previous period.

The features just described are in good overall agreement with the experimental observations of Nishioka et al. [20],[21]. From a comparison of the shape and the location of the high-shear layer we may relate the stages t ≃ 128 and t ≃ 134 of the simulation to the one- and two-spike stages of the experiment, respectively. However, a detailed comparison also shows some differences. For example, in the experiment the low-velocity regions and the kinks of the high-shear layer are more localized, and therefore the spike signals appearing around $x_3 \simeq -0.4$ are sharper than in the present simulation. Certainly our discretization is too low to resolve the 3-spike stage which should occur at $t \simeq 140$. Gradual lack of resolution is indicated by the "wiggles" appearing in some of the plots. We stress that in particular a high resolution in the normal direction is needed. With the present 30×60×90 discretization significantly better results have been obtained than in an earlier 36×80×54 simulation.

As a result from the final stage $t = 145$ of the 36×80×54 simulation we discuss the local mean velocity profiles obtained at the peak and the 8 nearest spanwise grid points, fig. 9. Each profile is normalized with its local friction velocity $u_\tau(x_2)$ and plotted over y_+, the distance from the wall measured in wall units. The profiles appear in two groups, one at the peak (j=0-2) and the other (j=3-8) on each side of the peak. This corresponds to the location of a low-speed streak at the peak and two high-speed streaks on both sides which are observed here in accordance with experimental results at the multi-spike stage [21]. The Reynolds number Re_τ, based on the local friction velocity u_τ and channel half width, rises to 176 in the high-speed region (j=8), compared with $Re_\tau = 100$ for the undisturbed laminar flow and $Re_\tau \simeq 210$ for the fully developed turbulent flow at the same flow rate [22]. For $y_+ \leq 60$ the profiles in the high-speed region clearly show the tendency to follow the wall law for the turbulent profile.

4. LONG-TIME INTEGRATION OF LOW-RESOLUTION EQUATIONS

Direct simulations of transition to fully developed turbulence require calculations over long time intervals with a high spatial resolution. Such numerical experiments are very expensive. In fact, starting from laminar parabolic flow, the mean velocity profile in the interior of the channel develops on a viscous time scale so that an integration time $t = O(Re)$ is needed [24]. The spatial resolution requirements for the simulation of turbulent channel flow are discussed in [17],[27]. They are derived from the necessary periodicity lengths and from the finest scales that must be resolved. If only the "large" coherent structures are resolved and the smaller ones are modelled the estimated total number of grid points is $N \simeq 0.05\, Re_{CL}^2$ (the Reynolds number Re_{CL} is based on the turbulent centerline velocity and channel half width). For a direct simulation with resolution of the dissipation (Kolmogorov) scales the estimated grid point number increases to $N \simeq$

$(5\ Re_{CL})^{9/4}$. However, this estimate may be too pessimistic, in particular for the present case of subcritical Reynolds numbers. Calculations of Rozhdestvensky & Simakin [28] indicated that some integral properties of turbulent channel flows at subcritical Reynolds numbers may be reflected already by direct simulations with very coarse resolution ($N_1=N_2=8$) in the horizontal directions. Certainly this resolution is too low to describe the real physics of turbulence even at these low Reynolds numbers. Nevertheless it seems to be worthwhile to study the statistical properties of such low-resolution models and to compare them with experimental data in more detail than has been done in [28].

In the following, we describe first results of our long-time integrations with a Fourier mode truncation as low as $|k_j| \leq 3$. In the normal direction we use either 32 or 64 Chebyshev polynomials. The streamwise wavenumber is $\alpha_1 = 1.12$ as before. The time step is $\Delta t = 0.05$. Further parameters of the four different calculations reported here are given in table 1. In this table Δx_{j+} is the grid spacing in wall units and N_{VS} is the number of grid points within the viscous sublayer $y_+ \leq 5$. t_{INT} and t_{AV} denote the total integration time and the averaging time, respectively. Because of the few grid points in the horizontal directions the vector lengths are relatively short and therefore the CPU time increases to 25 μs / (grid point × timestep). All calculations are done in the rest frame using the spanwise-symmetric code version. We impose alternatively a constant space-averaged pressure gradient of magnitude $2/Re_{lam}$ (where Re_{lam} is the Reynolds number based on laminar centerline velocity and channel half-width) or a constant mass flux. Most computations are done with the second formulation because it reduces the time until the integral characteristics become quasi-stationary. Starting from an initial velocity field of the same type as used in the transition simulations the Navier-Stokes equations are integrated in time until the total energy of the disturbances and the flow rate and/or the wall shear stress become quasi-stationary. Then the equations are further integrated in time, and a running time average of the mean velocity is calculated. The calculations are considered to be complete when the averaged turbulence quantities become stationary. The symmetry of the statistical data about the channel centerline has been checked. For example, the maxima of the rms fluctuations in the two channel halves agree within a few percent.

For comparison with experimental data we have chosen the experiment of Eckelmann [3] at Re_{CL} = 2800. The Reynolds number Re_τ, based on (mean turbulent) friction velocity u_τ and channel half width, is about 142. Using the laminar relation $Re_\tau = \sqrt{2\, Re_{lam}}$, the corresponding Reynolds number Re_{lam} is set to 10083 for the computation with fixed mean pressure gradient (case 1). For the computation with fixed mass flux (cases 2 - 4) we set Re_{lam} = 3585. We define the velocity fluctuation u_j as

$$u_j = v_j - \langle v_j \rangle$$

where $\langle \,\rangle$ denotes an average over the horizontal coordinates x_1 and x_2 and over the time interval t_{AV}.

Fig. 10a shows the development of the wall Reynolds number Re_τ and fig. 10b the change of the flow rate as a function of time in case 1. Re_τ initially rises on a convective (transitional) time scale and then, as necessary in this case, decays to a state where it fluctuates about its initial value. Nearly at the same time when the time-averaged Re_τ has reached its initial value the flow rate becomes constant. Fig. 10c shows the result for Re_τ obtained with constant mass flux (case 2). Here the initial transient time is much shorter. (Note that only every t = 20 time units data have been saved for plotting). The average value $\langle Re_\tau \rangle$ = 150 fits the experimental curve of Nishioka et al. (fig. 7 in [22]).

In figs. 11-13 we present statistical results obtained in case 2 and compare them with the experimental data of Eckelmann [3]. Fig. 11 shows the mean velocity data. The viscous sublayer and buffer region are well developed. The overall agreement is good except for a velocity defect near the channel centerline. Here the mean profile may still not be sufficiently developed on the viscous time scale. Due to this defect Re_{CL} attains the value 2700 only while it is 2800 in the experiment. The Reynolds stress $-\langle u_1 u_3 \rangle / u_\tau^2$ is shown in fig. 12. The maximum value and its location agree well with the experiment. Note also the approximate symmetry of the $|\langle u_1 u_3 \rangle|$ profile about the channel centerline. Fig. 13 shows the rms values $u_j' = \langle u_j^2 \rangle^{1/2}$ of the three fluctuating velocity components normalized with the friction velocity. Comparison with the data of Eckelmann at Re = 2800 shows reasonable qualitative and quantitative agreement, except for the u_3 component in the central part of the channel. The streamwise fluctuations have a maximum of 2.9 at y_+ = 11. The maxima of the u_2 and

u_3 components are located farther away from the wall. The spanwise fluctuations attain their maximum value of 1.5 at $y_+ = 18$. The fluctuations normal to the wall have a broad maximum of 1.0 at approximately $y_+ = 70$. The values near the centerline are overestimated by the computations.

Next we consider the skewness and flatness factors, which are defined as

$$S(u_j) = <u_j^3> / <u_j^2>^{3/2}$$
$$F(u_j) = <u_j^4> / <u_j^2>^2,$$

respectively. The results obtained in cases 2 and 4 are plotted in fig. 14. Because skewness and flatness factors were not given in [3] we include the experimental data obtained by Kreplin & Eckelmann [14] for $Re_{CL} = 3850$ ($Re_\tau = 194$). Away from the walls the overall agreement between our computational and the experimental data, except for $F(u_2)$, is not too bad. Towards the wall, however, larger discrepancies occur. The skewness and flatness factors of both the u_1 and u_3 components attain their highest values near the wall. The agreement with experimental data is better in case 4 which has twice the spanwise box length and resolution of case 2.

The statistical results of the other cases are similar. In particular, increasing the resolution in normal direction only (case 3) has no significant influence. In case 4 the averaged wall Reynolds number $<Re_\tau> \simeq 170$ is slightly higher than the experimental value. This quantity seems to be sensitive to the choice of the simulation parameters.

5. CONCLUSIONS

An accurate spectral scheme for the numerical integration of the 3-D Navier-Stokes equations has been implemented in an efficient code on a Cray-1S computer. The code has been applied to study the spike stages of the transition process with a resolution of $N_1 \times N_2 \times N_3 = 30 \times 60 \times 90$ grid points, and for long-time integrations of the equations of motion discretized with only $|k_j| \leq 3$ Fourier modes in the horizontal directions.

The results obtained for the 1- and 2-spike stages are basically in agreement with experimental observations. In the later stages the simulations suffer from insufficient spatial resolution. Further work with higher resolution is necessary to clarify the nature of the flow and the mechanisms active in these highly developed stages at the origin of turbulence.

The horizontal resolution used in the long-time integrations is certainly too low to resolve the scales present in a real transitional or turbulent flow, especially near the walls. Inspection of the energies of the Fourier modes $(k_1, k_2) \neq (0,0)$ reveals that they all have comparable magnitudes. Nevertheless it seems remarkable that the chaotic solutions of such "low-resolution models" attain a statistically steady state with some of the low-order statistical moments surprisingly close to the experimental data. It remains to be clarified how the results depend on the box lengths and on symmetries present in the initial conditions, and how they change with moderately increasing horizontal resolution.

REFERENCES

[1] AGARD Report No.709, Special course on stability and transition of laminar flow (1984).

[2] P. DRAZIN, W. REID: Hydrodynamic Stability. Cambridge University Press (1981).

[3] H. ECKELMANN: Experimentelle Untersuchungen in einer turbulenten Kanalströmung mit starken viskosen Wandschichten. Mitt. MPI f. Strömungsforschung und AVA Göttingen Nr. 48 (1970).

[4] R. EPPLER, H. FASEL (eds.): Laminar-Turbulent Transition. (Proc. IUTAM Symposium, Stuttgart 1979). Springer, Berlin (1980).

[5] N.A. FELISS, M.C. POTTER, M.C. SMITH: An experimental investigation of incompressible channel flow near transition. J. Fluids Eng. 99, 693-698 (1977).

[6] TH. HERBERT: Secondary instability of plane channel flow to subharmonic three-dimensional disturbances. Phys. Fluids 26, 871-874 (1983).

[7] TH. HERBERT: Secondary instability of shear flows, in AGARD-R-709 [1] (1984).

[8] Y.S. KACHANOV, V.Y. LEVCHENKO: The resonant interaction of disturbances at laminar-turbulent transition in a boundary layer. J. Fluid Mech. 138, 209-247 (1984).

[9] P.S. KLEBANOFF, K.D. TIDSTROM, L.M. SARGENT: The three-dimensional nature of boundary-layer instability. J. Fluid Mech 12, 1-34 (1962).

[10] L. KLEISER: Three-dimensional processes in laminar- turbulent transition, in H.L. Jordan et al. (eds.), Nonlinear Dynamics in Transcritical Flows, Springer Lecture Notes in Engineering 13, 123-154 (1985).

[11] L. KLEISER, E. LAURIEN: Three-dimensional numerical simulation of laminar-turbulent transition and its control by periodic disturbances, in [13], 29-37 (1985).

[12] L. KLEISER, U. SCHUMANN: Spectral simulations of the laminar-turbulent transition process in plane Poiseuille flow, in R.G. Voigt et al. (eds.): Spectral methods for partial differential equations. SIAM, Philadelphia, 141-163 (1984).

[13] V.V. KOZLOV: Laminar-Turbulent Transition. (Proc. IUTAM Symposium, Novosibirsk 1984). Springer, Berlin (1985).

[14] H.-P. KREPLIN, H. ECKELMANN: Behavior of the three fluctuating velocity components in the wall region of a turbulent channel flow. Phys. Fluids 22, 1233-1239 (1979).

[15] E. LAURIEN, L. KLEISER: Numerical simulation of transition control in boundary layers. Proc. 6th GAMM Conference on Numerical Methods in Fluid Mechanics, Göttingen, 1985. To be published by Vieweg-Verlag.

[16] C.L. MAY, L. KLEISER: Numerical simulation of subharmonic transition in plane Poiseuille flow. Bull. APS 30, 1748 (1985).

[17] P. MOIN, J. KIM: Numerical investigation of turbulent channel flow. J. Fluid Mech. 118, 341-377 (1982).

[18] M. NISHIOKA, S. IIDA, Y. ICHIKAWA: An experimental investigation of the stability of plane Poiseuille flow. J. Fluid Mech. 72, 731-751 (1975).

[19] M. NISHIOKA, M. ASAI, S. IIDA: An experimental investigation of the secondary instability, in [4], 37-46 (1980).

[20] M. NISHIOKA, M. ASAI, S. IIDA: Wall phenomena in the final stage of transition to turbulence, in R.E. Meyer (ed.): Transition and Turbulence. Academic Press, 113-126 (1981).

[21] M. NISHIOKA, M. ASAI: Evolution of Tollmien-Schlichting waves into wall turbulence, in T. Tatsumi (ed.): Turbulence and Chaotic Phenomena in Fluids, North-Holland, 87-92, (1984).

[22] M. NISHIOKA, M. ASAI: Some observations of the subcritical transition in plane Poiseuille flow. J. Fluid Mech. 150, 441-450 (1985).

[23] S.A. ORSZAG, A.T. PATERA: Secondary instability of wall-bounded shear flows. J. Fluid Mech. 128, 347-385 (1983).

[24] S.A. ORSZAG, A.T. PATERA: Calculation of von Karman's constant for turbulent channel flow. Phys. Rev. Lett. 47, 832-835 (1981).

[25] V.C. PATEL, M.R. HEAD: Some observations on skin friction and velocity profiles in fully developed pipe and channel flows. J. Fluid Mech. 38, 181-201 (1969).

[26] M.P. RAMAZANOV: Development of finite-amplitude disturbances in Poiseuille flow, in [13], 183-190 (1985).

[27] R.S. ROGALLO, P. MOIN: Numerical simulation of turbulent flows. Ann. Rev. Fluid Mech. 16, 99-137 (1984).

[28] B.L. ROZHDESTVENSKY, I.N. SIMAKIN: Secondary flows in a plane channel: their relationship and comparison with turbulent flows. J. Fluid Mech. 147, 261-289 (1984).

[29] P.R. SPALART: Numerical simulation of boundary-layer transition, Springer Lecture Notes in Physics 218, 531-535 (1985).

[30] W.S. SARIC, A.S.W. THOMAS: Experiments on the subharmonic route to turbulence in boundary layers, in T. Tatsumi (ed.): Turbulence and Chaotic Phenomena in Fluids, North-Holland, 117-122, (1984).

[31] A. WRAY, M.Y. HUSSAINI: Numerical experiments in boundary-layer stability. Proc. Roy. Soc. Lond. A 392, 373-389 (1984).

[32] T.A. ZANG, M.Y. HUSSAINI: Numerical experiments on subcritical transition mechanisms. AIAA Paper 85-0296 (1985).

Table 1. Parameters of long-time integrations

Case	α_2	N_1	N_2	N_3	Δx_{1+}	Δx_{2+}	N_{VS}	t_{INT}	t_{AV}	Re_{lam}
1	2.1	8	8	32	100	53	3	8300	2000	10083
2	2.1	8	8	32	107	56	3	6000	2000	3585
3	2.1	8	8	64	107	56	5	3000	1200	3585
4	1.05	8	16	32	107	56	3	6000	2000	3585

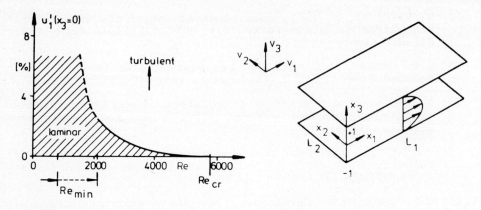

Fig. 1:
Regions of laminar and turbulent channel flow [5]

Fig. 2:
Integration domain and coordinates

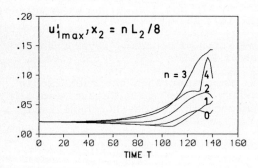

Fig. 3:
Development of maximum rms fluctuation at different spanwise positions

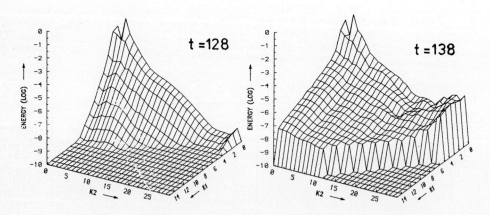

Fig. 4: Energy $E(k_1, k_2)$ of the Fourier modes

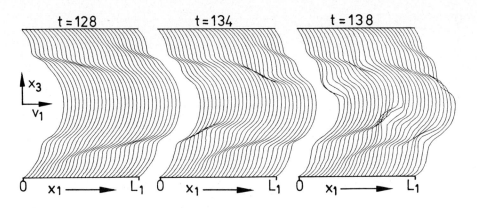

Fig. 5: Instantaneous velocity profiles in peak plane

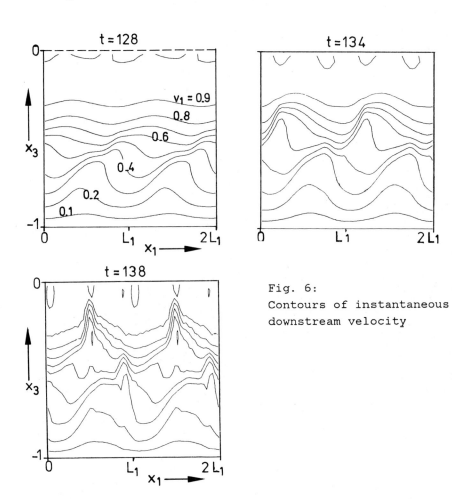

Fig. 6: Contours of instantaneous downstream velocity

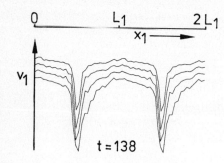

Fig. 7: Instantaneous velocity traces at $x_3 = -0.14$, -0.17, -0.21 and -0.24

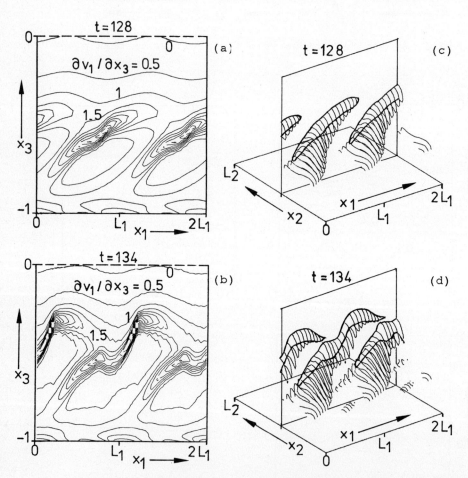

Fig. 8: Development of the high-shear layer.
(a),(b) contours of $\partial v_1/\partial x_3$ in peak plane,
(c),(d) iso-lines $\partial v_1/\partial x_3 = 1.65$

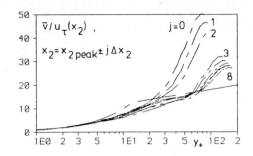

Fig. 9:
Local mean velocity profiles near peak at t=145

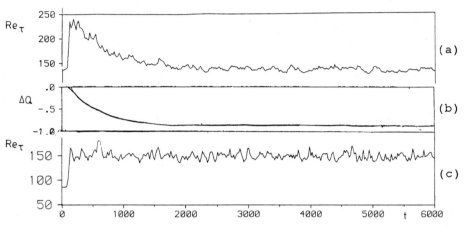

Fig. 10:
(a) evolution of wall Reynolds number Re_τ and
(b) change of flow rate for fixed pressure gradient (case 1),
(c) evolution of Re_τ for fixed flow rate (case 2)

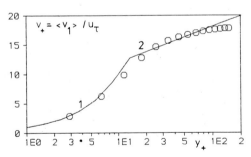

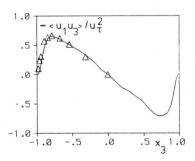

Fig. 11:
Mean velocity. o calculation;
curve 1, $v_+ = y_+$;
curve 2, $v_+ = 6.1 \log y_+ + 5.9$

Fig. 12:
Reynolds stress.
— calculation,
△ experiment [3]

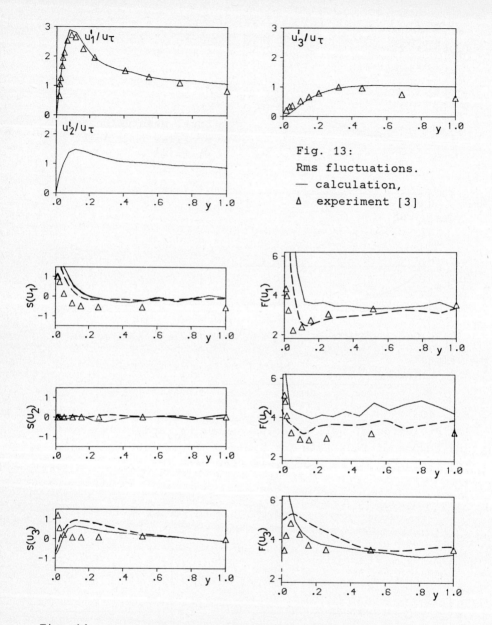

Fig. 13:
Rms fluctuations.
— calculation,
△ experiment [3]

Fig. 14:
Skewness and flatness factors. — calculation, case 2;
-- calculation, case 4; △ experiment [14]

Vortical Mechanisms in Shear Flow Transition

Thorwald Herbert
Department of Engineering Science and Mechanics
Virginia Polytechnic Institute and State University
Blacksburg, VA 24061, U.S.A.

ABSTRACT

The three-dimensional stages of early transition in plane shear flows originate from parametric instability of the streamwise (almost) periodic flow in the presence of finite-amplitude TS waves. The instability is governed by a Floquet system of linear disturbance equations that allows for a rich variety of generally three-dimensional disturbances. The theory predicts the quantitative properties of these disturbances in essential agreement with the experimental data. This capability is a major step towards predicting and ultimately controlling transition. Analysis of the flow field shows that the dramatic growth of three-dimensional disturbances results from the dynamics of an array of vorticity concentrations near the critical layer subject to the surrounding shear. Nonlinear analysis holds the promise of unraveling the feedback loop of self-sustained transition. The energy transfer between mean flow, two-dimensional, and three-dimensional disturbances has given a first lead in this direction.

1. Introduction

Although a wealth of experimental and computational data is now available on transitional and fully turbulent flows, there remains a great lack of understanding of the mechanisms involved. What really causes a laminar shear flow to break down into a small scale, high-frequency motion once the harmless two-dimensional TS waves exceed a certain amplitude? What really causes the well organized structures in the viscous sublayer to reproduce and burst into new supply of turbulent motion? What are the feedback loops that keep these processes going? To what extent are these phenomena governed by the same basic mechanisms?

In attempting to contribute some answers, we recognize the similarities between the phenomena in the viscous sublayer [1], [2], and those in transitional flow. The problem is therefore approached from the laminar beginnings through the cascade of weak and strong instabilities. We build on the classical results for shear-flow instability which show that primary instability occurs with respect to two-dimensional TS waves which develop on a slow viscous time scale. We concentrate on prototype flows such as plane Poiseuille flow (for clean mathematics), Blasius boundary layer and mixing layer (for the available experimental data base). We refrain, however, from using the concept of the weakly nonlinear theory that requires the intuitive choice of interacting modes. Plenty of resonant or non-resonant models have already been proposed but have failed to prove their power.

We recognize that finite-amplitude primary TS waves result in a streamwise (almost) periodic modulation of the steady laminar shear flow, and therefore may lead to parametric excitation of new classes of disturbances, which are in general three-dimensional [3]. Obviously, the linear stability analysis of this modulated flow rests on Floquet systems of stability equations with periodic coefficients. Mathematical properties of these systems provide the general form of secondary instability modes and permit a classification by the type of resonance. Primary resonance leads to *fundamental* modes with the streamwise wavelength λ_x of the TS wave. These modes are formally consistent with the observations of peak-valley splitting [4], [5], [6]. Principal parametric resonance provides *subharmonic* modes of streamwise wavelength $2\lambda_x$ as they have been discovered in recent experiments [6], [7], [8]. Subharmonic modes can also occur in a two-dimensional version, e.g. as vortex pairing in mixing layers [9]. Finally, subharmonic and fundamental modes can be considered as special cases of a more general class of three-dimensional disturbances that can grow in space and time.

Applications of Floquet theory in stability analysis are mostly concerned with time-periodic flows [10]. Time-periodicity, however, does not cause a spatial redistribution of vorticity, and therefore has no effect on the qualitative characteristics of the disturbances. Only a few applications to streamwise periodic flows have been reported. Most notable perhaps are the early studies on vortex pairing [11] and on the occurrence of three-dimensionality in the Blasius boundary layer [12]. This approach rapidly developed over the past five years, supported by numerical simulations of transition [13], [14] and new observations. Often, the concepts of secondary instability analysis are entangled with the computational method used [15]. In our work, we have attempted to develop the Floquet analysis of secondary instability as a stand-alone tool similar to the classical analysis of primary instability. Obviously, numerical methods are to be employed for obtaining quantitative results.

There are at least three exciting results of this effort. First, the formal results of the Floquet analysis are consistent with and provide understanding for the variety of observed phenomena. Second, the quantitative results such as growth rates and disturbance velocity distributions are in encouraging agreement with experimental data. Third, the mechanism of the secondary instability is found to be of a vortical nature and involves an intricate interlace of vortex stretching and convection. Energy transfer into secondary disturbances is highly localized in space. Interesting enough, nonlinear interaction of the three-dimensional disturbances reproduces a streamwise periodic vorticity distribution and consequently may lead to self-sustained and dramatic three-dimensional evolution.

In the following sections, we outline the concepts of the stability analysis, present some prototype results on growth rates and disturbance velocities, and discuss the energy transfer between the components of the flow in context with the vorticity distribution created by the primary wave.

2. Floquet analysis of secondary instability

The flow of an incompressible fluid along a solid boundary or between two walls is governed by the Navier-Stokes equations

$$\frac{\partial \mathbf{v}}{\partial t} + (\mathbf{v} \cdot \nabla)\mathbf{v} = -\nabla p + \frac{1}{R}\nabla^2 \mathbf{v}, \quad \nabla \cdot \mathbf{v} = 0 \tag{1}$$

with appropriate boundary conditions. All quantities are nondimensional, and R is the Reynolds number. The quantities $\mathbf{v}(x',y,z,t) = (u,v,w)$ and $p(x',y,z,t)$ are velocity and pressure fields, respectively, where x' is streamwise, y is normal to the wall(s), z is normal to the x',y plane and t is time. From (1) we derive the transport equation for the vorticity $\omega = \nabla \times \mathbf{v} = (\xi,\eta,\varsigma)$,

$$\frac{\partial \omega}{\partial t} + (\mathbf{v}\cdot\nabla)\omega - (\omega\cdot\nabla)\mathbf{v} = \frac{1}{R}\nabla^2\omega \qquad (2)$$

in order to eliminate the pressure. We consider a two-dimensional flow $\mathbf{v}_2 = (u_2,v_2,0)$ subject to small three-dimensional disturbances $\mathbf{v}_3 = (u_3,v_3,w_3)$ according to

$$\mathbf{v}(x',y,z,t) = \mathbf{v}_2(x',y,t) + B\,\mathbf{v}_3(x',y,z,t) \qquad (3)$$

where B is sufficiently small for linearization. Substitution into (2) and comparison in like powers of B provides two sets of equations for the basic flow and the disturbances, respectively. Terms of order $O(B^2)$ are neglected for an analysis of linear secondary instability.

2.1 The periodic basic flow

The basic flow $\mathbf{v}_2$ is required to satisfy the two-dimensional version of eq. (1) and the vorticity transport equation

$$(\frac{1}{R}\nabla^2 - \frac{\partial}{\partial t})\varsigma_2 - (\mathbf{v}_2\cdot\nabla)\varsigma_2 = 0 \qquad (4)$$

for the spanwise component of $\omega_2 = (0,0,\varsigma_2)$, and the prescribed boundary conditions. Note that the vortex stretching term $(\omega\cdot\nabla)\mathbf{v}$ vanishes for two-dimensional flow.

We consider (exact or approximate) solutions of eq. (4) in the form

$$\mathbf{v}_2(x',y,t) = \mathbf{v}_0(y) + A\,\mathbf{v}_1(x',y,t) \qquad (5)$$

where $\mathbf{v}_0 = (u_0,0,0)$ is the steady flow solution to equations (1), e.g. plane Poiseuille flow, or the Blasius boundary layer flow. The component $\mathbf{v}_1$ is considered in the form of a wave, i.e. periodic in t, periodic in x' with wavelength $\lambda_x = 2\pi/\alpha$, and traveling with phase velocity c_r in the x' direction. In a coordinate system x,y,z traveling with the wave, we have

$$\mathbf{v}_1(x',y,t) = \mathbf{v}_1(x,y) = \mathbf{v}_1(x+\lambda_x,y), \quad x = x' - c_r t, \qquad (6)$$

and consequently, $\mathbf{v}_1$ and $\mathbf{v}_2$ are steady and streamwise periodic. By proper normalization of $\mathbf{v}_1$, the amplitude A measures the maximum streamwise rms fluctuation (usually denoted as u'_m), and the streamwise fluctuation assumes this maximum at $x = 0$.

We express $\mathbf{v}_1 = (u_1,v_1,0)$ and the vorticity $\nabla \times \mathbf{v}_1 = (0,0,\varsigma_1)$ in terms of the stream function ψ_1 such that

$$u_1 = \partial\psi_1/\partial y, \quad v_1 = -\partial\psi_1/\partial x, \quad \varsigma_1 = -\nabla^2\psi_1. \qquad (7)$$

The periodic component of the basic flow is then governed by the nonlinear partial differential equation

$$[\frac{1}{R}\nabla^2 - (u_0 - c_r)\frac{\partial}{\partial x}]\varsigma_1 + \frac{\partial\varsigma_0}{\partial y}\frac{\partial\psi_1}{\partial x} = A[\frac{\partial\psi_1}{\partial y}\frac{\partial}{\partial x} - \frac{\partial\psi_1}{\partial x}\frac{\partial}{\partial y}]\varsigma_1 \qquad (8)$$

with homogeneous boundary conditions on $\partial \psi_1 / \partial x$ and $\partial \psi_1 / \partial y$. Since ψ_1 is periodic in x, it can be represented by the Fourier series

$$\psi_1(x,y) = \sum_{n=-\infty}^{\infty} \phi_n(y) e^{in\alpha x}, \qquad (9)$$

where $\phi_{-n} = \phi_n^{\dagger}$ is necessary for a real solution, and $\dagger$ denotes the complex conjugate.

In some cases like plane Poiseuille flow, strictly periodic solutions of the Navier-Stokes equations can be found by numerically solving a truncated version of the nonlinear system for ϕ_n, $0 \leq n \leq N$ [25]. Such solutions exist for certain parameter combinations that satisfy a nonlinear dispersion relation $F(R, \alpha, A, c_r) = 0$. These equilibrium states of constant amplitude allow cleaner mathematics in formulating the secondary instability theory. On the other hand, their use severely restricts the scope of the theory especially with respect to flows like the Blasius boundary layer where such equilibrium states are not established.

An alternative way of constructing the periodic basic flow rests on the shape assumption and a generalization of the parallel-flow assumption frequently used in the classical stability theory. The shape assumption is equivalent to discarding the nonlinear terms on the right hand side of eq. (8) and considering the amplitude A as a parameter. This assumption is justified by the observations that secondary instability occurs at small amplitudes and originates from the redistribution of vorticity, not from the amplitude-sensitive Reynolds stresses. With the shape assumption, the streamfunction is given by

$$\psi_1(x,y) = \phi_1(y) e^{i\alpha x} + \phi_1^{\dagger}(y) e^{-i\alpha x}, \qquad (10)$$

where ϕ_1 is the eigenfunction associated with the principal mode of the Orr-Sommerfeld equation for given R and α,

$$\{\frac{1}{R}D^2 - i\alpha[(u_0 - c)D^2 - u''_0]\}\phi_1 = 0, \qquad (11)$$

where $D^2 = d^2/dy^2 - \alpha^2$, and the prime denotes d/dy. The eigenvalue problem associated with eq. (11) provides complex eigenvalues $c = c_r + ic_i$ or $\alpha = \alpha_r + i\alpha_i$ depending on whether the temporal or spatial growth concept is chosen. The non-vanishing imaginary part describes exponential growth or decay of the amplitude. Generalizing the parallel-flow assumption, we consider the local (or instantaneous) amplitude as independent of x (or t). Obviously, this neglect of the slow amplitude variation is justified only for sufficiently large growth rates of the three-dimensional disturbances.

2.2. Three-dimensional disturbances

The analysis of three-dimensional disturbances $\mathbf{v}_3$ rests on the vorticity equation

$$(\frac{1}{R}\nabla^2 - \frac{\partial}{\partial t})\boldsymbol{\omega}_3 - (\mathbf{v}_2 \cdot \nabla)\boldsymbol{\omega}_3 - (\mathbf{v}_3 \cdot \nabla)\boldsymbol{\omega}_2 + (\boldsymbol{\omega}_2 \cdot \nabla)\mathbf{v}_3 + (\boldsymbol{\omega}_3 \cdot \nabla)\mathbf{v}_2 = 0. \qquad (12)$$

This equation is linear in $\mathbf{v}_3$ due to neglecting terms of order $O(B^2)$. As in the classical work, we eliminate the spanwise disturbance velocity w_3 by exploiting continuity. Since $\partial \eta_3/\partial z$ and $\partial \xi_3/\partial z$ can be expressed in terms of u_3, v_3, we obtain two coupled differential equations for u_3 and v_3,

$$[\frac{1}{R}\nabla^2 - (u_0 - c)\frac{\partial}{\partial x} - \frac{\partial}{\partial t}]\frac{\partial \eta_3}{\partial z} + \varsigma_0 \frac{\partial^2 v_3}{\partial z^2} + A\{(-\frac{\partial \psi_1}{\partial y}\frac{\partial}{\partial x} + \frac{\partial \psi_1}{\partial x}\frac{\partial}{\partial y}$$
$$- \frac{\partial^2 \psi_1}{\partial x \partial y})\frac{\partial \eta_3}{\partial z} + \frac{\partial^2 \psi_1}{\partial x^2}(\frac{\partial^2 u_3}{\partial x \partial y} + \frac{\partial^2 v_3}{\partial y^2}) - \frac{\partial^2 \psi_1}{\partial y^2}\frac{\partial^2 v_3}{\partial z^2}\} = 0, \quad (13)$$

$$[\frac{1}{R}\nabla^2 - (u_0 - c)\frac{\partial}{\partial x} - \frac{\partial}{\partial t}]\nabla^2 v_3 - \frac{d\varsigma_0}{dy}\frac{\partial v_3}{\partial x} + A\{(-\frac{\partial \psi_1}{\partial y}\frac{\partial}{\partial x} + \frac{\partial \psi_1}{\partial x}\frac{\partial}{\partial y})\nabla^2 v_3$$
$$+ \frac{\partial^2 \psi_1}{\partial x^2}(\frac{\partial \varsigma_3}{\partial y} + \frac{\partial \eta_3}{\partial z}) - \frac{\partial^2 \psi_1}{\partial x \partial y}(\frac{\partial \varsigma_3}{\partial x} + \frac{\partial \xi_3}{\partial z}) - \frac{\partial \varsigma_1}{\partial x}(2\frac{\partial u_3}{\partial x} + \frac{\partial v_3}{\partial y}) - \frac{\partial \varsigma_1}{\partial y}\frac{\partial v_3}{\partial x}$$
$$- (u_3\frac{\partial}{\partial x} + v_3\frac{\partial}{\partial y})\frac{\partial \varsigma_1}{\partial x}\} = 0, \quad (14)$$

with homogeneous boundary conditions on u_3, v_3, and $\partial v_3/\partial y$.

In the limit $A \to 0$, the coefficients in (13), (14) are independent of x, z, and t, and normal modes can be introduced in these variables. Equation (13) reduces to Squire's equation for the y-component of vorticity whereas eq. (14) provides the Orr-Sommerfeld equation for oblique waves in the basic flow $\mathbf{v}_0$, written in a moving coordinate system.

For $A \neq 0$, the x-periodic streamfunction ψ_1 appears in the coefficients. The normal mode concept can still be applied with respect to z and t and the disturbances can be written in the form

$$\mathbf{v}_3(x,y,z,t) = e^{\sigma t} e^{i\beta z} \mathbf{V}(x,y). \quad (15)$$

We consider the spanwise wave number $\beta = 2\pi/\lambda_z$ as real, whereas $\sigma = \sigma_r + i\sigma_i$ is in general complex. Insight into classes of solutions and their streamwise structure can be obtained from the Floquet theory of ordinary differential equations with periodic coefficients.

Although $\mathbf{v}_3$ in the form (15) is in general complex, we note that the disturbance equations provide real physical solutions including the complex conjugate $\mathbf{v}_3^\dagger$. Consequently, the system of equations can be written in real form. In this case, Floquet theory suggests solutions in the form

$$\mathbf{V}(x,y) = e^{\gamma x} \tilde{\mathbf{V}}(x,y), \quad \tilde{\mathbf{V}}(x + 2\lambda_z, y) = \tilde{\mathbf{V}}(x,y) \quad (16)$$

where $\gamma = \gamma_r + i\gamma_i$ is a characteristic exponent, and $\tilde{\mathbf{V}}$ is periodic in x with wavelength $2\lambda_z$. Hence, we can write $\mathbf{v}_3$ in the form

$$\mathbf{v}_3 = e^{\sigma t} e^{\gamma x} e^{i\beta z} \sum_{m=-\infty}^{\infty} \hat{\mathbf{v}}_m(y) e^{im\hat{\alpha}x}, \quad \hat{\alpha} = \alpha/2. \quad (17)$$

The functions $\hat{u}_m$, $\hat{v}_m$ are governed by an infinite system of ordinary differential equations.

2.3 Classification of modes

Since the basic flow $\mathbf{v}_2$ with wavenumber $\alpha = 2\hat{\alpha}$ provides coupling only between components $\hat{\mathbf{v}}_m$ and $\hat{\mathbf{v}}_{m\pm 2}$, this system splits into two separate systems for even and odd m that describe two classes of solutions

$$\mathbf{v}_f = e^{\sigma t} e^{\gamma x} e^{i\beta z} \tilde{\mathbf{v}}_f(x,y), \quad \tilde{\mathbf{v}}_f = \sum_{m \text{ even}} \hat{\mathbf{v}}_m(y) e^{im\hat{\alpha}x}, \quad (18)$$

$$\mathbf{v}_s = e^{\sigma t} e^{\gamma x} e^{i\beta z} \tilde{\mathbf{v}}_s(x,y), \quad \tilde{\mathbf{v}}_s = \sum_{m \text{ odd}} \hat{\mathbf{v}}_m(y) e^{im\hat{a}x}. \tag{19}$$

The periodic functions $\tilde{\mathbf{v}}_f$ and $\tilde{\mathbf{v}}_s$ obviously satisfy

$$\tilde{\mathbf{v}}_f(x + \lambda_x, y) = \tilde{\mathbf{v}}_f(x,y), \quad \tilde{\mathbf{v}}_s(x + 2\lambda_x, y) = \tilde{\mathbf{v}}_s(x,y). \tag{20}$$

Therefore, we denote $\mathbf{v}_f$ as the fundamental mode, associated with primary resonance, and $\mathbf{v}_s$ as the subharmonic mode, originating from principal parametric resonance.

The occurrence of two complex quantities, σ and γ, in the eigenvalue problem for secondary disturbances leads to an ambiguity similar to that associated with the Orr-Sommerfeld equation. Only two of the four real quantities $\sigma_r, \sigma_i, \gamma_r, \gamma_i$ can be determined; the other two must be chosen. In this way, we can distinguish various physically relevant classes of solutions, e.g.

(1) Temporally growing modes. With $\gamma_r = 0$, the temporal growth rate is given by σ_r, while σ_i can be interpreted as frequency shift. Modes with $\sigma_i = \gamma_i c_r$ travel synchronous with the basic flow.

(2) Spatially growing modes. Since spatial growth is implicitly related to the laboratory frame x', y, z, we choose $\sigma = \gamma c_r$. The spatial growth rate is given by γ_r while γ_i is the shift in the streamwise wavenumber.

It is interesting to note that the distinction of fundamental and subharmonic modes in (18) and (19) is redundant, if $\gamma_i \neq 0$. Replacing γ in (18) by $\gamma = \gamma' \pm \hat{a}$ and renumbering the terms in the Fourier series leads to an expression equivalent to (19). Nevertheless, it is convenient to distinguish these two classes for temporally growing modes with $\gamma_i = 0$ and spatially growing modes with $\sigma_i = \gamma_i c_r$. In fact, we will restrict the following discussion to these cases that allow for synchronization between two-dimensional wave and three-dimensional disturbances. Moreover, we will mostly use the (simpler) temporal growth concept.

3. Results

For the numerical work, truncated series (9), (18), (19) are substituted into (13), (14). It has been shown [24] that the lowest Fourier approximations provide results of satisfactory accuracy. The rather lengthy disturbance equations (in real form) are given in [19] and [21] for subharmonic and fundamental modes, respectively. We solve these equations by a spectral collocation method, using an algebraic transformation in the case of a semi-infinite interval. Treatment as a boundary value problem allows easy access to the eigenvalue spectrum and identification of the principal (most unstable) modes in the absence of any prior knowledge.

In the symmetric Poiseuille flow, the fundamental mode can appear in symmetric ($\mathbf{v}_{fs}$) or antisymmetric ($\mathbf{v}_{fa}$) form. A typical eigenvalue spectrum is shown in fig. 1. The spectrum is symmetric to the σ_r axis as expected for a problem associated with a real matrix. The parameters are chosen according to Nishioka's [5] experimental conditions. Three eigenvalues are located in the right half-plane, $\sigma_r > 0$, indicating instability of the basic flow to three-dimensional modes. The principal mode is associated with a real eigenvalue, i.e. this mode is phase-locked, and travels synchronous with the primary wave. At relevant amplitudes, this characteristic is shared by subharmonic and fundamental modes in the Blasius flow. Synchronization of two-dimensional wave and three-dimensional disturbances has also been found in the experiments [4], [7].

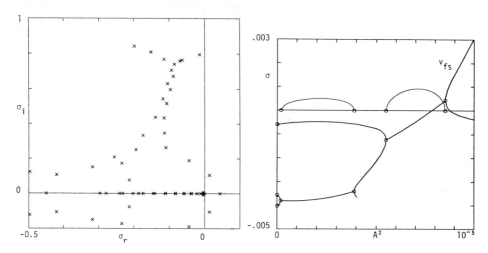

Figure 1. Spectrum of eigenvalues $\sigma = \sigma_r + i\sigma_i$ for the anti-symmetric fundamental mode $\mathbf{v}_{f_a}$ in plane channel flow (shape assumption). $Re = 5000$, $A = 0.025$, $\alpha = 1.12$, $\beta = 2.0$.

Figure 2. Analytical connection of the principal mode $\mathbf{v}_{f_s}$ of secondary instability in plane channel flow (equilibrium state) to three modes of primary instability. $\alpha = 1.02$, $\beta = 1.0$. (———) σ_r, (———) $|\sigma_i|$.

The parametric dependence of the principal modes is traced by a local (Wielandt iteration) procedure. The trace to low amplitudes, $A \to 0$, reveals the intricate analytical connection to modes of primary instability. Figure 2 shows an example for the symmetric fundamental mode in plane Poiseuille flow. The modes at $A = 0$ can be identified as eigenmodes of Squire's equation (that consist only of a streamwise velocity component) and longitudinal vortex modes of the Orr-Sommerfeld equation for $\alpha = 0$. Interesting to note is the collapse of two real eigenvalues into a complex conjugate pair, and the ultimate conversion of this pair into a strongly growing principal mode and a strongly decaying mode as the amplitude increases. The details of this metamorphosis, and the number of modes involved may change with the parameters. Knowledge of these analytical connections suggests new weakly nonlinear models that are free from biased intuition.

The dependence of the temporal growth rate σ_r on the spanwise wavenumber β is shown in figures 3 and 4 for plane Poiseuille flow and Blasius flow, respectively. The common feature of these figures are the large values of σ_r - much larger than the growth rates for TS waves, the broad band of the instability, and the rather sharp cutoff at low wavenumbers. The exceptional behaviour of the symmetric fundamental mode with $\sigma_r > 0$ at $\beta = 0$ in fig. 3 is associated with the instability of the two-dimensional equilibrium state that has been used as basic flow (an interesting by-product of the analysis). We also note that growth rates for subharmonic and fundamental modes are very similar. At lower amplitudes, however, the subharmonic instability is stronger, and must be considered the main cause of three-dimensional effects in low-disturbance environments.

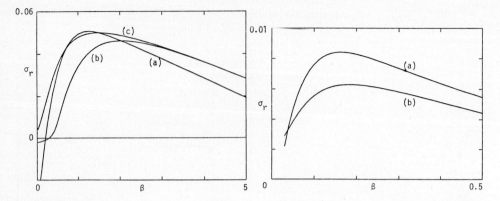

Figure 3. Growth rates σ_r vs. spanwise wave number β for principal modes in plane channel flow (equilibrium state). (a) subharmonic, (b) anti-symmetric, and (c) symmetric fundamental mode. $Re = 5000$, $A = 0.0248$, $\alpha = 1.12$.

Figure 4. Growth rates σ_r vs. spanwise wave number β for principal modes in Blasius boundary layer flow. (a) subharmonic, (b) fundamental (peak-valley splitting) mode. $Re = 950$, $A = 0.01408$, $F = 58.8$.

Figure 5 compares the theoretical predictions of spatial amplitude growth for TS wave and subharmonic disturbances with experimental data [7] for the Blasius flow. The discrepancy at low values of the Reynolds number is due to transient effects near the vibrating ribbon in the experiments. In earlier work [19], the amplitude growth was obtained from temporal growth rates using the crude (and suspicious) transformation $\gamma_r \approx \sigma_r/c_r$ and

$$\ln\frac{B(R)}{B_0} = 2 \int_{R_0}^{R} \frac{\sigma_r}{c_r} dr, \qquad (21)$$

where R_0 and B_0 are Reynolds number and initial amplitude, respectively, at onset of secondary instability, $\sigma_r = 0$. In order to answer various criticism on the ground that the conversion of temporal to spatial growth rates is justified only for sufficiently small (primary TS wave) growth rates [26], we repeated this analysis based on the spatial growth concept [23], [24]. Surprisingly, there is little difference in the results.

Figure 6 shows the comparison of spatial growth rates obtained by transformation of temporal data and by direct calculation. Considering the substantial increase in computational effort when using the spatial growth concept (where the eigenvalue α occurs nonlinearly in the equations), and the approximate character of our linear stability analysis, the small difference in the growth rates appears irrelevant. The reason for this agreement is obscure. The fact in itself, however, may answer a question that often arises in evaluating computer simulations of transition. These simulations depend on the temporal growth concept. Their results, however, agree very well with (spatial growth) observations. Figure 6 shows that spatial and temporal evolution in the three-dimensional stage of transition are very similar. More insight is needed, however, to really understand this similarity in physical terms.

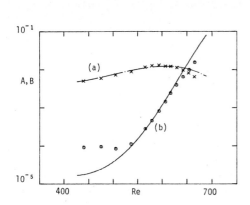
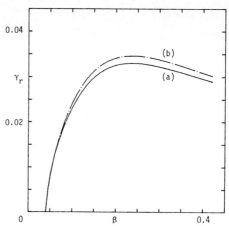

Figure 5. Amplitude variation with Reynolds number Re for (a) the TS wave and (b) the subharmonic mode in Blasius boundary layer flow. $F = 124$, $\beta = 0.2$. Comparison of theoretical results (———) with experimental data (x,o) of Kachanov & Levchenko [7].

Figure 6. Spatial growth rate γ_r vs. spanwise wave number β for the principal subharmonic mode in Blasius boundary layer flow. Results of (a) direct calculation and (b) transformation of temporal data, eq. (21). $Re = 826$, $F = 83$, $A = 0.01$.

Beyond the growth rates (eigenvalues), our analysis provides the disturbance velocity distributions (eigenvectors/functions). Figure 7 shows the comparison of experimental and theoretical results for the principal fundamental mode in Poiseuille flow [27]. Figure 8 gives analogue results for the subharmonic mode in the Blasius flow [7]. Similar agreement extends to the phase distribution and to other measured quantities. The streamwise rms fluctuation in figures 7, 8 assumes a pronounced maximum near the critical layer at y_c, where $u_0(y_c) = c_r$. Figure 8 also shows that the secondary disturbance velocity is confined in a narrow region within the boundary layer (see also [24] for fundamental modes). These properties are essentially different from primary disturbances. The striking similarity of figures 7 and 8 - for different mean flows and different modes - indicates the generic nature of the secondary instability mechanism. Variation of the parameters has little effect on these velocity distributions.

4. Nonlinear effects

All the results obtained to date are consistent and in encouraging agreement with the scarce experimental data base. The Floquet analysis of linear secondary instability appears capable of accurately predicting the characteristics of the early three-dimensional stages of transition. Further development, however, demands accounting for nonlinear effects. This need is obvious because of the strong growth of three-dimensional disturbances and is clearly shown in fig. 9 taken from the experimental study of Kachanov & Levchenko [7]. At low levels of the amplitude A, secondary disturbances may grow but ultimately decay with the two-dimensional wave. At higher amplitude levels, however, the stronger growth leads to three-

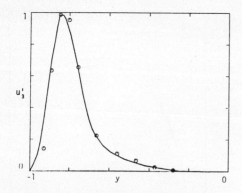

Figure 7. Normalized streamwise rms fluctuation of the fundamental (peak-valley splitting) mode in plane channel flow. $Re = 5000$, $A \approx 0.025$, $\alpha = 1.12$, $\beta = 2.0$. Comparison of theoretical result (———) and experiment (o). From Nishioka & Asai [27].

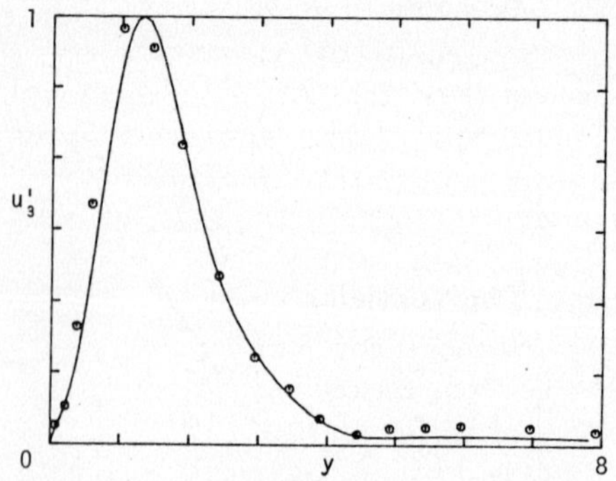

Figure 8. Normalized streamwise rms fluctuation of the subharmonic mode in Blasius boundary layer flow. $Re = 608$, $F = 124$, $A = 0.0122$, $\beta = 2.0$. Comparison of theoretical result (———) and experimental data (o) of Kachanov & Levchenko [7].

dimensional amplitudes B large enough to affect the two-dimensional wave development. Figure 9 shows that the nonlinear interaction prevents the decay of the TS wave. Primary and secondary disturbances team up to generate a rapid evolution towards breakdown. Current efforts aim at a qualitative and quantitative description of this self-sustained growth mechanism.

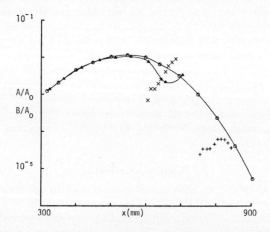

Figure 9. Amplitude growth curves for disturbances at different levels of the TS amplitude. (o) $A_0 = 0.00163$, fundamental frequency f, (+) $A_0 = 0.00163$, subharmonic frequency $f/2$, (Δ) $A_0 = 0.00654$, f, (x) $A_0 = 0.00645$, $f/2$. From Kachanov & Levchenko [7].

As a first step, we have analyzed the vorticity field and the energy transfer between mean flow v_0, two-dimensional wave $A\,v_1$, and three-dimensional disturbances $B\,v_3$ [28]. A similar analysis based on numerical transition simulations for fundamental modes has been performed by Orszag & Patera [29]. Their work clearly reveals the generic nature of the secondary instability mechanism. The results reported here are for plane Poiseuille flow.

4.1 The vorticity field

Without any stability calculation, the origin of secondary instability can be found from a close look at the periodic basic flow v_2. Figure 10 shows the streamlines of v_2 for an amplitude $A = 2.5\%$ in the wave-fixed coordinate system. The centers of the cats eyes indicate extrema of the stream function that are located just outside the critical layer at $\pm y_c$. The associated vorticity contours are given in fig. 11. Viscous effects are restricted to the close neigborhood of the wall: the high levels of vorticity near the wall that diffuse into the flow extend only to the critical layer. For $|y| < y_c$, streamlines and isolines of vorticity are nearly parallel, indicating essentially inviscid flow [29]. A weak extremum of vorticity occurs at positions close to the centers of the cats eyes. Consequently, the flow in this neighborhood resembles two arrays of distributed vortices at the edge of the viscous layers near the walls. As the amplitude increases, the vortices strengthen and move away from the critical layers into the inviscid region.

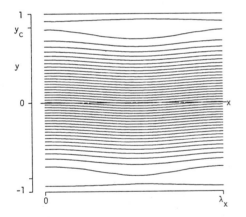

Figure 10. Streamlines of the periodic flow in a plane channel (shape assumption). $Re = 5000$, $A = 0.025$, $\alpha = 1.12$.

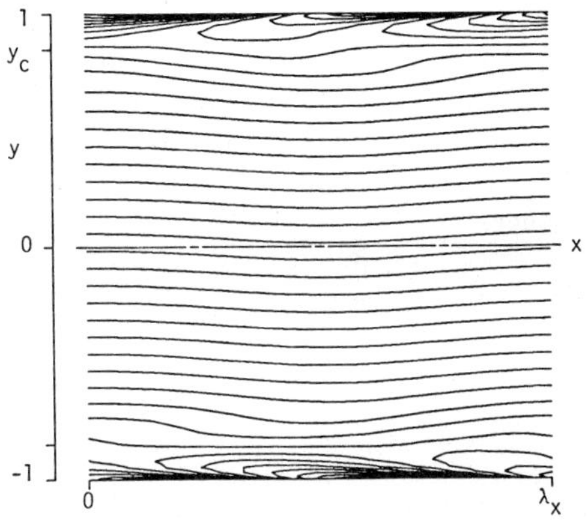

Figure 11. Contour lines of the spanwise vorticity component for the periodic flow in a plane channel (shape assumption). $Re = 5000$, $A = 0.025$, $\alpha = 1.12$.

The dynamics of a single vortex tube in a shear layer has been analyzed by Stuart [30], and in more physical terms by Wortmann [31] in context with flow visualizations of peak-valley splitting. The distributed vortex tube in a two-dimensional flow is a salient feature only in presence of a spanwise, three-dimensional perturbation. Owing to the surrounding shear, any deformation of the tube initiates transport processes which in turn enhance the deformation. Wortmann concludes: "we have to expect a strong, exponential growth of any three-dimensional perturbation once the Reynolds number and wave amplitude of the Tollmien wave establish a local vorticity peak of sufficient strength near the critical layer." His arguments clearly indicate the necessity of combined vortex-stretching and the retrograde rotation of a bent vortex tube for the exponential growth of three-dimensionality. Wortmann's model has been verified by the more detailed and formal analysis of the vorticity dynamics by Orszag & Patera [29]. This work also provides quantitative data for the fundamental mode of secondary instability at large amplitude.

4.2 Energy transfer

Our analysis of the energy transfer between mean flow $\mathbf{v}_0$, TS wave $A\,\mathbf{v}_1$, and three-dimensional disturbances $B\,\mathbf{v}_3$ differs from earlier work [29] in three aspects. First, the basic flow is constructed under the shape and parallel-flow assumptions. Second, the energy balance is considered at a fixed time, i.e. for fixed A,B, and the associated velocity distributions obtained from linear stability analysis. Finally, we consider not only the global energy balance but also the spatial distribution.

We consider a domain Ω between the walls $|y|=1$ which is $\lambda_z = 2\pi/\beta$ wide and one or two TS wavelengths $\lambda_x = 2\pi/\alpha$ long for fundamental or subharmonic modes, respectively. For given amplitudes $a_0 = 1$, $a_1 = A$, $a_3 = B$, the analysis of the global energy balance rests on the equations

$$\frac{dE_0}{dt} = -T_{01} - T_{03} + D_0 \tag{22}$$

$$\frac{dE_1}{dt} = T_{01} - T_{13} + D_1 \tag{23}$$

$$\frac{dE_3}{dt} = T_{03} + T_{13} + D_3 \tag{24}$$

where

$$E_k = \frac{1}{2} a_k^2 \int_\Omega \mathbf{v}_k \cdot \mathbf{v}_k \, d\Omega \tag{25}$$

$$D_k = \frac{1}{Re} a_k^2 \int_\Omega \nabla^2 \mathbf{v}_k \cdot \mathbf{v}_k \, d\Omega - \delta_{k0} \int_\Omega \nabla p_0 \cdot \mathbf{v}_k \, d\Omega \tag{26}$$

$$T_{jk} = -a_j a_k^2 \int_\Omega \mathbf{v}_k \nabla \mathbf{v}_j \cdot \mathbf{v}_k \, d\Omega \tag{27}$$

for $k = 0, 1, 3$, and δ_{k0} is the Kronecker symbol. The term T_{jk} denotes the transfer from the jth to the kth component and hence $T_{jk} = -T_{kj}$. For $B = 0$ ($T_{j3} = 0$), eqs. (22), (23) describe the energetics of the primary instability. In particular, we obtain the growth rate c_i of the TS wave from

$$\alpha c_i = \frac{dE_1}{2E_1 dt} = \frac{T_{01} + D_1}{2E_1}. \tag{28}$$

The growth rate is independent of $a_1 = A$ as long as the shape assumption is applied, i.e. the nonlinear distortion of $\mathbf{v}_0$ and $\mathbf{v}_1$ and the generation of harmonics is neglected. For $B \neq 0$, the system shows that the mean flow may exchange energy directly with the TS wave and the three-dimensional mode. In addition, there is also an indirect way of transferring energy through the term T_{31}.

For the spatial distributions, the integration is only carried out in one or two variables, providing functions of the remaining variables. Figure 12 shows the distribution of the power for a TS wave at $R = 5000$, $\alpha = 1.12$, where positive values are indicated by the heavy line width. Except in the immediate neighborhood of the wall, this figure is very similar to the distribution of the Reynolds stress. Although activity is distributed across the flow, integration over x cancels most of the contributions and leads to just some little bumps in the distribution of power over y. These bumps integrate into a multiple (see eq. (28)) of αc_i, the small (positive or negative) growth rate of TS waves on a viscous time scale.

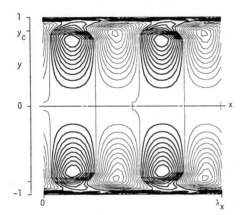

Figure 12. Distribution of the energy transfer dE_1/dt from mean flow into the TS wave in the (x,y)-plane. $Re = 5000$, $\alpha = 1.12$. Heavy lines indicate values ≥ 0.

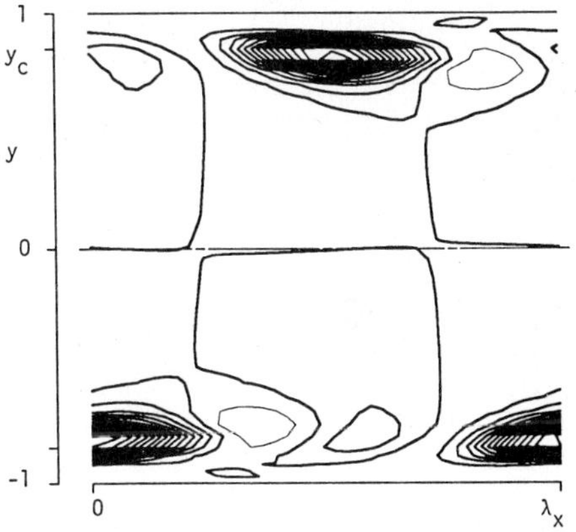

Figure 13. Distribution of the energy transfer dE_3/dt into the symmetric fundamental mode $\mathbf{v}_{fs}$ in the (x,y)-plane at the peak position, $z = 0$. $Re = 5000$, $\alpha = 1.12$, $A = 0.025$, $\beta = 2.0$.

Figure 12 is in strong contrast to the power distribution for the symmetric fundamental mode of secondary instability shown in fig. 13. This picture is almost identical with the distribution of the energy transfer T_{03} from mean flow into the fundamental mode. The densely spaced contours clearly show the peak power at the location of the vorticity extremum in fig. 11. Figure 14 shows the distribution of the energy transfer T_{13} from the two-dimensional periodic component into the fundamental mode. Remarkable is the negative transfer in the downstream part of the vorticity concentration that overweights the gain in the upstream region. Integration of the power distribution in fig. 13 over x provides the y-distribution of dE_3/dt shown in fig. 15a. The sharp peaks near the critical layer reflect the energy transfer T_{03} from $\mathbf{v}_0$ into $\mathbf{v}_3$ shown in fig. 15b, slightly modified by transfer between $\mathbf{v}_1$ and $\mathbf{v}_3$, and viscous dissipation near the wall.

The pictures analogue to fig. 13, 14, and 15 for the antisymmetric fundamental mode are very similar, while each of the subharmonic modes shows maximum power coincident with the vorticity extrema located near one of the walls only, as shown in fig. 16. This one-sided energy transfer is consistent with the theoretical prediction that a single mode of subharmonic instability may initiate transition in the

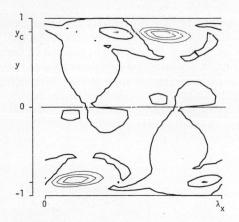
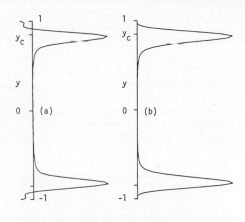

Figure 14. Distribution of the energy transfer T_{13} from the TS wave into the fundamental mode $\mathbf{v}_{fs}$ in the (x,y)-plane at the peak position, $z = 0$. $Re = 5000$, $\alpha = 1.12$, $A = 0.025$, $\beta = 2.0$.

Figure 15. Energy transfer (a) dE_3/dt, (b) T_{03} as a function of y for the symmetric fundamental mode $\mathbf{v}_{fs}$. $Re = 5000$, $\alpha = 1.12$, $A = 0.025$, $\beta = 2.0$.

neighborhood of only one of the channel walls [17]. This curious phenomenon has been observed in the transition simulations of May & Kleiser [32].

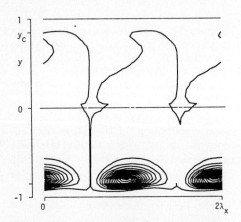
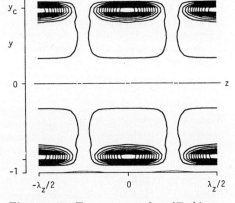

Figure 16. Distribution of the energy transfer dE_3/dt into the subharmonic mode $\mathbf{v}_s$ in the (x,y)-plane at $z = 0$. $Re = 5000$, $\alpha = 1.12$, $A = 0.025$, $\beta = 2.0$.

Figure 17. Energy transfer dE_3/dt as a function of y, z for the symmetric fundamental mode $\mathbf{v}_{fs}$. $Re = 5000$, $\alpha = 1.12$, $A = 0.025$, $\beta = 2.0$. Heavy lines indicate values ≥ 0.

The highly localized energy transfer in the (x,y)-plane is another indicator for the close association between vorticity redistribution and secondary instability which is, after all, not very surprising. A similar, but less obvious concentration of the power can bee seen in the spanwise direction, as shown in fig. 17 for the symmetric fundamental mode. Energy transfer is concentrated at the positions of peak ($z = 0$) and valley, i.e. where the deflection of the vortices is strongest. The pattern for the antisymmetric fundamental mode is almost identical, while for the subharmonic mode the concentrations at opposite walls are shifted by $\lambda_z/4$.

For large amplitudes A, Orszag & Patera [29] found that the growth of the fundamental mode originates from direct and strong energy transfer through T_{03}. The TS wave, therefore, plays only a catalyst role in mediating the energy transfer between $\mathbf{v}_0$ and $\mathbf{v}_3$, consistent with the parametric character of secondary instability. This conclusion is fully supported by our results.

Table 1 shows the values of the various terms for amplitudes $A = B = 0.025$, and for the symmetric fundamental mode. The growth rates in the last row were obtained from linear analysis of primary and secondary instability. Reconstruction of these growth rates from eq. (28) and

$$\sigma_r = \frac{dE_3}{2E_3 dt} = \frac{T_{03} + T_{13} + D_3}{2E_3} \tag{29}$$

provides not only a useful check, but gives some appreciation for the different scales of primary (viscous) and secondary (vortical) instability. The energy transfer T_{03} from the mean flow into the three-dimensional mode is about 20 times larger than the transfer T_{01} into the TS wave. Only 1/3 of this energy is lost by dissipation. Most interesting, however, is the additional loss T_{13} that amounts to less than 1/10th of T_{03}, but boosts the energy transfer into the TS wave considerably. In fact, T_{31} is about 1.5 times T_{01} and leads to a growth rate of $dE_1/(2E_1 dt) = 0.0143$ for the two-dimensional, streamwise periodic component of the flow. The situation is very similar for the other modes.

Table 1. Energy transfer terms in eqs. (23), (24) for the symmetric fundamental mode $\mathbf{v}_{fs}$ in plane channel flow (shape assumption). $Re = 5000$, $\alpha = 1.12$, $A = 0.025$, $\beta = 2.0$, $B = 0.025$.

Two-dimensional, $A\mathbf{v}_1$		Three-dimensional, $B\mathbf{v}_3$ (symmetric fundamental)	
E_1	$2.593097 \cdot 10^{-4}$	E_3	$4.703205 \cdot 10^{-4}$
dE_1/dt	$3.700424 \cdot 10^{-6}$	dE_3/dt	$4.357592 \cdot 10^{-5}$
T_{01}	$3.397289 \cdot 10^{-6}$	T_{03}	$7.108149 \cdot 10^{-5}$
T_{31}	$5.144587 \cdot 10^{-6}$	T_{13}	$-5.144587 \cdot 10^{-6}$
D_1	$-4.841452 \cdot 10^{-6}$	D_3	$-2.236098 \cdot 10^{-5}$
αc_i	$-2.784618 \cdot 10^{-3}$	σ_r	$4.632590 \cdot 10^{-2}$

The observation of the positive transfer term T_{31} is a first lead to understanding the self-sustained growth of three-dimensional disturbances. As the amplitude A exceeds a certain threshold, the increasing strength of the distributed vortices leads to parametric growth of the amplitude B of three-dimensional modes. Since all terms in eq. (24) are of order $O(B^2)$, the growth rate σ_r changes only due to the relatively weak distortion of the velocity profiles with increasing B. For the two-dimensional periodic component, however, T_{31} contributes to the growth rate as a term of order $O(B^2/A)$. Even for negative αc_i, the two-dimensional component will grow as B exceeds some threshold. The simultaneous growth of two- and three-dimensional components establishes a feedback loop, that will rapidly lead to breakdown of the laminar flow. Although our energy analysis neglects nonlinear effects such as the distortion of the velocity profiles and the generation of harmonics, the results for varying amplitudes are consistent with computer simulations [13], [14], and measurements [7] (see fig. 9) of transition.

5. Conclusions

The Floquet analysis of secondary instability appears as a natural and very powerful concept. The results obtained to date indicate that linearization in itself is no major drawback if the basic state is properly chosen. The choice of the streamwise periodic basic state with embedded vortex arrays provides the essentials underlying the mechanism of secondary instability. Formal analysis yields various distinct classes of disturbances that may grow in time or space. Numerical results show that temporal and spatial growth rates are closely related by a simple transformation. Although this result awaits deeper understanding, it explains the surprising, and often disputed agreement of temporal computer simulations and spatial experiments on transition. Numerical results on the qualitative characteristics of three-dimensional disturbances in plane Poiseuille flow and Blasius boundary layer are fully consistent with observations. In spite of approximations and truncations introduced, the theory reproduces experimental data to within acceptable error bounds. Our results have also stimulated new experiments and transition simulations, and helped to understand previously obscure phenomena.

The theory of linear secondary instability, and the understanding of its mechanism have reached a mature state. Work may be directed toward extensions accounting for TS amplitude growth or non-parallelism of the mean flow. In view of the rapid growth of secondary modes, however, the study of nonlinear effects appears as the priority task. The analysis of the energetics of the early stage of three-dimensional development is a first step in this direction. This analysis shows not only the very localized source of the feeding mechanism for three-dimensional modes, but indicates a simple feedback loop for self-sustained evolution of the three-dimensional disturbance field. This feedback loop, and the associated threshold amplitudes are now under more detailed investigation based on the nonlinear disturbance equations. We expect this work to predict more accurate lower limits for the onset of turbulence in terms of Reynolds number, amplitudes and spatial scales of three-dimensional disturbances. We also expect insight into feedback mechanisms that may be common to transition and the bursting phenomenon in the laminar sublayer.

ACKNOWLEDGEMENT

I am grateful to my graduate students Fabio P. Bertolotti, Joseph W. Croswell, and German R. Santos who contributed valuable ideas and results to this study. This work has been supported by the National Science Foundation under Contract MEA-8120935, by the Office of Naval Research under Contract N00014-84-K-0093, and by the Air Force Office of Scientific Research under Contract F49620-84-K-0002.

REFERENCES

[1] Smith, C. R. & Metzler, S. P. 1983 *The characteristics of low-speed streaks in the near-wall region of a turbulent boundary layer*, J. Fluid Mech. 129, 27-54.

[2] Moin, P. & Kim J. 1985 *The structure of the vorticity field in turbulent channel flow. Part 1. Analysis of instantaneous fields and statistical correlations.* J. Fluid Mech. 155, 441-464.

[3] Herbert, Th. & Morkovin, M. V. 1980 *Dialogue on bridging some gaps in stability and transition research,* in: Laminar-Turbulent Transition (eds. R. Eppler & H. Fasel), 47-72, Springer-Verlag.

[4] Klebanoff, P. S., Tidstrom, K. D. & Sargent, L. M. 1962 *The three-dimensional nature of boundary-layer instability,* J. Fluid Mech. 12, 1-34.

[5] Nishioka, M., Asai, M. & Iida, S. 1980 *An experimental investigation of secondary instability,* in: Laminar-Turbulent Transition (eds. R. Eppler & H. Fasel), 37-46, Springer-Verlag.

[6] Saric, W. S. & Thomas, A. S. W. 1983 *Experiments on the subharmonic route to turbulence in boundary layers,* in: Turbulence and Chaotic Phenomena in Fluids (ed. T. Tatsumi), 117-122, North-Holland.

[7] Kachanov, Yu. S. & Levchenko, V. Ya. 1984 *The resonant interaction of disturbances at laminar-turbulent transition in a boundary layer,* J. Fluid Mech. 138, 209-247.

[8] Kozlov, V. V. & Ramazanov, M. P. 1984 *Development of finite-amplitude disturbances in Poiseuille flow,* J. Fluid Mech. 145, 149-157.

[9] Winant, C. D. & Browand, F. K. 1974 *Vortex pairing: the mechanism of turbulent mixing-layer growth at moderate Reynolds number,* J. Fluid Mech. 63, 237-255.

[10] Davis, S. H. 1976 *The stability of time-periodic flows,* Ann. Rev. Fluid Mech. 8, 57-74.

[11] Kelly, R. E. 1967 *On the stability of an inviscid shear layer which is periodic in space and time,* J. Fluid Mech. 27, 657-689.

[12] Maseev, L. M. 1968 *Occurrence of three-dimensional perturbations in a boundary layer,* Fluid Dyn. 3, 23-24.

[13] Orszag, S. A. & Patera, A. T. 1981 *Subcritical transition to turbulence in plane shear flows,* in: Transition and Turbulence (ed. R. E. Meyer), 127-146, Academic Press.

[14] Kleiser, L. 1982 *Numerische Simulationen zum laminar-turbulenten Umschlagsprozess der ebenen Poiseuille-Strömung,* Dissertation, Universität Karlsruhe. See also: *Spectral simulations of laminar-turbulent transition in plane Poiseuille flow and comparison with experiments,* Springer Lecture Notes in Physics 170, 280-287 (1982).

[15] Pierrehumbert, R. T. & Widnall, S. E. 1982 *The two- and three-dimensional instabilities of a spatially periodic shear layer,* J. Fluid Mech. 114, 59-82.

[16] Herbert, Th. 1981 *Stability of plane Poiseuille flow - theory and experiment,* VPI-E-81-35, Blacksburg, VA. Published in: Fluid Dyn. Trans. 11, 77-126 (1983).

[17] Herbert, Th. 1983 *Secondary instability of plane channel flow to subharmonic three-dimensional disturbances,* Phys. Fluids. 26, 871-874.

[18] Herbert, Th. 1983 *Subharmonic three-dimensional disturbances in unstable shear flows,* AIAA Paper No. 83-1759.

[19] Herbert, Th. 1984 *Analysis of the subharmonic route to transition in boundary layers,* AIAA Paper No. 84-0009.

[20] Herbert, Th. 1984 *Modes of secondary instability in plane Poiseuille flow,* in: *Turbulence and Chaotic Phenomena in Fluids* (ed. T. Tatsumi), 53-58, North-Holland.

[21] Herbert, Th. 1985 *Three-dimensional phenomena in the transitional flat-plate boundary layer,* AIAA Paper No. 85-0489.

[22] Herbert, Th. 1985 *Secondary instability of plane shear flows - theory and applications,* in: *Laminar-Turbulent Transition* (ed. V. V. Kozlov), 9-20, Springer-Verlag.

[23] Herbert, Th. & Bertolotti, F. P. 1985 *Effect of pressure gradients on the growth of subharmonic disturbances in boundary layers,* Proc. Conference on Low Reynolds Number Airfoil Aerodynamics, (ed. T. J. Mueller), 65-77.

[24] Herbert, Th., Bertolotti, F. P. & Santos, G. R. 1985 *Floquet analysis of secondary instability in shear flows,* ICASE/NASA Workshop on Stability of Time-Dependent and Spatially Varying Flows, to be published by Springer-Verlag.

[25] Herbert, Th. 1977 *Finite amplitude stability of plane parallel flow,* AGARD CP 244, 3/1.

[26] Gaster, M. 1962 *A note on the relation between temporally-increasing and spatially-increasing disturbances in hydrodynamic stability,* J. Fluid Mech. 14, 222-224.

[27] Nishioka, M. & Asai, M. 1985 *Three-dimensional wave-disturbances in plane Poiseuille flow,* in: Laminar-Turbulent Transition (ed. V. V. Kozlov), 173-182, Springer-Verlag.

[28] Croswell, J. W. 1985 *On the energetics of primary and secondary instabilities in plane Poiseuille flow,* M.S. Thesis, VPI & SU.

[29] Orszag, S. A. & Patera, A. T. 1983 *Secondary instability of wall-bounded shear flows,* J. Fluid Mech. 128, 347-385.

[30] Stuart, J. T. 1965 *The production of intense shear layers by vortex stretching and convection,* AGARD Report 514.

[31] Wortmann, F. X. 1981 *Boundary-layer waves and transition,* in: *Advances in Fluid Mechanics* (ed. E. Krause), Lecture Notes in Physics 148, 268-279, Springer-Verlag.

[32] May, C. L. & Kleiser, L. 1985 *Numerical simulation of subharmonic transition in plane Poiseuille flow,* Bull. Amer. Phys. Soc. 30, 1748.

SUBGRID SCALE MODELS FOR HOMOGENEOUS ANISOTROPIC TURBULENCE

B. AUPOIX

ONERA/CERT
Department of Aerothermodynamics
2 avenue Edouard Belin - 31055 TOULOUSE Cedex (FRANCE)

SUMMARY

A statistical model is used to study the subgrid scale modelling problem for homogeneous turbulence. Pressure and transfer effects of the subgrid scales are modelled for wavenumbers small relative to the filter cut. The model is then extended to the complete wavenumber range from very large scales to the filter cut. As these pressure and transfer effects are defined as integrals over the subgrid scales, representation of subgrid energy and anisotropy spectra are studied. Lastly, a subgrid scale model has to be extracted from this representation of the subgrid pressure and transfer terms.

1 - INTRODUCTION

Direct resolutions of the NAVIER equations are restricted to small Reynolds number flows due to limitation of computer speed and storage. As proposed by LEONARD [1973], the NAVIER equations can be filtered to resolve only the energy-containing eddies while the smaller eddies, in inertial and dissipative ranges, must be modelled. The usual assumption is that these small eddies, or subgrid scales, are governed by the energy cascade process ; they are then less flow-dependent and have a more universal character. Standard models try to express these small scales in terms of the large eddies. We have previously shown that, for isotropic turbulence, at least some statistical information must be retained about the subgrid scales to correctly model the effects of the subgrid scales on the large scales (AUPOIX and COUSTEIX [1982], AUPOIX [1984]).

This paper is devoted to subgrid scale modelling in homogeneous, anisotropic turbulence. We will show how the basic concepts previously developed for subgrid scale modelling of isotropic turbulence extend to the anisotropic case.

2 - FILTERED NAVIER EQUATIONS

FOURIER transforms are convenient to write the NAVIER

equations for homogeneous turbulence :

$$k_i \hat{u}_i(\underline{k}) = 0$$

$$\left[\frac{\partial}{\partial t} + \nu k^2\right] \hat{u}_i(\underline{k}) + \frac{\partial U_i}{\partial x_l} \hat{u}_l(\underline{k}) - 2 \frac{\partial U_m}{\partial x_l} \frac{k_i k_m}{k^2} \hat{u}_l(\underline{k})$$

$$- \frac{\partial U_l}{\partial x_m} k_l \frac{\partial \hat{u}_i(\underline{k})}{\partial k_m} = - i k_l \left[\delta_{im} - \frac{k_i k_m}{k^2}\right]$$

$$\times \iint \delta(\underline{k} - \underline{p} - \underline{q}) \hat{u}_m(\underline{p}) \hat{u}_l(\underline{q}) d\underline{p}^3 d\underline{q}^3$$

where $\partial U_l/\partial x_m$ are the mean velocity gradients which are imposed, $\hat{u}(\underline{k})$ the FOURIER transform of the velocity field, $\delta(\underline{k})$ the DIRAC function ($\delta(\underline{k}) = 0$ when $||k|| \neq 0$) and δ_{ij} the KRONECKER tensor.

As the pressure has been eliminated from the momentum equation with the help of the POISSON equation, continuity is automatically ensured by solutions of the momentum equation. The LHS of the momentum equation includes time derivative, viscous effects and linear effects due to the action of the mean velocity gradients at wavevector $\underline{k}$. The RHS corresponds to non-linear effects due to interactions with all wavevectors $\underline{p}$ and $\underline{q}$ such as $\underline{k} = \underline{p} + \underline{q}$.

Following LEONARD's ideas, the NAVIER equations are filtered to only resolve for the most energetic wavenumbers. We have chosen to use low-pass filter to capture the maximum of energy in the simulations. For wavevectors $\underline{k}$ inside the filter, the RHS can be written as the sum of two terms : a term $T_i^R(\underline{k})$ called resolvable term which corresponds to interactions with wavevectors $\underline{p}$ and $\underline{q}$ both inside the filter and can be computed, and a term $T_i^S(\underline{k})$ called subgrid scale term which corresponds to interactions with wavevectors outside the filter and must be modelled.

The role of the subgrid scale term, i.e. the drain of energy to the small scales and the random injection of noise from the small scales, has been brought into evidence in previous papers (AUPOIX-COUSTEIX [1982], AUPOIX [1984]).

Various techniques can be used to model these subgrid terms. Dimensional analysis and arguments based upon KOLMOGOROV energy cascade were first used to propose subgrid scale models. Statistical models and renormalization group theory have been widely used to obtain information about the subgrid terms. Lastly, direct resolution of the NAVIER equations can be used to obtain "experimental" values of the subgrid terms and to check the models.

As for isotropic turbulence, we shall use statistical models -also called two-point closures as they are dealing with FOURIER transforms of the two-point correlation tensor- to derive subgrid scale models. Statistical models are convenient to study the role of the subgrid terms as they give a fine description of non-linear interactions. The subgrid contribution to the non-linear interactions can easily be obtained and modelled and gives statistical information about the subgrid term. Such an approach has already been fruitful for isotropic turbulence. To extend this approach to anisotropic turbulence, a statistical model for homogeneous anisotropic turbulence is needed.

3 - CAMBON'S EDQNM MODEL

In isotropic turbulence, one has only to deal with the energy spectrum E(k) which is a function of the wavenumber. In anisotropic turbulence, the amount of information to be considered is much more important ; the unknowns are now the components of the tensor :

$$\phi_{ij}(\underline{k}) = \langle \hat{u}_i(\underline{k})\hat{u}_j(-\underline{k})\rangle = \langle \hat{u}_i(\underline{k})\hat{u}_j^*(\underline{k})\rangle$$

where * denotes the complex conjugate and $\langle . \rangle$ a statistical average. This tensor is now a function of the wavevector and can be interpreted either as the FOURIER transform of the two-point correlation $\langle u_i(\underline{x}) u_j(\underline{x} + \underline{r})\rangle$ or as the spectrum of the Reynolds stresses since :

$$\langle u_i u_j \rangle = \int \phi_{ij}(\underline{k}) \, d^3\underline{k} .$$

The evolution equation for this tensor can be deduced from the NAVIER equation for $\hat{u}_i$:

$$\left[\frac{\partial}{\partial t} + 2\nu k^2 \right] \phi_{ij}(\underline{k}) + \frac{\partial u_i}{\partial x_l} \phi_{jl}(\underline{k}) + \frac{\partial u_j}{\partial x_l} \phi_{il}(\underline{k})$$

$$- 2 \frac{\partial u_l}{\partial x_m} (k_i \phi_{mj}(\underline{k}) + k_j \phi_{mi}(\underline{k})) - \frac{\partial u_l}{\partial x_m} \frac{\partial}{\partial k_m} (k_l \phi_{ij}(\underline{k})) =$$

$$\left[\delta_{li} - \frac{k_i k_l}{k^2} \right] T_{lj}(\underline{k}) + \left[\delta_{lj} - \frac{k_j k_l}{k^2} \right] T_{li}^*(\underline{k})$$

with $T_{ij}(\underline{k}) = k_l \int \langle \hat{u}_i(\underline{k}) \hat{u}_l(\underline{p}) \hat{u}_j(-\underline{k}-\underline{p}) \rangle \, d^3\underline{p}$.

The LHS corresponds to time derivative, viscous dissipation, production and linear effects by action of the mean velocity gradients while the RHS corresponds to non-linear effects due to interactions between wavevectors and must be modelled. BERTOGLIO [1981] proposed a way to solve this equation in the framework of the EDQNM theory. We shall focus our attention on a simplified model proposed by CAMBON [1981] which is easier to handle and better suited for analytical treatments.

CAMBON's basic simplification is to use spherical averages. He defines the spherically averaged spectrum tensor as the integral over the sphere of radius k :

$$\varphi_{ij}(k) = \int \phi_{ij}(\underline{k}) \, dA(\underline{k})$$

the evolution equation of which reads :

$$\left[\frac{\partial}{\partial t} + 2\nu k^2\right] \varphi_{ij}(k) = -\frac{\partial U_i}{\partial x_l}\varphi_{jl}(k) - \frac{\partial U_j}{\partial x_l}\varphi_{il}(k)$$

$$+ P^L_{ij}(k) + S^L_{ij}(k) + P^{NL}_{ij}(k) + S^{NL}_{ij}(k)$$

with

$$P^L_{ij}(k) = 2\frac{\partial U_l}{\partial x_m}\int \frac{k_l}{k^2}[k_i\,\phi_{mj}(\underline{k}) + k_j\,\phi_{mi}(\underline{k})]\,dA(\underline{k})$$

$$S^L_{ij}(k) = \frac{\partial U_l}{\partial x_m}\int \frac{\partial}{\partial k_m}[k_l\,\phi_{ij}(\underline{k})]\,dA(\underline{k})$$

$$P^{NL}_{ij}(k) = -\int \frac{k_l}{k^2}[k_i\,T_{lj}(\underline{k}) + k_j\,T^*_{li}(\underline{k})]\,dA(\underline{k})$$

$$S^{NL}_{ij}(k) = \int [T_{ij}(\underline{k}) + T^*_{ji}(\underline{k})]\,dA(\underline{k}).$$

The symbols P and S refer to pressure and transfer terms. Pressure terms are trace-free and redistribute the energy at wavevector $\underline{k}$ while the transfer terms correspond to exchange of energy with other wavevectors. The superscript L refers to linear terms which represent the action of the mean velocity gradient on the turbulence at wavevector $\underline{k}$ while the superscript NL refers to non-linear terms due to the action of the turbulence upon itself, i.e. between wavevectors.

With this averaged model, the information about the distribution of the Reynolds stress spectra on the spheres of radius k is lost and has to be restored. CAMBON proposed to model it as :

$$\phi_{ij}(\underline{k}) = F_{ij}[\varphi_{lm}(k), k_j/k, a(k)]$$

which corresponds to a decomposition of $\phi_{ij}(\underline{k})$ in terms of spherical harmonics. The coefficient $a(k)$ is called the structural parameter and controls the Reynolds stress distribution over the sphere of radius k. With this representation, the linear terms which needed no modelling in the ϕ_{ij} equation can be closed. The EDQNM theory gives the expression of the non-linear terms as function of the ϕ_{ij} and then of the φ_{ij}.

The evolution equation for the spherically averaged spectrum

$$\left[\frac{\partial}{\partial t} + 2\nu k^2\right] \varphi_{ij}(k) = -\frac{\partial U_i}{\partial x_l}\varphi_{jl}(k) - \frac{\partial U_j}{\partial x_l}\varphi_{il}(k)$$

$$+ P^L_{ij}(k) + S^L_{ij}(k) + P^{NL}_{ij}(k) + S^{NL}_{ij}(k)$$

is now closed as :

$$P^L_{ij} = 2E\left[\frac{2}{5}d_{ij} - 3D\left[d_{lj}H_{li} + d_{li}H_{lj} - \frac{2}{3}\delta_{ij}d_{lm}H_{lm}\right]\right.$$
$$\left. + \frac{14}{3}\left[D + \frac{4}{7}\right](\omega_{il}H_{lj} + \omega_{jl}H_{li})\right]$$

$$S^L_{ij} = -\frac{2}{15}d_{ij}\frac{\partial}{\partial k}(kE) + 2d_{il}\frac{\partial}{\partial k}[kDEH_{jl}]$$
$$+ 2d_{jl}\frac{\partial}{\partial k}[kDEH_{il}] - \frac{\delta_{ij}}{3}d_{lm}\frac{\partial}{\partial k}[(2+11D)kEH_{lm}]$$

$$P^{NL}_{ij} = \iint \theta_{kpq}\frac{2}{pq}(x+yz)H_{ij}(q)\left[k^2 p\, E(p)E(q)\right.$$
$$[y(z^2-y^2)(a(q)+3) + (y+xz)\frac{a(q)}{5}]$$
$$\left. - p^3 E(k)E(q)y(z^2-x^2)(a(q)+3)\right]dp\,dq$$

$$S^{NL}_{ij} = \iint \theta_{kpq}\frac{2}{pq}\left[(xy+z^3)\left[k^2 p E(p)E(q)\left[\frac{\delta_{ij}}{3} + H_{ij}(p) + H_{ij}(q)\right]\right.\right.$$
$$\left. - p^3 E(k)E(q)\left[\frac{\delta_{ij}}{3} + H_{ij}(k) + H_{ij}(q)\right]\right]$$
$$\left. + H_{ij}(q)(k^2 p\, E(p)E(q) C_{kpq} - p^3 E(k)E(q) C_{pkq})\right]dp\,dq$$

where

$$d_{ij} = \frac{1}{2}\left[\frac{\partial u_i}{\partial x_j} + \frac{\partial u_j}{\partial x_i}\right]$$

$$\omega_{ij} = \frac{1}{2}\left[\frac{\partial u_i}{\partial x_j} - \frac{\partial u_j}{\partial x_i}\right]$$

$$D = \frac{2}{7}(1 + \frac{4}{5}a)$$

$$E = \frac{1}{2}\varphi_{11} \text{ is the energy spectrum}$$

$$H_{ij} = \frac{\varphi_{ij}}{2E} - \frac{\delta_{ij}}{3} \text{ is the anisotropy spectrum}$$

$$C_{kpq} = \frac{1}{2}(xy+z)\left[(y^2-z^2)(a(q)+3) + \frac{2}{5}a(q)(1+z^2)\right].$$

The EDQNM integrals extend over all wavenumbers k, p, q

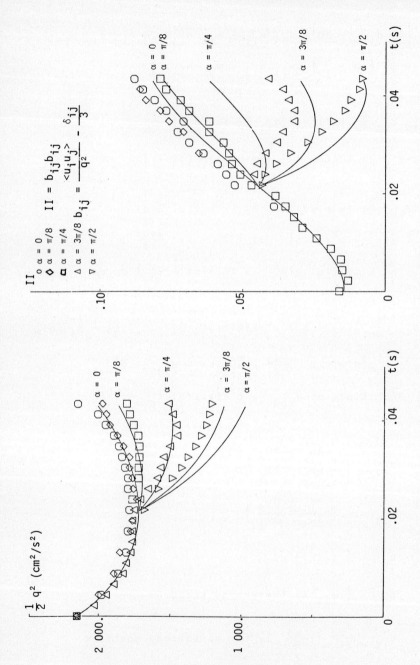

Fig. 1 - GENCE's experiment - Comparison between prediction and experiment with the CAMBON's anisotropic EDQNM model

such as the wavevectors $\underline{k}$, $\underline{p}$, $\underline{q}$ form a triangle. x, y and z are the cosine of the interior angles respectively opposite to the wavevectors $\underline{k}$, $\underline{p}$ and $\underline{q}$.

Two adjustable parameters are on the one hand the structural parameter a(k) which represents the repartition of each Reynolds stress on the sphere of radius k and is obtained from rapid distortion theory and, on the other hand, the eddy damping θ_{kpq} which is modelled according to ANDRE and LESIEUR [1977] and CAMBON et al [1981] :

$$\theta_{kpq} \approx \frac{1}{\mu(k) + \mu(p) + \mu(q)}$$

$$\mu(k) = \nu k^2 + \lambda \left[\int_0^k p^2 E(p) \, dp + \omega_{ij} \omega_{ij} \right]^{1/2}$$

to account for mean flow rotation effects (CAMBON et al [1981], AUPOIX et al [1983] and papers referenced in). The constant λ is evaluated from the KOLMOGOROV constant ($\lambda \approx 0.36$).

The adjustable coefficients are deduced from theory and no adjustment with reference to experiments is needed.

This spherically averaged EDQNM model has proved to give good predictions of strained or weakly sheared turbulence (CAMBON et al [1981]). On figure 1, prediction of the time evolution of kinetic energy and of the second invariant of the anisotropy tensor are plotted in the case of the GENCE and MATHIEU [1979] experiment of grid turbulence submitted to two successive plane strains of different axis of relative angle α. The prediction agrees with experiment only when the two strains have identical axis ($\alpha=0$). Similar predictions have been obtained with various models such as one-point closures or large eddy simulations. This seems to be due to differences between the theoretical and the experimental strains.

4 - APPLICATION OF CAMBON'S MODEL TO SUBGRID MODELLING

In this statistical model, the non-linear pressure and transfer terms are statistical estimates of the non-linear interactions between wavenumbers. The attractive feature of statistical models such as the ones derived from the EDQNM theory or related theories (DIA, TFM ...) is that these models express the non-linear term in a detailed manner, i.e. they do not only give some global expression for the non-linear terms at, let say, wavenumber k but they express these non-linear terms as the integrals of all non-linear interactions with all the pairs of wavenumbers p and q such as $\underline{k} = \underline{p} + \underline{q}$. With this detailed expression for the non-linear terms, it is then possible to separate them between large scale and subgrid scale contributions. In the framework of CAMBON's spherically averaged model, only isotropic filters can be used. We shall restrict ourself to an isotropic low-pass filter at wavenumber k_c.

We shall then focus our attention to the non-linear terms at wavenumber $k \leq k_c$ involving at least one wavenumber p or q greater than the filter cut. Such a term represents a large scale/subgrid scale interaction and corresponds to the subgrid term T^S in the filtered NAVIER equation.

The approach will be similar to the one used for isotropic turbulence. The subgrid contribution to the non-linear pressure and transfer terms at wavenumber $k \leq k_c$ is a priori an intractable integral. Therefore, some simple approximation for these non-linear subgrid terms will be looked for. In fact, we are really interested not in the subgrid pressure and transfer terms in a statistical approach but in the subgrid term T^S in the filtered NAVIER equation. As the subgrid pressure and transfer terms are statistical averages involving the subgrid term T^S, the statistical approach is just a way to get information about the subgrid scale term T^S. The final step is to derive some expression or condition for the subgrid term T^S from the subgrid pressure and transfer terms.

4.1. Non local modelling of the subgrid terms

Non-linear pressure and transfer terms are expressed in CAMBON's EDQNM model as complicated integrals over all wavenumbers such as corresponding wavevectors can form a triangle. When an isotropic filter is applied, the subgrid contributions to the non-linear pressure and transfer terms for the large scales are still complicated integrals. However, following ideas introduced by KRAICHNAN [1976] and developed by LESIEUR and SCHERTZER [1979], these integrals can be reduced to simpler formulations for wavenumbers very small relative to the filter cut k_c. Triadic interaction rule $\underline{k} - \underline{p} - \underline{q} = 0$ then imposes $k << k_c \approx p, q$, i.e. only non local interactions are considered. The non local development technique introduced by KRAICHNAN then gives the expression for the subgrid pressure and transfer terms :

$$P_{ij}^{NL}(k) = - \frac{32}{175} k^4 \int_{k_c}^{\infty} \theta_{0pp} [10 + a(p)] \frac{E^2(p) H_{ij}(p)}{p^2} dp$$

$$+ \frac{16}{105} k^2 E(k) \int_{k_c}^{\infty} \theta_{0pp} \Big[[a(p)+3] p \frac{\partial}{\partial p} [E(p) H_{ij}(p)]$$

$$+ E(p) H_{ij}(p) \Big[5(a(p)+3) + p \frac{\partial a(p)}{\partial p} \Big] \Big] dp$$

$$S_{ij}^{NL}(k) = + 2k^4 \int_{k_c}^{\infty} \theta_{0pp} \frac{E^2(p)}{p^2} \left[\frac{14}{15} \left(\frac{\delta_{ij}}{3} + 2 H_{ij}(p) \right) \right.$$

$$\left. + \frac{8}{25} a(p) H_{ij}(p) \right] dp$$

$$- 2k^2 \varphi_{ij}(k) \frac{1}{15} \int_{k_c}^{\infty} \theta_{0pp} \left[5 E(p) + p \frac{\partial E(p)}{\partial p} \right] dp$$

$$- 2k^2 E(k) \int_{k_c}^{\infty} \theta_{0pp} \left[\frac{2}{15} \left[5 E(p) H_{ij}(p) + p \frac{\partial}{\partial p} (E(p)H_{ij}(p)) \right] \right.$$

$$\left. + E(p) H_{ij}(p) \left[\frac{8}{15} (a(p) + 3) + \frac{8}{25} a(p) \right] \right] dp .$$

It must be noticed that these subgrid pressure and transfer terms are obtained in terms of integrals over the subgrid scales.

4.2. Isotropic turbulence

If the subgrid scales are assumed isotropic, these terms reduce to :

$$P_{ij}^{NL}(k) = 0$$

$$S_{ij}^{NL}(k) = \frac{28 \delta_{ij}}{45} k^4 \int_{k_c}^{\infty} \theta_{0pp} \frac{E^2(p)}{p^2} dp$$

$$- 2 v_e k^2 \varphi_{ij}(k)$$

$$v_e = \frac{1}{15} \int_{k_c}^{\infty} \theta_{0pp} \left[5 E(p) + p \frac{\partial E}{\partial p} \right] dp .$$

The pressure term is zero as no redistribution of energy at a given wavevector occurs.
The transfer term is decomposed into two terms : a k^4 beating term which inputs energy from the subgrid scales into the large scales and a drain term of eddy viscosity form. It has been previously shown that the k^4-beating term is negligible in most practical cases. The drain term corresponds to an effective viscosity for wavenumbers very small relative to the cut as scale separation occurs. Moreover, this effective viscosity gives a good approximation for the energy transfer between large and subgrid scales when the filter cut is located at the beginning of the inertial range (AUPOIX [1984]).

Readers interested in more details about the models developed for isotropic turbulence could refer to AUPOIX [1984] and references in it.

4.3. Anisotropic turbulence

When the subgrid scales are anisotropic, the subgrid pressure and transfer terms reflect the anisotropy of the subgrid scales. The complete form given previously is too complex to easily derive information on the role of these terms. However, the structural parameter $a(k)$ rapidly reaches an asymptotic value - 4.5 in the inertial range. If this value is assumed to be valid at the filter cut, the subgrid terms read :

$$P_{ij}^{NL}(k) = - \frac{176}{175} k^4 \int_{k_c}^{\infty} \theta_{0pp} \frac{E^2(p) H_{ij}(p)}{p^2} dp$$

$$- \frac{8}{35} k^2 E \int_{k_c}^{\infty} \theta_{0pp} \left[5 E(p) H_{ij}(p) + p \frac{\partial}{\partial p} [E(p) H_{ij}(p)] \right] dp$$

$$S_{ij}^{NL}(k) = \frac{4}{15} k \int_{k_c}^{\infty} \theta_{0pp} \frac{E(p)}{p^2} \left[7 \frac{\delta_{ij}}{3} - \frac{16}{5} H_{ij}(P) \right] dp$$

$$- 2 \nu_e k^2 \varphi_{ij}(k)$$

$$- 2 k^2 E(k) \int_{k_c}^{\infty} \theta_{0pp} \left[- \frac{118}{75} E(p) H_{ij}(p) + \frac{2}{15} p \frac{\partial}{\partial p} [E(p) H_{ij}(p)] \right] dp$$

or, globally :

$$P_{ij}^{NL}(k) + S_{ij}^{NL}(k) = k^4 \int_{k_c}^{\infty} \theta_{0pp} \frac{E^2(p)}{p^2} \left[\frac{28}{45} \delta_{ij} - \frac{368}{175} H_{ij}(p) \right] dp$$

$$- 2 \nu_e k^2 \varphi_{ij}(k) + k^2 E(k) \int_{k_c}^{\infty} \theta_{0pp} \left[\frac{1052}{525} E(p) H_{ij}(p) - \frac{52}{105} \frac{\partial}{\partial p} [E(p) H_{ij}(p)] \right] dp .$$

The shape of the anisotropy spectrum $H_{ij}(k)$ is a priori unknown and reflects the time history of the turbulence. The anisotropy spectrum can then exhibit sign reversal in complicated cases. If we restrict ourself to simple cases such as initial isotropic turbulence submitted to only one strain or one shear, the anisotropy spectrum for each Reynolds stress has the same sign over the large scales and the inertial range but exhibits sign reversal in the dissipative range. However the main contribution to the above integrals comes from the inertial range where the anisotropy spectrum has the same sign as in the large scales. Then, the above integrals show that the pressure terms are draining the anisotropy of the large scales and so play a classical return to isotropy role. It must be stressed again that these pressure terms are trace-free and redistribute the energy at each wavenumber.

For the transfer term, the k^4 beating term injects some energy in an isotropic way and also redistributes energy in a return to isotropy way. The effective viscosity term drains all the components of the Reynolds stress spectra in an isotropic viscosity way. But the most surprising contribution comes from the last $k^2 E(k)$ term. As the energy spectrum decreases, $\partial[E(p) H_{ij}(p)]/\partial p$ can be assumed negative and the integral has the sign of the anisotropy. This corresponds to an input of anisotropy from the subgrid scales into the large scales ; the subgrid scales "talk back" to the large scales to impose on them their anisotropy. As the subgrid scale anisotropy is imposed by the large scales, there is a self-enforcement process for the large scales anisotropy.

The global balance between pressure and transfer terms shows that the k^4 terms still play a return to isotropy role while the $k^2 E(k)$ terms globally increase the large scale anisotropy.

4.4. Extension of the non local model

These non local formulations have been derived for wavenumbers very small relative to the filter cut and are theoretically valid only for these wavenumbers. For isotropic turbulence, the effective viscosity obtained by non local hypothesis has shown to be a good approximation for the subgrid transfer term for all wavenumbers from the very large scales to the filter cut when this filter cut lies in the beginning of the inertial range. It thus seems interesting to try to use the non local expression for the subgrid pressure and transfer terms all over the large scales. Tests have been performed on various strained and sheared experiments, for different times and various locations of the filter cut. Some results are given on the figures 2.

Figure 2a shows the energy and Reynolds stress spectra at the initial station of the GENCE experiment. As these spectra have not been measured, they have been generated from Reynolds stress and dissipation rate data with ad hoc formulae given by CAMBON et al [1981]. The location of the filter cut at $k_c = 4$ cm^{-1} is also plotted. As φ_{22} and φ_{33} are of the same order of magnitude, only the subgrid terms for φ_{11} and φ_{22} have been plotted on figures 2c to 2f. To limit their range of variations, these subgrid terms are given in viscosity form as divided by $k^2 \varphi$, versus the reduced wavenumber k/k_c. The continuous line is the "exact" subgrid term, i.e. the term obtained with the complete EDQNM

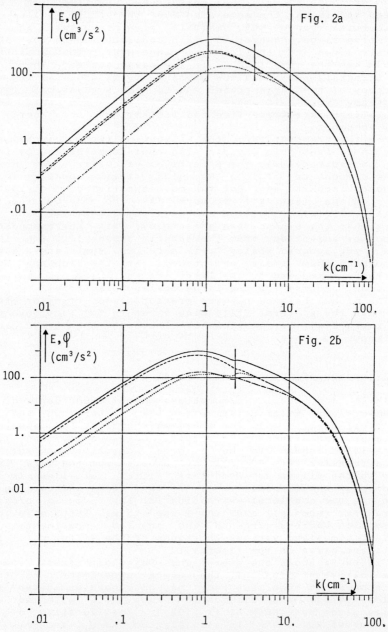

Fig. 2 - GENCE's experiment - Initial evaluated (2a) and final computed (2b) spectra and location of the cut

—— E ... φ_{11} —·— φ_{22} ----- φ_{23}

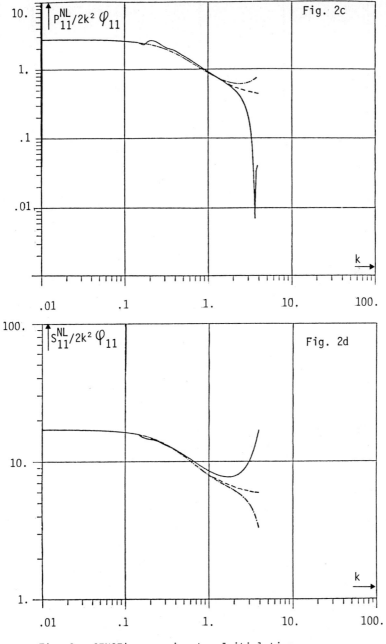

Fig. 2 - GENCE's experiment - Initial time
Pressure (2c) and transfer (2d) subgrid terms for φ_{11}
——— "exact" subgrid term   —·—·— modelled subgrid term
---- modelled subgrid term with k^4 terms removed

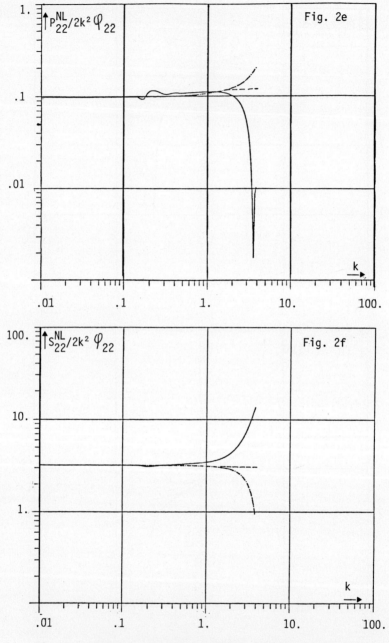

Fig. 2 - GENCE's experiment - Initial time
Pressure (2e) and transfer (2f) subgrid terms for φ_{22}
——— "exact" subgrid term   —·—·— modelled subgrid term
———— modelled subgrid term with k^4 terms removed

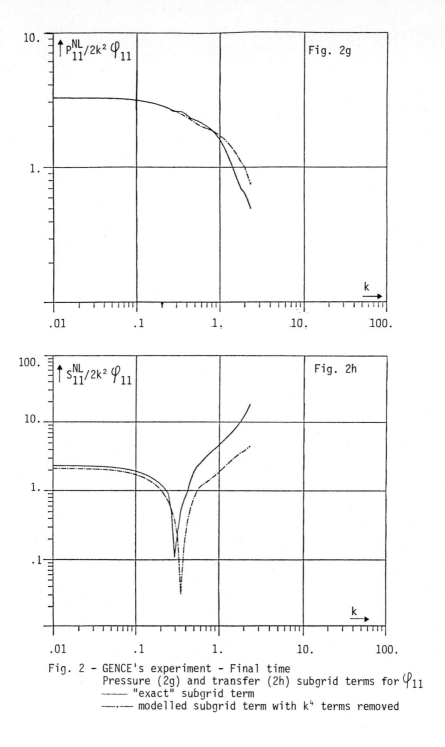

Fig. 2 - GENCE's experiment - Final time
Pressure (2g) and transfer (2h) subgrid terms for φ_{11}
—— "exact" subgrid term
—·— modelled subgrid term with k^4 terms removed

integrals. The oscillations at about k ≈ 0.2 are due to numerical problems in the treatment of these integrals. The dashed lines are the evaluations of the subgrid terms obtained with the non local formulae. For small wavenumbers, the two curves collapse as expected but diverge in the vicinity of the filter cut. This divergence seems natural as no local interactions are correctly accounted for in the formulae. However the main source of divergence is the k^4 terms which were very small in the large eddies and are increasing in the vicinity of the cut. As these terms play a return to isotropy role, they are giving the wrong tendency and increase the discrepancy. When the k^4 terms are removed, the dotted line is obtained. The agreement is enforced, the "exact" and the model curves only diverge in the last octave where local effects are predominant. In this case, with k^4 terms removed, the non local formulae give a good approximation for the subgrid pressure and transfer terms.

Figure 2b gives the computed energy and Reynolds stresses spectra at the end of the first strain in the GENCE's experiment. The filter cut is now at k_c = 2.5 cm^{-1}. Figures 2g and 2h show the subgrid terms for φ_{11}. As absolute values are plotted, the kink on the transfer curve corresponds to a sign reversal. Due to the complicated anisotropy spectra, there is a backscatter on the φ_{11} component only which is grossly reproduced by the non local model.

The study over a large set of test cases for strained or weakly sheared turbulence has shown that the non local formulae, with k^4 terms removed, give good approximations for the subgrid pressure and transfer terms all over the large scales when the filter cut lies in the beginning of the inertial range.

From these figures, we can also see that subgrid pressure and transfer terms can have different relative orders of magnitude. On figures 2c and 2d or 2e and 2f, the pressure term is an order of magnitude smaller than the transfer term. On figures 2g and 2h, the pressure term is greater than the transfer term. Figure 2h shows a case where the removal of the k^4 terms introduced some error in the very large eddies, however this error always remains small.

Lastly, the subgrid scale terms are highly anisotropic and eddy viscosity models do not hold as pointed out by BERTOGLIO and MATHIEU [1983]. If the eddy viscosity term is the main term in the subgrid transfer, then :

$$\frac{S_{11}^{NL}(k)}{k^2 \varphi_{11}(k)} \sim 2\nu_e \sim \frac{S_{22}^{NL}(k)}{k^2 \varphi_{22}(k)}.$$

Figures 2d and 2f show that this is not so as the ratio between these two terms is about five.

4.5. Application of the non local formulae to subgrid modelling

The non local evaluations of the subgrid pressure and transfer terms give good estimates all over the wavenumber range. However, these terms are integrals over the subgrid scales of terms including energy and anisotropy spectra. In a large eddy simulation, only the large scales are known. To be able to compute the subgrid pressure and transfer terms and, then, the subgrid term T^s in the filtered NAVIER equation, some statistical information about the subgrid scales must be known.

A first approach to obtain the required statistical quantities for the subgrid scales is to couple the evolution of one realisation of the large scales with an evolution of the small scales in a statistical sense. This technique has been introduced by CHOLLET [1983] and AUPOIX et al [1983] for isotropic turbulence and extended to anisotropic turbulence by BERTOGLIO [1984]. This leads to intricate and lengthy computation. Moreover, extension to inhomogeneous flow presently seems out of reach.

Another strategy, developed for isotropic turbulence, is to assume a priori shapes for the various subgrid scale spectra and to couple the large scales and the subgrid scales evolution in some statistical way to adjust the levels of the subgrid scale spectra.

4.5.a. Subgrid scale representation

A first set of simplifications is to assume, as mentioned before, the structural parameter a(k) to be equal to its asymptotic value - 4.5. Moreover, the eddy damping time can also be assumed equal to its inertial value :

$$\mu(k) = \lambda \ [(3/4) \ k^3 \ E(k) + \omega_{ij}\omega_{ij} \ ]^{1/2}$$

so that assumptions are only needed for the energy and anisotropy spectra.

Various forms have been proposed to model the small scales energy spectrum. Their common characteristic is that they express this spectrum in terms of only one adjustable parameter ε which represents the energy flux transferred along the inertial range and dissipated by the smaller eddies. The simplest model is the celebrated KOLMOGOROV law :

$$E(k) = K_o \ \varepsilon^{2/3} \ k^{-5/3} \qquad K_o \sim 1.4 \ .$$

To be able to simulate low Reynolds effects, some viscous cut-off has to be introduced in the energy spectrum. A simple model is the one proposed by PAO :

$$E(k) = K_o \ \varepsilon^{2/3} \ k^{-5/3} \ \exp \left[\ - \frac{3K_o}{2} \left[\frac{k}{k_D} \right]^{4/3} \right]$$

$$k_D = \left[\frac{\varepsilon}{\nu^3} \right]^{1/4} .$$

We have proposed the intricate formula :

$$E(k) = K_o \varepsilon^{2/3} k^{-5/3} \exp[-3.5\eta^2 (1 - \exp[6\eta + 1.2 - [196\eta^2 - 33.6\eta + 1.4532]^{1/2}])]$$

$$\eta = k/k_D$$

which is an approximation for the solution of the EDQNM equations at high Reynolds number in the inertial and dissipative range. Such a formula exhibits the bump of the compensated spectrum ($E(k)/K_o \varepsilon^{2/3} k^{-5/3}$) at the transition between inertial and dissipative ranges pointed out by MESTAYER et al [1984] and is more suited than PAO's spectrum for EDQNM computations.

As concerns the anisotropy, experiments performed by MESTAYER [1982] have shown that anisotropy remains in the inertial range and local isotropy is only found in the dissipative range. EDQNM computations have been performed to try to define some universal shape for the anisotropy spectrum over the subgrid scales.

These computations showed no universal equilibrium solution and very complex spectra with sign reversals in the dissipative range. We turned back to rapid distortion theory and empiricism to develop a simple model able to mimic the subgrid scale anisotropy.

For isotropic turbulence submitted to mean velocity gradient on a time interval short enough to neglect non-linear and viscosity effects, CAMBON's model gives :

$$H_{ij}(k) = b_{ij} \frac{1}{4} \left[5 + \frac{k}{E} \frac{\partial E}{\partial k} \right] \quad \text{where} \quad b_{ij} = \frac{\langle u_i u_j \rangle}{q^2} - \frac{\delta_{ij}}{3}.$$

This expression for the anisotropy spectrum is of course the solution also obtained with the help of rapid distortion theory for short time intervals.

The anisotropy spectrum is expressed in terms of the Reynolds stress anisotropy and of the slope of the energy spectrum in log-log coordinates. For a KOLMOGOROV spectrum, this model reduces to :

$$H_{ij}(k) = \frac{5}{6} b_{ij}.$$

For PAO's or EDQNM spectra, the rapid distortion solution exhibits a sign reversal in the part of dissipative range where the energy spectrum is steeper than k^{-5}. But this simple model a priori suffers from two severe lacks to correctly represent subgrid anisotropy.

First, for sheared turbulence, computations performed with CAMBON's model showed that off-diagonal stresses exhibit a $k^{-7/3}$ spectrum in the inertial range, as expected. A first a priori modification was introduced to account for this phenomenon, as :

$$H_{ij}(k) \sim b_{ij} \left[5 + \frac{k}{E} \frac{\partial E}{\partial k} \right]$$

$$* \left[1 + H \left[\frac{k}{k_{max}} - 1 \right] H \left[|\frac{\partial u_i}{\partial x_j} + \frac{\partial u_j}{\partial x_i}| \right] \left[\left[\frac{k}{k_{max}} \right]^{-2/3} - 1 \right] \right]$$

where k_{max} is the wavenumber corresponding to the maximum of the energy spectrum and H is the HEAVISIDE function defined as :

$$H(x) = 0 \quad \text{when} \quad x \leq 0$$
$$H(x) = 1 \quad \text{when} \quad x > 0.$$

The proportionality constant is then adjusted to still satisfy the relation :

$$\langle u_i u_j \rangle = \int_0^\infty 2E(k) \left[H_{ij}(k) + \frac{\delta_{ij}}{3} \right] dk .$$

Secondly, the model is obtained in a rapid distortion approach which neglects all non-linear effects and emphasizes linear effects on the small scales. Non-linear effects tend to reduce small scale anisotropy and dissipation anisotropy. With the basic model, the dissipation anisotropy is equal to half the Reynolds stress anisotropy. Some viscous damping has to be introduced in the model to reduce the dissipation anisotropy. This was done by adding an a priori damping of the PAO form :

$$\exp \left[- \beta \left[\frac{k}{k_D} \right]^{4/3} \right]$$

where β was adjusted to reduce the dissipation anisotropy as proposed by CAMBON et al [1981]. Two values have been tested for β, i.e. β = 16.5 which gives a dissipation anisotropy of about 11 % of the Reynolds stress anisotropy and a value of 100 which gives a ratio of 1 %.

At last, spectra such as the one given on figure 2b show that there is no universal law for the anisotropy spectra over the large scales. So there should exist no real connection between the small scale anisotropy and the global anisotropy and some parameter related to the small scales should replace the Reynolds stress anisotropy in the above formulation.

4.5.b. Coupling of large and subgrid scales

Shapes have been defined for the energy and anisotropy spectra over the subgrid scales. However, these laws include adjustable parameters such as the dissipation rate ε in the energy spectrum law. To determine such values, the statistical evolution of the subgrid scales has to be linked to the evolution of the large scales as shown in AUPOIX et al [1982].

The dissipation rate ε which adjusts the energy spectrum level for the subgrid scale is linked to the kinetic energy of the subgrid scales as :

$$\langle \tfrac{1}{2} q'^2 \rangle = \int_{k_c}^{\infty} E(k)\, dk .$$

The knowledge of this kinetic energy is then enough to determine the subgrid energy spectrum. Some transport equation for this subgrid kinetic energy can be handled as in SCHUMANN's model. We chose to use the transport equation for the total kinetic energy to avoid the flux term. This equation reads :

$$\frac{d}{dt} \langle \tfrac{1}{2} q^2 \rangle = - \langle u_i u_j \rangle \frac{\partial u_i}{\partial x_j} - \varepsilon$$

where the Reynolds stresses include large and subgrid scales contribution and the dissipation rate ε is given by the subgrid scale spectrum. The subgrid kinetic energy is then the difference between total and large scale kinetic energies.

The crudest models for the subgrid anisotropy express it in terms of the global anisotropy. The subgrid anisotropy can then be deduced from the knowledge of the large scale anisotropy and the large and subgrid scale kinetic energies. More elaborate models assume that the anisotropy spectrum shape only holds for the small scales, without reference to the global anisotropy. In the anisotropy spectrum law, the global anisotropy is then replaced by a coefficient to be adjusted. By analogy with the energy spectrum, it seems natural to write transport equations for either global or subgrid Reynolds stresses. Unfortunately, while for isotropic turbulence, the kinetic energy transport equation was exact, the Reynolds stress transport equations are no longer exact and modelling is required.

If the equations are written for the global Reynolds stresses, the main trouble is the modelling of the pressure-strain term which is the integral of P^L and P^{NL} terms. If equations are written for the subgrid Reynolds stresses, the assumption $a = -4.5$ discards one pressure term but other trouble arises mainly from linear transfer through the filter cut. The cure was to assume the anisotropy law to be valid over the highest modes of the large scales and to express the subgrid anisotropy in terms of the anisotropy of the modes near the filter cut.

To link the subgrid anisotropy to the anisotropy of the large scales near the filter cut has some drawbacks as it may introduce numerical noise, mainly for low resolution codes.

4.5.c. <u>Disaveraging</u>

With the coupling between large and subgrid scales and

the assumption about the form of the subgrid energy spectrum, the subgrid energy spectrum is completely determined. Moreover, the subgrid anisotropy spectra can be expressed in terms of the large scale anisotropy. So the integrals over the subgrid scales can be calculated to evaluate the subgrid pressure and transfer terms with the non local formulae.

But we are really interested in the subgrid term $T_i^S(\underline{k})$ in the filtered NAVIER equation. The evaluation of the subgrid pressure and transfer terms is just a means to get information about this transfer term as they are connected by :

$$\text{subgrid part of } (P_{ij}^{NL}(k) + S_{ij}^{NL}(k)) = \int \langle \hat{u}_i(\underline{k}) \; T_j^S(-\underline{k}) + \hat{u}_j(-\underline{k}) \; T_i^S(\underline{k}) \rangle \; dA(k).$$

The information included in the subgrid pressure and transfer terms is averaged both in a statistical sense and over spheres of radius k. Desaveraging is needed to derive a model for the subgrid term T^S which satisfies both the above relation and the continuity constraint $k_i \; T_i^S(\underline{k}) = 0$.

When the subgrid scales are assumed isotropic, the subgrid pressure and transfer terms reduce to an eddy viscosity model. The constraints are then :

$$\int \langle \hat{u}_i(\underline{k}) \; T_j^S(-\underline{k}) + \hat{u}_j(-\underline{k}) \; T_i^S(\underline{k}) \rangle \; dA(k) = -2\nu_e k^2 \varphi_{ij}(k)$$

$$k_i \; T_i^S(\underline{k}) = 0$$

and as

$$\varphi_{ij}(k) = \int \langle \hat{u}_i(\underline{k}) \; \hat{u}_j(-\underline{k}) \rangle \; dA(k)$$

a solution for the subgrid term is :

$$T_i^S(\underline{k}) = -\nu_e \; k^2 \; \hat{u}_i(\underline{k}).$$

It must be pointed out that various solutions can be proposed to satisfy the continuity and statistical constraints as information has been lost when dealing with statistics.

When the subgrid scale anisotropy is taken into account, the subgrid pressure and transfer terms read :

$$-2 \; \nu_e \; k^2 \; \varphi_{ij}(k) + k^2 \; E(k) \; \psi_{ij}$$

where ψ_{ij} is linked to the anisotropy of either all the large scales or the large scales near the filter cut according to the subgrid anisotropy model used. The first term can still give an eddy viscosity contribution to the subgrid term $T_i^S(\underline{k})$. The contribution of the second term is more tricky to model as there is generally no direct connection between ψ_{ij} and the anisotropy at wavenumber k. BERTOGLIO [1984] proposed a way to generate subgrid terms $T_i^S(\underline{k})$ submitted to similar constraints. Simpler solutions are looked for but will not be discussed herein as they have not been tested in large eddy simulations.

5 - LARGE EDDY SIMULATIONS OF STRAINED TURBULENCE

As we do not have enough computing facilities, we cannot perform direct resolution of the NAVIER equations, extract the subgrid scale stresses from these simulations and compare with the models. So the only way to check the validity of our subgrid scale model will be to perform simulations of homogeneous flow experiments and to compare computation and experiments. Unfortunately, very little information is available about experimental spectra for homogeneous anisotropic turbulence, so the comparison must be done at the Reynolds stresses level.

5.1. Extension of the effective viscosity model

Large eddy simulations of homogeneous turbulence submitted to constant velocity gradient are performed following ROGALLO's method [1981]. To get advantage of FOURIER's methods, the filtered NAVIER equations are solved in a domain convected by the mean flow. As the domain is distorted by the mean flow, the filter no longer remains a sphere but an evolving ellipsoid.

Subgrid scale models previously developed for isotropic (i.e. spherical) filters have to be extended to these configurations. Our first attempt deals with a very simple model of the effective viscosity kind. The effective viscosity is now to be evaluated as the integral outside the filter :

$$\nu_e = \frac{1}{15} \int \theta_{opp} \left[7\, U(p) + p\, \frac{\partial U}{\partial p} \right] d^3\underline{p}$$

where $U(p) = \frac{E(p)}{4\pi p^2}$. This simple model assumes an equipartition of the energy over the sphere of radius k. The kinetic energy of the subgrid scales which is computed to couple large and subgrid scales is also obtained as :

$$\langle \tfrac{1}{2} q'^2 \rangle = \int U(p)\, d^3\underline{p}\,.$$

To have a simple model, PAO's law is assumed for the subgrid energy spectrum. While the subgrid anisotropy is omitted in the subgrid modelling, trials showed that it must be taken into account to evaluate the global Reynolds stresses and the kinetic energy production. The simple 5/6 law was used for subgrid anisotropy.

5.2. Generation of the initial velocity field

As pointed out previously, the subgrid scale model can only be tested by comparing experimental and computed Reynolds stresses. So the initial velocity field must have the experimental anisotropy. Standard methods to generate initial velocity fields, i.e. MONTE CARLO's method, only give isotropic turbulent fields without skewness. When these

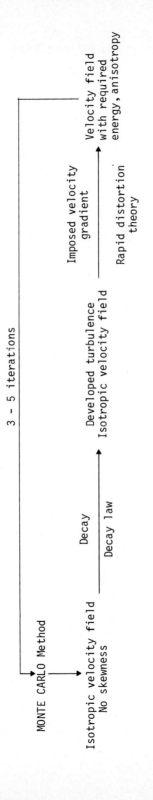

Fig. 3 - Generation of the initial velocity field

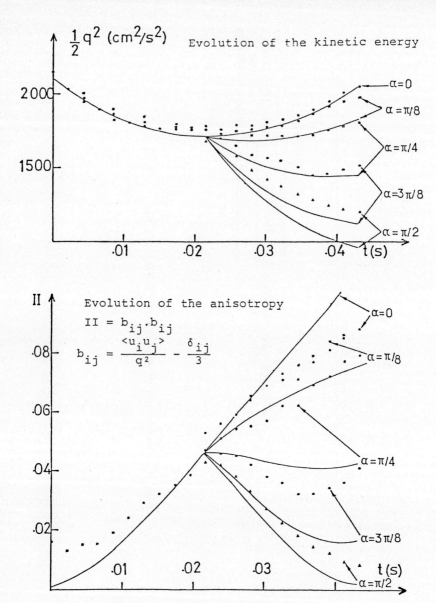

Fig. 4 - GENCE experiment - Turbulence submitted to two plane strains of relative angle α

☐ o △   Experiment
───   LES

fields are allowed to decay, third-order moments build up and skewness reaches an asymptotic value corresponding to fully developed turbulence. Strain can then be imposed upon this developed isotropic turbulence to obtain the required anisotropy. Rapid distortion theory gives the estimates for the isotropic turbulent field and the strain to apply to obtain the desired energy and anisotropy. Decay law for isotropic turbulence gives an evaluation of the initial skewness-free isotropic field to generate to obtain the isotropic field corresponding to developed turbulence.

However, neither decay law nor rapid distortion theory really hold. The decaying isotropic velocity field is initially skewness-free and is building up its skewness. The decay law for fully developed turbulence does not exactly predict the decay of such a developing turbulent field. Moreover, the distortion is applied over a time interval which is of the order of magnitude of the eddy turnover time. This is done to avoid spurious linear effects such as high anisotropy of the dissipation which occurs when a high strain is applied over a very short time, i.e. when rapid distortion theory is valid. Consequently, several computations are needed to obtain the desired energy and anisotropy (figure 3).

This method is very expensive as the initial field generation is longer than the experiment computation. Moreover, as the subgrid scale model will control the evolution of energy and anisotropy, this procedure has to be repeated for each subgrid scale model to generate velocity fields consistent with the subgrid scale model.

5.3. LES of the GENCE's experiment

Large eddy simulation of the GENCE's experiment of grid turbulence submitted to two successive plane strains have been performed. Figure 4 shows the evolution of both energy and anisotropy for the various directions of the second strain axis. As with other models, agreement is not obtained when the two strains have different axis.

Even for the no-rotation case ($\alpha = 0$), while the prediction of the kinetic energy is correct, the evolution of anisotropy is puzzling. Figure 4 corresponds to developed, quasi-isotropic initial turbulence. In this case, the final anisotropy is overestimated. Similar computations have been performed with an initial field having the experimental anisotropy. The kinetic energy prediction is slightly modified while the anisotropy evolution is directly shifted with the initial value.

Various causes can be suggested for this overestimation of the anisotropy. They are :

- numerical problems due to the choice of the size of the computational domain ;
- the eddy viscosity subgrid scale model ;
- the 5/6 anisotropy law ;
- the extension of the model to non isotropic filtering.

6 - TEST OF VARIOUS SUBGRID MODELS

6.1. Use of the filtered EDQNM equations

A lot of parameters can influence the behaviour of the subgrid scale models and have to be checked. Moreover, a model can behave correctly for a test case while it gives erroneous predictions for others, so several test cases must be used. With the initial velocity field generation procedure presented above, tests of various subgrid scale models on several experiments will represent a huge amount of computation. To reduce this computational time, we decided to use CAMBON's model to check the various assumptions for the subgrid model. An isotropic filter is introduced in the computation so that about 50-60 % of the kinetic energy is captured in the large scales. Various subgrid scale models corresponding to the hypotheses presented in Chapter 4 are then introduced to account for the subgrid scales. Strained turbulence experiments of TOWNSEND, TUCKER and REYNOLDS, GENCE and MARECHAL and sheared experiments of ROSE and CHAMPAGNE, HARRIS and CORRSIN were used to check the subgrid models.

As the large scales are computed with CAMBON's EDQNM model and the small scales are replaced by a subgrid scale model, a good subgrid model should be a model which gives the correct prediction not with reference to the experiment but with reference to the complete EDQNM computation.

6.2. Computations of the GENCE's experiment

The various model predictions will be presented only for the GENCE's experiment. We shall restrict our attention to the no rotation ($\alpha = 0$) case.

The first subgrid model tested is of eddy-viscosity form. When the energy spectrum is assumed a KOLMOGOROV's law and the anisotropy the corresponding 5/6 law, the whole model simplifies to easy-to-handle algebraic relations. Unfortunately, the KOLMOGOROV's spectrum has shown to be unable to predict the decay of low Reynolds number turbulence (AUPOIX [1984]) and this model also fails to predict the GENCE's experiment. The energy drain from the large to the subgrid scales is overestimated and the kinetic energy evolution is underestimated (figure 5a).

A first correction is to use a subgrid energy spectrum shape which accounts for low Reynolds number effects. The intricate form derived from EDQNM computation has been selected. The subgrid model is now analogous to the one used in the LES presented above. Figure 5a shows similar trends with a correct estimate of the energy and an overprediction of anisotropy. Let us stress again that the reference is not the experiment but the complete EDQNM computation, the results of which are given on figure 1. These complete EDQNM computations underestimate the anisotropy with reference to experiment.

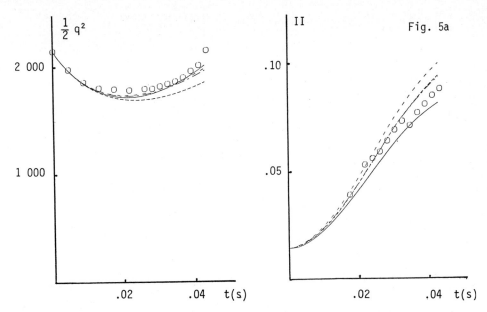

Fig. 5 - GENCE's experiment - Computation with the filtered EDQNM model
——— complete EDQNM model
------ eddy viscosity model-KOLMOGOROV's spectrum-5/6 anisotropy law
— — eddy viscosity model-"EDQNM" spectrum-5/6 anisotropy law
—·—·— eddy viscosity model-"EDQNM" spectrum-anisotropy law derived from rapid distortion theory

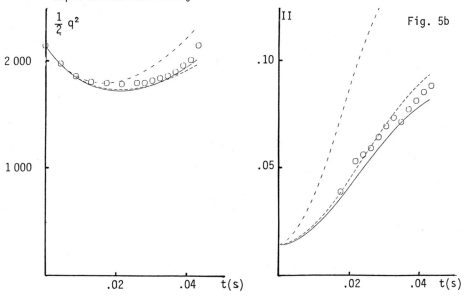

——— complete EDQNM model
----- eddy viscosity model-"EDQNM" spectrum-anisotropy law from R.D.T.
—·— complete pressure and transfer terms-"EDQNM" spectrum-anisotropy law from R.D.T.

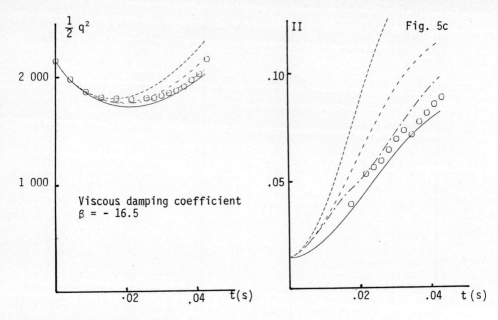

——— complete EDQNM model
Complete pressure and transfer terms - "EDQNM" spectrum -
------ anisotropy derived from rapid distortion theory
— — viscous damping in the anisotropy law
-·-·-· viscous damping - subgrid anisotropy linked to the modes near the cut

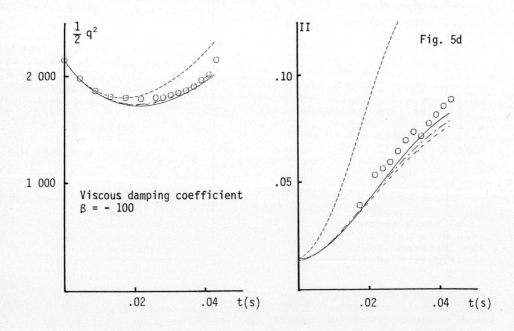

The 5/6 law for the anisotropy is then replaced by the law derived from rapid distortion theory without any viscous cut-off. The prediction of anisotropy is improved but the anisotropy is still overpredicted as shown on figure 5a.

The fact that this simple anisotropy law overestimates the subgrid anisotropy is brought into evidence on figure 5b. The eddy viscosity model has been replaced by the non local formulation for the subgrid pressure and transfer terms with k^4 terms omitted, hereafter labelled as complete model. As this model reinjects the subgrid anisotropy into the large scales, the anisotropy is now drastically overestimated. Consequently, the kinetic energy production is overestimated and so for the kinetic energy.

The introduction of a viscous damping term greatly modified the results. For a first value of the damping coefficient ($\beta = -16.5$), both energy and anisotropy are still overestimated but the discrepancy is reduced (figure 5c). A possible source of error is that the anisotropy feedback from the subgrid scales to the large scales is amplified by the fact that the subgrid anisotropy is expressed in terms of the large scales. When the subgrid anisotropy is linked to the anisotropy of the higher modes of the large scales, the agreement is improved (figure 5c).

Lastly, a stronger viscous damping corresponding to $\beta = -100$ was tested. This model gives good prediction, not only for the GENCE's case but for all the test cases. For sheared flow experiments, the correction to predict a $k^{-7/3}$ spectrum for the off-diagonal term proved to be necessary.

7 - CONCLUSION

CAMBON's spherically averaged EDQNM model has proved to be an efficient tool to study subgrid scale models for homogeneous anisotropic turbulence. The non local hypothesis leads to analytical formulae for the subgrid pressure and transfer terms which can be validated over the large scales. These subgrid terms have two roles : they drain energy via an effective viscosity term and they reinject the subgrid scale anisotropy into the large scales. As these subgrid terms are defined as integrals over the subgrid scales, large and subgrid scales evolution must be computed and coupled. Representation formulae are needed for the small scales energy and anisotropy spectra. Tests with the filtered EDQNM model showed that the energy spectrum law must account for low Reynolds number effects and that sophisticated formulae are required for the subgrid anisotropy. More work is needed to correctly model the subgrid anisotropy but a satisfactory model has already been obtained. Lastly, extensive tests have been done only on the subgrid pressure and transfer terms in the filtered EDQNM model and must be continued for the subgrid term in the filtered NAVIER equations.

REFERENCES

- ANDRE J.C., LESIEUR M. [1977] "Influence of helicity on the evolution of isotropic turbulence at high Reynolds number" - JFM Vol. 81, pp. 187-207.
- AUPOIX B., COUSTEIX J. [1982] "Subgrid scale models for isotropic turbulence" - Proceedings of the Symposium on Refined Modelling of Flows - PARIS, Sept. 7-10, 1982.
- AUPOIX B., COUSTEIX J. [1982] "Simple subgrid scale stresses models for homogeneous isotropic turbulence" - La Recherche Aérospatiale 1982-4.
- AUPOIX B., COUSTEIX J., LIANDRAT J. [1983] "Effects of rotation on isotropic turbulence" - Turbulent Shear Flow 4 - KARLSRUHE.
- AUPOIX B. [1984] "Eddy viscosity subgrid scale models for homogeneous turbulence" - Lecture Notes in Physics N° 230 - Springer Verlag pp. 45-64.
- BERTOGLIO J.P. [1981] "A model of threedimensional transfer in non isotropic homogeneous turbulence" - Turbulent Shear Flow 3 - DAVIS.
- BERTOGLIO J.P., MATHIEU J. [1983] "Study of subgrid models for sheared turbulence" - Turbulent Shear Flow 4 - KARLSRUHE.
- BERTOGLIO J.P. [1984] "A stochastic subgrid model for sheared turbulence" - Lectures Notes in Physics N° 230 pp. 100-149.
- CAMBON C., JEANDEL D., MATHIEU J. [1981] "Spectral modelling of homogeneous non isotropic turbulence" JFM Vol. 104, pp. 247-262.
- CAMBON C., BERTOGLIO J.P., JEANDEL D. [1981] "Spectral closures for homogeneous turbulence" - The 1980-81 AFOSR-HTTM-STANFORD Conference on Complex Turbulent Flow, Vol. III, pp. 1307-1311.
- CHOLLET J.P. [1983] "Two-point closures as a subgrid scale modelling for large eddy simulations" - Turbulent Shear Flows 4 - KARLSRUHE.
- GENCE J.N., MATHIEU J. [1979] "On the application of successive plane strains to grid generated turbulence" - JFM Vol. 93, Part 3, pp; 501-513.
- KRAICHNAN R.H. [1976] "Eddy viscosity in two and three dimensions" - JAS Vol. 33, pp. 1521-1536.
- LEONARD A. [1973] "Energy cascade in large eddy simulation of turbulent fluid flow" - Advances in Geophysics, Vol. 18A, pp. 237-248.
- LESIEUR M., SCHERTZER D. [1978] "Amortissement autosimilaire d'une turbulence à grand nombre de Reynolds" - Journal de Mécanique, Vol. 17, N° 4, pp. 609-646.
- MESTAYER P. [1982] "Local isotropy and anisotropy in a high Reynolds number turbulent boundary layer" - JFM, Vol. 125, pp. 475-505.
- MESTAYER P., CHOLLET J.P., LESIEUR M. [1984] "Inertial subrange of velocity and scales variance spectra in high Reynolds number threedimensional turbulence" - Proceedings of Turbulence and Chaotic Phenomena in Fluids - T. TATSUMI Editor Elsevier Science Publisher (North Holland) - pp. 285-288.
- ROGALLO R.S. [1981] "Numerical experiments in homogeneous turbulence" - NASA TM 81315.

APPLICATION OF RENORMALISATION GROUP (RG) METHODS

TO THE SUBGRID MODELLING PROBLEM

W.D. McComb

Department of Physics, University of Edinburgh

King's Buildings, Edinburgh. EH9 3JZ. U.K.

SUMMARY

The technique of 'iterative averaging' has previously been used to formulate equations for the large-eddy simulation of the velocity field in k-space, with the mean effect of subgrid scales being represented by an effective viscosity $\nu(k)$, which is the fixed-point value of the RG transformations. Here we present some preliminary aspects of a more general and systematic development of the theory in terms of moment expansion. At the lowest nontrivial level this leads to a slightly modified form of the effective viscosity previously reported. The theory is also extended by the derivation of an equation for the dissipation rate. The solution for the inertial range spectrum is shown to be the Kolmogorov form $E(k) \sim k^{-5/3}$ and this power law is found to hold to all orders of the moment expansion.

INTRODUCTION

It is well known that turbulent flows of any practical significance lie far beyond the scope of full numerical simulation. The main problem is the large number of degrees of freedom, as measured (for example) by the number of independently excited modes in wavenumber space. Thus the problem resolves itself into the need to eliminate modes, in some statistical sense, in order to bring the reduced number of degrees of freedom within the range of existing (or, even, envisaged) computers.

A frontal attack on this problem is to filter the modes at $k = k_c$, where k_c stands for cut-off wavenumber. Then modes $k \lesssim k_c$ are dealt with by direct numerical simulation of the Navier-Stokes equations (NSE) on the interval $0 \lesssim k \lesssim k_c$; whereas the effect of the subgrid modes ($k \gtrsim k_c$) is represented by an enhanced viscosity acting on the truncated NSE. This approach is known as large eddy simulation (LES) and has attracted much attention in recent years [1,2]. However, as well as overcoming problems in numerical methods and computer software, the prospective large-eddy simulator has to find an analytical form for the effective viscosity which represents the drain of energy from modes $k \lesssim k_c$ to modes $k \gtrsim k_c$, by inertial transfer. This is known as the 'subgrid modelling problem', and has mostly been tackled at a phenomenological level - although there has been quite a lot of activity in applying renormalised perturbation theory to the calculation of the subgrid effective viscosity for homogeneous turbulence [3-5].

However, since the early 1970s, we have had in RG a powerful and systematic method of mode elimination for problems involving many degrees of freedom [6]. In the context of fluid turbulence RG may be seen as a way of eliminating the effect of eddies progressively; with first the smallest eddies; then the next smallest eddies; and so on: eventually replacing their mean effect by an effective turbulent viscosity. In other words, the molecular kinematic viscosity of the fluid ν_o becomes renormalised by the collective action of the turbulent eddies.

Of course this method cannot be taken over from (say) critical phenomena and applied in some prescriptive fashion to turbulence. As Wilson [6] has pointed out, the difficulties faced initially in applying RG to a new problem may seem as formidable as those involved with applying any other method. And indeed, the pioneering application of RG to turbulence [7], was only valid in the asymptotic case $k \to 0$. This, and other low-order perturbation theories have been given a critical examination in the review by Kraichnan [8], so we shall not give a general discussion here. However, we should mention the work of Rose [9], who was apparently the first person to recognise the relevance of RG to the subgrid problem, and who applied the method to the subgrid modelling of passive scalar convection.

Our own approach has been reported at various stages of its development in the literature [10-12]. It began as an iterative time-averaging of the equations of motion and was not restricted to isotropic or homogeneous turbulence [10]. Connection with RG was made by (a) Fourier-transforming into ω space, and (b) invoking the Taylor hypothesis of frozen convection to take the analysis into k space. This was followed by the calculation of a Heisenberg-type effective viscosity [10,11] and later the analysis was carried out directly in k space, and some of the underlying assumptions exposed [12]. We have also discussed the extension of the method to the formal derivation of the LES equations of motion [13,14]. Our objective here is to present some preliminary aspects of a more general treatment which, among other things, leads to a slight modification to the previously reported form of the effective viscosity.

THE CONCEPT OF RG

Consider the velocity field as represented by the discrete set of components $U_\alpha(\underline{k},t)$ on the interval $0 \lesssim k \lesssim k_o$, where k_o is the largest wavenumber present and may be defined through the dissipation integral

$$\varepsilon = \int_0^\infty 2\nu_o k^2 E(k) \, dk = \int_0^{k_o} 2\nu_o k^2 E(k) \, dk, \qquad (1)$$

where ε is the dissipation rate and $E(k)$ is the energy spectrum.

We now choose a wavenumber cut-off k_1 such that $k_1 < k_o$ and, in practice,

$$k_1 \sim k_o \sim (\varepsilon/\nu_o^3)^{1/4}: \text{ the dissipation wavenumber} . \qquad (2)$$

In principle, RG then involves two stages:

A. Solve the NSE on $k_1 \lesssim k \lesssim k_o$. Substitute that solution for the mean effect of high-k modes into the NSE on $0 \lesssim k \lesssim k_1$. This results in an increment to the viscosity: $\nu_o \to \nu_1 = \nu_o + \delta\nu_o$.

B. Re-scale the basic variables so that the NSE on $0 \lesssim k \lesssim k_1$ looks like the original Navier-Stokes equation on $0 \lesssim k \lesssim k_o$.

This procedure is repeated for a cut-off wavenumber $k_2 < k_1 < k_0$, and so on.

The underlying physics of RG may be summarised as follows. In the viscous range of wavenumbers, it is reasonable to suppose that the turbulence is critically damped. That is, any mode k in the band $k_1 \lesssim k \lesssim k_0$ is driven by energy transfer from modes $k < k_1$, and the injected energy is dissipated locally by the effects of the molecular viscosity. Thus we can solve the NSE for $U_\alpha(\underline{k},t)$, in the band $k_1 \lesssim k \lesssim k_0$, in terms of the <u>bilinear</u> energy transfer from $k \lesssim k_1$; with the quadratic nonlinear term involving energy transfers within $k_1 \lesssim k \lesssim k_0$ being neglcted (although as we shall see later, this term may be treated systematically to all orders). As a result, the new NSE has an effective viscosity $\nu_1 = \nu_0 + \delta\nu_0$ in the range of wavenumbers $0 \lesssim k \lesssim k_1$.

If we then define an <u>effective</u> dissipation wavenumber $k_d^{(1)}$ for the new NSE,

$$k_d^{(1)} = (\varepsilon/\nu_1^3)^{1/4} < k_d^{(0)} = (\varepsilon/\nu_0^3)^{1/4}, \qquad (3)$$

then we may repeat the whole procedure for $k_2 \sim k_d^{(1)}$ and we may again linearise the (suitably scaled) NSE in the range $k_2 \lesssim k \lesssim k_1$.

NAVIER-STOKES EQUATIONS IN $\underline{k}$ SPACE

The Fourier representation (in a cubical box of side L) for the velocity field $U_\alpha(\underline{x},t)$ may be introduced through the relationship

$$U_\alpha(\underline{x},t) = \sum_{|\underline{k}| \leq k_0} U_\alpha(\underline{k},t) e^{i\underline{k}\cdot\underline{x}}. \qquad (4)$$

Then the NSE takes the form [15]:

$$(\frac{\partial}{\partial t} + \nu_0 k^2) U_\alpha(\underline{k},t) = \sum_{|\underline{j}| \leq k_0} M_{\alpha\beta\gamma}(\underline{k}) U_\beta(\underline{j},t) U_\gamma(\underline{k}-\underline{j},t), \qquad (5)$$

where the inertial transfer operator is given by

$$M_{\alpha\beta\gamma}(\underline{k}) = (2i)^{-1} \{k_\beta D_{\alpha\gamma}(\underline{k}) + k_\gamma D_{\alpha\beta}(\underline{k})\} \qquad (6)$$

and the projection operator by:

$$D_{\alpha\beta}(\underline{k}) = \delta_{\alpha\beta} - k_\alpha k_\beta |\underline{k}|^{-2}, \qquad (7)$$

and $\delta_{\alpha\beta}$ is the Kronecker delta.

In addition to the initial condition that $U_\alpha(\underline{k},0)$ be prescribed, we also have the boundary condition

$$U_\alpha(\underline{k},t) = 0 \text{ on } k = 0. \qquad (8)$$

For isotropic turbulence, the pair-correlation of velocities takes the form

$$\left(\frac{L}{2\pi}\right)^3 <U_\alpha(\underline{k},t)U_\beta(-\underline{k},t')> = D_{\alpha\beta}(\underline{k})Q(k;t,t'), \tag{9}$$

where $<>$ means the conventional averaging operation, and the expression for the energy spectrum follows in the usual way:

$$E(k,t) = 4\pi k^2 Q(k;t,t). \tag{10}$$

In the work that follows, we shall restrict our attention to turbulence which is both isotropic and homogeneous in space, and which is stationary in time. The last requirement implies that the energy spectrum is independent of time.

THE MOMENT HIERARCHY FROM PARTIAL AVERAGING

We begin by dividing up the velocity field in k space at $k = k_1$, thus:

$$\begin{aligned} U_\alpha(\underline{k},t) &= U_\alpha^-(\underline{k},t) & 0 \leq k \leq k_1 \\ &= U_\alpha^+(\underline{k},t) & k_1 \leq k \leq k_o \end{aligned} \tag{11}$$

We note that with this definition, U^- and U^+ are statistically independent variables [16].

We now introduce the partial average over the small scales. We denote it by $<>_o$ and define it by:

$$\begin{aligned} <U_\alpha(\underline{k},t)>_o &= U_\alpha^-(\underline{k},t) \\ <U_\alpha^-(\underline{k},t)>_o &= U_\alpha^-(\underline{k},t) \\ <U_\alpha^+(\underline{k},t)>_o &= 0 . \end{aligned} \tag{12}$$

For what follows, it is helpful to have a simpler notation, so we combine the index α, the wavenumber $\underline{k}$ and (where appropriate) frequency ω into a single subscript. Thus $U_\alpha(\underline{k},t) \to U_{\underline{k}}$ etc.

Then the NSE may be written as

$$\left(\frac{\partial}{\partial t} + \nu_o k^2\right)U_{\underline{k}} = \sum_{\underline{j}} M_{\underline{k}} U_{\underline{j}} U_{\underline{k}-\underline{j}} \tag{13}$$

and our immediate objective is to follow a procedure like Reynolds averaging, but using (11) and (12) on which to base our statistical treatment of the NSE, as given by (13).

Averaging both sides of (13) according to (12) we obtain for the explicit scales ($0 \leq k \leq k_1$) the result

$$\left(\frac{\partial}{\partial t} + \nu_o k^2\right)U_{\underline{k}}^- = \sum_{\underline{j}} M_{\underline{k}} <U_{\underline{j}} U_{\underline{k}-\underline{j}}>_o . \tag{14}$$

Then the equation for the implicit scales ($k_1 \leq k \leq k_o$) is found by subtracting (14) from (13) to obtain

$$\left(\frac{\partial}{\partial t} + \nu_o k^2\right) U_{\underline{k}}^+ = \sum_{\underline{j}} M_{\underline{k}} \{U_{\underline{j}} U_{\underline{k}-\underline{j}} - \langle U_{\underline{j}} U_{\underline{k}-\underline{j}}\rangle_o\}. \tag{15}$$

Although $U_{\underline{k}}^+$ has been eliminated from the explicit-scales NSE, we must still take into account the coupling of $U_{\underline{j}}^+$ and $U_{\underline{k}-\underline{j}}^+$ to cause fluctuations in the band $0 \leq k \leq k_1$.

We may obtain an equation for the mean effect of this coupling in the following way. Rewrite (15) as two equations for $U_{\underline{j}}^+$ and $U_{\underline{k}-\underline{j}}^+$, multiply through by $U_{\underline{k}-\underline{j}}^+$ and $U_{\underline{j}}^+$ respectively, add the two equations and average over the small scales to yield:

$$\{\frac{\partial}{\partial t} + \nu_o j^2 + \nu_o |\underline{k}-\underline{j}|^2\} \langle U_{\underline{j}}^+ U_{\underline{k}-\underline{j}}^+\rangle_o$$
$$= 2\sum_{\underline{p}} M_{\underline{j}} \langle \{U_{\underline{p}} U_{\underline{j}-\underline{p}} - \langle U_{\underline{p}} U_{\underline{j}-\underline{p}}\rangle_o\} U_{\underline{k}-\underline{j}}^+\rangle_o , \tag{16}$$

where we have used invariance under interchange of $\underline{j}$ and $\underline{k}-\underline{j}$ to simplify the RHS of (16).

Alternatively, we may write the formal solution of (16) as

$$\langle U_{\underline{j}}^+ U_{\underline{k}-\underline{j}}^+\rangle_o = 2G_o(\underline{j},\underline{k}-\underline{j};\omega) \sum_{\underline{p}} M_{\underline{j}} \langle\{U_{\underline{p}} U_{\underline{j}-\underline{p}} - \langle U_{\underline{p}} U_{\underline{j}-\underline{p}}\rangle_o\} U_{\underline{k}-\underline{j}}^+\rangle_o , \tag{17}$$

the Green's function being given by

$$G_o(\underline{j},\underline{k}-\underline{j};\omega) = \frac{1}{i\omega + \nu_o j^2 + \nu_o |\underline{k}-\underline{j}|^2} , \tag{18}$$

where ω is the frequency. The convolution sum on the RHS of (17) must therefore be regarded as being over frequency as well as wavenumber, although we have not shown this explicitly.

Evidently equation (17) contains triple moments on the RHS and so we have not eluded the closure problem. In fact we can carry on generating equations for moments of all orders, just as in the conventional NSE hierarchy.

Our problem at this stage is to relate our partial averaging operation $\langle \rangle_o$ to the conventional (global) average $\langle \rangle$. We have already encountered the latter in connection with equation (9). It will shortly be seen that the way in which we deal with this problem involves the major underlying assumption of this work.

On the basis of global averaging, homogeneity demands that

$$\langle U_{\underline{j}}\rangle = 0 \;;\; \langle U_{\underline{j}} U_{\underline{j}'}\rangle = Q(\underline{j})\delta(\underline{j}+\underline{j}') \tag{19}$$

and, in general,

$$\langle U_{\underline{j}} U_{\underline{\ell}} U_{\underline{p}} \ldots U_{\underline{r}}\rangle = 0 \tag{20}$$

unless $\underline{j} + \underline{l} + \underline{p} + \ldots + \underline{r} = 0$.

It should be emphasised that we have deliberately avoided using $\underline{k}$ for any of the above illustrations. From here on, $\underline{k}$ is <u>always</u> the labelling wavevector of the RG iteration, <u>always</u> lies in the band $0 \leq k \leq k_n$, and <u>always</u> appears as the resultant of a set of dummy wavevectors which all lie in the band $k_n \leq k \leq k_{n-1}$.

If we now substitute (11) into the RHS of (17) and average according to (12); and carry on the same process to higher orders, we see that the resulting hierarchy can contain two types of moments. Referring to these as Type A and Type B, we have:

$$\text{Type A} = \langle U^+_{\underline{j}} U^+_{\underline{l}} U^+_{\underline{p}} \ldots U^+_{\underline{r}} \rangle_0 \qquad (21)$$

where $|\underline{j} + \underline{l} + \underline{p} + \ldots + \underline{r}|$ lies in the band $0 \leq k \leq k_1$.

$$\text{Type B} = \langle U^-_{\underline{j}} U^+_{\underline{l}} U^+_{\underline{p}} \ldots U^+_{\underline{r}} \rangle_0 \qquad (22)$$

where $|\underline{j} + \underline{l} + \underline{p} + \ldots + \underline{r}|$ lies in the band $0 \leq k \leq k_1$.

We cannot evaluate Type A moments directly, but we can solve the NSE equation to express Type A moments in terms of Type B moments. This brings us to our main assumption. We <u>assume</u> that the low-k dependence of Type B moments can be factored out as follows:

$$\langle U^-_{\underline{j}} U^+_{\underline{l}} U^+_{\underline{p}} \ldots U^+_{\underline{r}} \rangle_0 = U^-_{\underline{j}} \langle U^+_{\underline{l}} U^+_{\underline{p}} \ldots U^+_{\underline{r}} \rangle$$

$$= 0 \text{ unless } \underline{l} + \underline{p} + \ldots + \underline{r} = 0. \qquad (23)$$

Let us see the effect of this by considering two examples. At first order we have:

$$\sum_{\underline{j}} M_{\underline{k}} \langle U^-_{\underline{j}} U^+_{\underline{k}-\underline{j}} \rangle_0 = \sum_{\underline{j}} M_{\underline{k}} \langle U^-_{\underline{j}} U^+_{\underline{k}-\underline{j}} \rangle$$

$$= 0, \qquad (24)$$

and at second order we have:

$$\sum_{\underline{p}\underline{j}} M_{\underline{j}} \langle U^-_{\underline{p}} U^+_{\underline{j}-\underline{p}} U^+_{\underline{k}-\underline{j}} \rangle_0 = \sum_{\underline{p}} M_{\underline{j}} U^-_{\underline{p}} \langle U^+_{\underline{j}-\underline{p}} U^+_{\underline{k}-\underline{j}} \rangle$$

$$= \sum_{\underline{p}} M_{\underline{j}} U^-_{\underline{p}} \delta(\underline{k}-\underline{p}) Q^+(|\underline{k}-\underline{j}|)$$

$$= M_{\underline{j}} Q^+(|\underline{k}-\underline{j}|) U^-_{\underline{k}} , \qquad (25)$$

where we have used (19), and $Q^+(|\underline{k}-\underline{j}|)$ is just $Q(|\underline{k}-\underline{j}|)$ with $k_1 \leq |\underline{k}-\underline{j}| \leq k_0$. In this way, we can generate a moment hierarchy which is linear in $U^-_{\underline{k}}$.

THE RG EQUATIONS

With the assumption just discussed, equation (14) for the explicit scales becomes

$$(\frac{\partial}{\partial t} + \nu_o k^2)U_{\underline{k}}^- - \sum_{\underline{j}} M_{\underline{k}} \langle U_{\underline{j}}^+ U_{\underline{k}-\underline{j}}^+ \rangle_o = \sum_{\underline{j}} M_{\underline{k}} U_{\underline{j}}^- U_{\underline{k}-\underline{j}}^- \qquad (26)$$

for $0 \leq k \leq k_1$.

The coupling to implicit scales (now written on the LHS) is obtained as the sum of a moment expansion. In view of the complexity of this, we shall only sketch out its general form. Let us put $G_o(\underline{j}, |\underline{k}-\underline{j}|; \omega) = G_{o2}$. Then G_o with three wavenumber arguments becomes G_{o3}; and so on. Also put $\langle U_{\underline{j}}^+ U_{\underline{k}-\underline{j}}^+ \rangle = Q_2^+$. Then the global average of the product of four U^+ becomes Q_4^+; and so on. Finally, dropping the subscripts on summations and inertial transfer operators, we write the moment expansion representing the coupling between explicit and implicit scales in the highly symbolic form:

$$\langle U_{\underline{j}}^+ U_{\underline{k}-\underline{j}}^+ \rangle_o = 2G_{o2} MQ_2^+ U_{\underline{k}}^-$$

$$+ \{2G_{o2} G_{o3} G_{o3} \Sigma\Sigma MMMQ_4^+ + 2G_{o2} G_{o3} G_{o4} \Sigma\Sigma MMMQ_4^+\} U_{\underline{k}}^-$$

$$+ \{2G_{o2} G_{o3} G_{o4} G_{o5} G_{o5} \Sigma\Sigma\Sigma\Sigma MMMMMQ_6^+$$

$$+ 2G_{o2} G_{o3} G_{o4} G_{o5} G_{o6} \Sigma\Sigma\Sigma\Sigma MMMMMQ_6^+\} U_{\underline{k}}^-$$

$$+ \ldots\ldots \qquad (27)$$

Two points about this should be noted. First, renormalisation does not involve summing the series on the RHS of (27), in contrast to perturbation methods. Instead, we calculate the effective viscosity by substituting (27) into (26). Then we have $\nu_1 = \nu_o + \delta\nu_o$, $G_o(\underline{k},\omega) \to G_1(\underline{k},\omega)$ and repeat for $k_{n+1} < k_n < \ldots\ldots k_1 < k_o$ until

$$G_{n+1} = G_n = G^*, \qquad (28)$$

with appropriate rescaling of basic variables. Renormalisation is therefore achieved at the fixed point, as defined by (28).

The second point is that the general physics of turbulence implies that the RHS of (27) should be rapidly convergent. As explained earlier, for physical reasons, we may expect such convergence to be retained (in terms of rescaled variables) during the RG iteration. Hence, it seems physically plausible to truncate the above moment expansion at low order.

In fact we shall truncate the expansion at second order. Then the calculation of correction terms will be found to justify this approximation provided the wavenumber bands $k_n \leq k \leq k_{n-1}$ are not too wide.

SECOND-ORDER CALCULATION OF THE EFFECTIVE VISCOSITY

We may recover the equations previously reported [10-14] for the effective viscosity by retaining only the first term on the RHS of (27). (There is one difference and we shall draw attention to this shortly.) This gives a second-order theory and, interestingly, is the same order as employed in renormalised perturbation theories (e.g. [17]), although the method of renormalisation is different.

We shall also make the algebra simpler by approximating the Green's function as given by equation (18), thus:

$$G_o(\underline{j}, \underline{k}-\underline{j}; \omega) \simeq \frac{1}{\nu_o j^2 + \nu_o |\underline{k}-\underline{j}|^2} \, . \tag{29}$$

Referring back to the original differential equation, we see that the ω arises as the inversion of the $\partial/\partial t$ on the LHS of equation (16). This appears despite the stationarity of the field because $<U_{\underline{j}}^+ U_{\underline{k}-\underline{1}}^+>_o$ is a random variable on $0 \leq k \leq k_1$. However, the frequency is bounded $\omega \leq \nu_o k^2$, and as $\nu_o k^2 < \nu_o j^2 + \nu_o |\underline{k}-\underline{j}|^2$ for $0 \leq k \leq k_1$ and $k_1 \leq j$, $|\underline{k}-\underline{j}| \leq k_o$, neglect of ω in (29) seems justified, particularly for the asymptotic case $k \to 0$. In practice, it turns out that the numerical effect of this term is small.

We now choose our wavenumber bands by putting

$$k_n = h^n k_o \, ; \quad 0 < h < 1, \tag{30}$$

where the scaling parameter h is arbitrarily chosen within the given limits. Then following the procedure outlined in the previous section, we obtain the recursion relation

$$\nu_{n+1}(k) = \nu_n(k) + \delta\nu_n(k), \tag{31}$$

where

$$\delta\nu_n(k) = \frac{2}{k^2} \int d^3j \, \frac{L_{kj} Q(|\underline{k}-\underline{j}|)}{\nu_n(j)j^2 + \nu_n(|\underline{k}-\underline{j}|)|\underline{k}-\underline{j}|^2} \tag{32}$$

$$0 \leq k \leq k_{n+1} \, ; \quad k_{n+1} \leq j, \, |\underline{k}-\underline{j}| \leq k_n,$$

and

$$L_{kj} = -2M_{\alpha\beta\gamma}(\underline{k}) \, M_{\beta\rho\delta}(\underline{j}) \, D_{\delta\gamma}(\underline{k}-\underline{j}) D_{\rho\alpha}(\underline{k}). \tag{33}$$

At the same time, equation (26) for the explicit scales may now be written (with full notation restored) as

$$\left\{ \frac{\partial}{\partial t} + \nu_{n+1}(k)k^2 \right\} U_\alpha^-(\underline{k},t) = \sum_{\underline{j}} M_{\alpha\beta\gamma}(\underline{k}) U_\beta^-(\underline{j},t) U_\gamma^-(\underline{k}-\underline{j},t), \tag{34}$$

$$0 \leq k, j, |\underline{k}-\underline{j}| \leq k_{n+1}.$$

where $\nu_{n+1}(k)$ is given by equations (31)-(33).

We may form the energy equation for the explicit scales by multiplying each term in (34) by $U_\alpha^-(-\underline{k},t)$ and averaging. If we then integrate each term with respect ot k over $0 \leq k \leq k_{n+1}$, the result is the renormalised dissipation equation

$$\int_0^{k_{n+1}} 2\nu_{n+1}(k)k^2 E(k)\, dk = \varepsilon. \tag{35}$$

which may be compared with equation (1).

Before going further, we should note that equation (32) for the increment to the effective viscosity differs from the form previously reported [10-14] by the presence of $\nu_n(|\underline{k}-\underline{j}|)\,|\underline{k}-\underline{j}|^2$ in the denominator and the factor of two in the numerator. However, the numerical difference between the old and new calculations of effective viscosity is no more than about 25%.

If we now assume that $E(k) = \alpha\, \varepsilon^r\, k^s$, and make the scaling transformation

$$k = k_n k', \tag{36}$$

it follows from (31) and (32) that the effective viscosity may be written as

$$\nu_n(k_n k') = \alpha^{1/2}\, \varepsilon^{r/2}\, k_n^{(s-1)/2}\, \tilde{\nu}_n(k'), \tag{37}$$

where α is the constant of proportionality in the assumed spectrum.

Substitution of (37) into (35) fixes the exponents as $r = 2/3$, $s = -5/3$: the well known Kolmogorov spectrum. With these results, equations (37), (31) and (32) become

$$\nu_n(k_n k') = \alpha^{1/2}\, \varepsilon^{1/3}\, k_n^{-4/3}\, \tilde{\nu}_n(k'), \tag{38}$$

$$\tilde{\nu}_{n+1}(k') = h^{4/3}\{\tilde{\nu}_n(hk') + \delta\tilde{\nu}_n(hk')\} \tag{39}$$

and

$$\delta\tilde{\nu}_n(k') = \frac{1}{2\pi k'^2}\int d^3j'\, \frac{L_{k'j'}|\underline{k}'-\underline{j}'|^{-11/3}}{\tilde{\nu}_n(j')j'^2 + \tilde{\nu}_n(|\underline{k}'-\underline{j}'|)\,|\underline{k}'-\underline{j}'|^2} \tag{40}$$

for the wavenumber bands $0 \leq k' \leq 1$ and $1 \leq j'$, $|\underline{k}'-\underline{j}'| \leq h^{-1}$.

Iteration of equations (39) and (40) reaches a fixed point with $\tilde{\nu}_{n+1} = \tilde{\nu}_n = \nu^*$, and this is shown in Fig. 1 for three different choices of $\tilde{\nu}_0$. Thus the new formula for the increment - as given by equation (40) - also leads to a fixed-point value ν^* which is independent of both n and $\tilde{\nu}_0$. However, it should be noted that the value of ν^* is about 25% lower than the result previously reported [10-14].

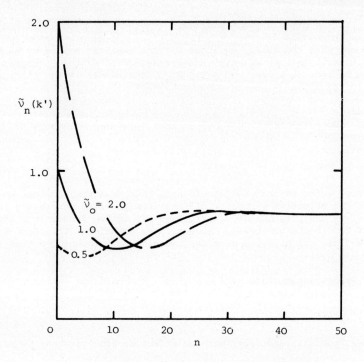

Fig. 1. Demonstration of the scaled eddy viscosity reaching a fixed point, independent of its initial value : h = 0.9; k' = 0.01.

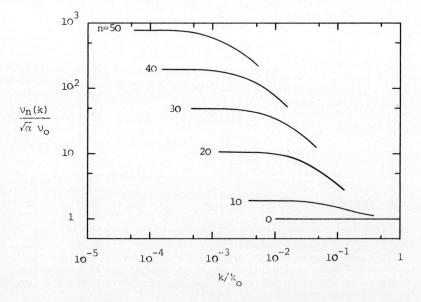

Fig. 2. Evolution of the actual eddy viscosity.

Fig. 2 shows the evolution of the actual eddy viscosity $\nu(k)$ as computed from $\tilde{\nu}$ and equation (38). As before [10-14], this becomes constant for the asymptotic case $k \to 0$, but shows the effect of non-local interactions near the cut-off (i.e. where $k' = 1$).

The above results show ν^* to be independent of two of the three RG parameters, ν_0 and n. The situation concerning the third parameter - the scaling factor h - is more problematical and is bound up with both the functional form of the dependence of ν^* on k' and the order at which we truncate the moment expansion on the RHS of equation (27). We shall only give brief details here. We begin by illustrating the problem.

Equations (35) and (38) can be solved to obtain a value for the Kolmogorov constant α. The result is

$$\alpha = \left\{ 2 \int_0^1 \nu^*(k') k'^{1/3} \, dk' \right\}^{-2/3}. \tag{41}$$

In Fig. 3 we plot the calculated value of α from equation (41) against the scaling parameter h. One merit of taking α as a test is that it does have known experimental values, albeit scattered in the range $1.2 \lesssim \alpha \lesssim 2.2$. We have plotted these bounds on the experimental values as dotted lines, and it can be seen that the theoretical value of the Kolmogorov constant - although showing some dependence on h - lies comfortably in between them.

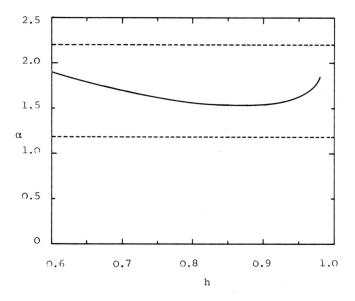

Fig. 3. Variation of the Kolmogorov constant α with the scaling parameter h (Dotted lines indicate the upper and lower bounds of the spread of experimental values).

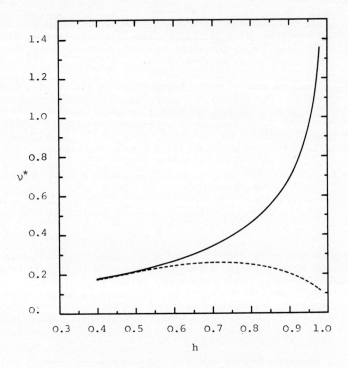

Fig. 4. Variation of the effective viscosity at the fixed point with the scaling parameter h. ———, k'=0; - - -, k'=1.

This dependence on h arises of course because ν^* depends on h. Clearly we cannot expect the iteration leading to ν^* to hold for $h \to 0$ (wavenumber bands too coarse) or for $h \to 1$ (wavenumber bands too fine). But somewhere in between we may hope for a region in which ν^* is relatively insensitive to the value of h. In Fig. 4 we plot both the asymptotic value $\nu^*(o)$ and the cut-off value $\nu^*(1)$ against h. Clearly the two forms of dependence are quite different - a fact that can be demonstrated analytically, although we shall not pursue that here. As we have shown previously [13,14] the actual effective viscosity $\nu(k)$ is less sensitive to changes in h; largely because the value of k where the fixed point is reached also depends on h.

This problem is, at least partially, related to the overall accuracy of truncating the moment expansion at a given order. The effect of including higher-order moments has been investigated. This is rather complicated, but we can summarise the result as follows. If we relabel the increment given by equation (32) as $\delta\nu_n^{(o)}$ then the correction due to fourth- and sixth-order moments can be indicated schematically as

$$\delta\nu_n^{(2)} \sim \delta\nu_n^{(o)}\{\delta\nu_n^{(o)}/\nu_n\}, \quad \delta\nu_n^{(4)} \sim \delta\nu_n^{(o)}\{\delta\nu_n^{(o)}/\nu_n\}^2,$$

and, by induction

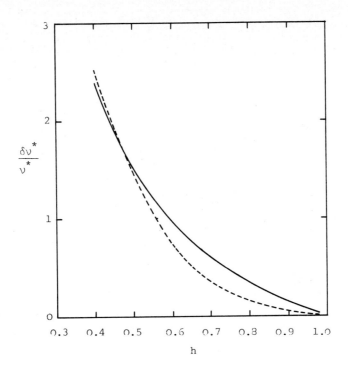

Fig. 5. Ratio of incremental change to effective viscosity at the fixed point, as a function of h. ———, k'=0; - - -, k'=1.

$$\delta v_n^{(2m)} \sim \delta v_n^{(o)} \{\delta v_n^{(o)}/v_n\}^m$$

for the moment of order 2m + 2.

In Fig. 5 we show a plot of $\delta v/v$ at the fixed point for k' = 0 and k' = 1, as a function of h. Clearly the neglect of correction terms of any order can only be justified for values of h greater than about h = 0.6.

At this stage, we are carrying this work further on two fronts. First, we are investigating the accuracy of these approximations in conjunction with a numerical simulation now being carried out on an array processor. Second we are exploring the analytical formulation further. Particularly - in view of certain aspects of the application of RG in critical phenomena - the nature of the theory for dimension above 3.

Much still has to be done. But, while the results presented here are rather preliminary in nature, we feel justified in claiming that the RG approach - implemented by iterative averaging - holds out the prospect of a subgrid theory based on rational approximations.

I would like to thank Dr. V. Shanmugasundaram for his help with the numerical work and with the preparation of this paper.

REFERENCES

[1] VOKE, P.R., COLLINS, M.W.: "Large eddy simulation: retrospect and prospect", PCH Physicochem. Hydrodyn., 4 (1983) pp. 119-161.

[2] ROGALLO, R.S., MOIN, P.: "Numerical simulation of turbulent flows", Ann. Rev. Fluid Mech., 16 (1984) pp. 99-137.

[3] KRAICHNAN, R.H.: "Eddy viscosity in two and three dimensions", J. Atm. Sci., 33 (1976) pp. 1521-1536.

[4] LESLIE, D.C., QUARINI, G.L.: "The application of turbulence theory to the formulation of subgrid modelling procedures", J. Fluid Mech., 79 (1979) pp. 65-91.

[5] CHOLLET, J.-P., LESIEUR, M.: "Parameterization of small scales of three dimensional isotropic turbulence using spectral closures", J. Atm. Sci., 38 (1981) pp. 2747-2757.

[6] WILSON, K.G.: "The renormalisation group: critical phenomena and the Kondo problem", Rev. Mod. Phys., 47 (1975) pp. 773-840.

[7] FORSTER, D., NELSON, D.R., STEPHEN, M.J.: "Long-time tails and the large-eddy behaviour of a randomly stirred fluid", Phys. Rev. Lett., 36 (1976) pp. 867-870.

[8] KRAICHNAN, R.H.: "Hydrodynamic turbulence and the renormalisation group", Phys. Rev. A., 25 (1982) pp. 3281-3289.

[9] ROSE, H.A.: "Eddy diffusivity, eddy noise and subgrid-scale modelling", J. Fluid Mech., 81 (1977) pp. 719-734.

[10] McCOMB, W.D.: "Reformulation of the statistical equations for turbulent shear flows", Phys. Rev. A., 26, (1982) pp. 1078-1094.

[11] McCOMB, W.D., SHANMUGASUNDARAM, V.: "Fluid turbulence and the renormalisation group: a preliminary calculation of the eddy viscosity", Phys. Rev. A., 28 (1983) pp. 2588-2590.

[12] McCOMB, W.D., SHANMUGASUNDARAM, V.: "Renormalisation group calculation of the eddy viscosity for isotropic turbulence", J. Phys. A.: Math. Gen., 18 (1985) pp. 2191-2198.

[13] McCOMB, W.D., SHANMUGASUNDARAM, V.: "Renormalisation methods applied to the calculation of the subgrid-scale eddy viscosity for isotropic turbulence", Paper presented to the XVIth Int. Conf. on Theor. Appl. Mech. (IUTAM), Lyngby, Denmark (1984).

[14] McCOMB, W.D.: "Renormalisation Group methods applied to the numerical simulation of fluid turbulence", Proc. NASA - ICASE Workshop on Theoretical Approaches to Turbulence, Springer-Verlag, Berlin (1986: to be published).

[15] LESLIE, D.C.: "Developments in the theory of turbulence", Oxford Univ. Press, London.(1973).

[16] MONIN, A.S., YAGLOM, A.M.: "Statistical fluid mechanics - Vol. II",

M.I.T. Press, Cambridge, Mass., U.S.A. (1975) pp. 18-21.

[17] McCOMB, W.D., SHANMUGASUNDARAM, V.: "Numerical calculation of decaying turbulence using the LET theory", J. Fluid Mech., <u>143</u> (1984) pp. 95-123.

HELICAL FLUCTUATIONS, FRACTAL DIMENSIONS AND PATH INTEGRAL IN THE THEORY OF TURBULENCE

E. Levich

The Institute of Applied Chemical Physics
and Physics Department
City College of the City University of New York

SUMMARY

The concept of helical fluctuations, fractal dimensions and path integral method are brought together in an attempt to formulate an adequate model for fine scales of turbulence. It is hoped that this model may be useful for the development of LES models.

INTRODUCTION

The Navier-Stokes (NS) equation is far too complex to be effectively used in practical problems of turbulent flows. Neither are the direct simulations, at least in the foreseeable future, capable of dealing with the high Reynolds (Re) number flows. Though the usefulness of direct simulations as an exponential tool on a par with laboratory experiments is quite apparent, still it would be highly desirable to have much simplified models of turbulence. Such models should be a subject of two obvious basic requirements. They should be able to describe adequately the large-scale properties of turbulence and be amenable to computer simulations more efficient than direct simulations of the NS equation. The basic approach of Large Eddy formulations (LES) is to represent the action of fine scales of turbulence on large scales in a certain averaged way. Thus a great economy in the number of independent modes can, in principle, be achieved. Unfortunately early hopes that the large-scale properties of turbulence are only weakly dependent on fine scales have been to a large extent shattered. We believe that, on the contrary, the interaction between different scales of turbulence is strong. Thus if we have no realistic model of fine scales of turbulence, it is doubtful that a realistic LES model can be constructed. On the other hand, LES models have until how been based on quite unrealistic models of fine-scale turbulence. Indeed, this is considered either in the framework of closures, such as direct interaction approximation (DIA), or by means of so-called infrared renormalization group perturbation theory. Both approaches neglect and as a matter of principle cannot account for the effects of intermittence of fine-scale turbulence, thereby, in our view, yielding highly distorted values of turbulent viscosity. On the other hand, should we have a more realistic model of fine-scale turbulence, not based on closures or related methods, this would probably allow the advancement of a basically sound concept behind the LES approach. Here we report on a certain attempt to develop such a model.

I. THE HELICAL NATURE OF TURBULENCE

Important for our vision of turbulence is a conjecture that the topological properties of inviscid flows may bear a direct relation to turbulence. In other words, it is our belief that the topological properties of streamlines and vorticity lines retain their importance even in viscous fluids in the limit of large Reynolds numbers [1-6].

Consider a deformable Lagrangian subdomain of inviscid fluid with the volume $V(L) \propto L^3$ and bounded by the streamlines having the kinetic energy

$$E(L) = \int_V \vec{v}^2 d^3r = \text{const.}$$

A simplified interpretation of a remarkable theorem [7] expanding on the results of [8,9] says that there is an uncountable infinity of Euler flows, distinguishable by their helicity and having the same kinetic energy $E(L)$. The maximal helicity is achieved in Beltrami flows. If $|\vec{\omega}|$ is bounded, then $|\lambda| < \lambda_{max}$. Hence, the maximal and minimal helicity is $\pm \lambda_{max} E$. There is an uncountable number of other Beltrami flows with different values of helicity H_i such that $-\lambda_{max} E \leq H_i \leq \lambda_{max} E$. Using the language of physics one can say that for a given energy the flows are infinitely degenerate by their topology. However, as the topology is conserved there is no transition from one flow to another. The topology of each of them serves as an infinite potential barrier. When viscosity, even infinitesimal, is introduced, the situation basically changes. The vorticity lines and streamlines can break and reconnect in a multitude of ways even in the limit $\nu \to 0 (\text{Re} \to \infty)$, since there would always be a set of points at which $\nu \nabla^2 \vec{v}$ is big, and thus one flow can tunnel into another. In other words, viscosity mixes up the topologies and, in this sense, creates an ensemble of all Euler flows (uncountable variety) and all transient flows. For an isotropic reflectionally symmetric "mixed" flow, the average helicity is zero, $\langle H \rangle = 0$.[1] However, each particular member of the ensemble can have any helicity, within the limits set above. The developed turbulence is a system extremely far from equilibrium. Therefore, it is only natural to expect that generally helicity will fluctuate strongly. Consider a volume V_0 of the mixed flow having $\langle H(L_0) \rangle = 0$. We argue that a part of the volume $V_1(\ell_1)$ has a non-zero helicity H_1 and the remaining part of the volume $V_2 \leq V_0 - V_1$ has helicity $-H_1$, so that $\langle H \rangle = 0$. The conjecture is that the most probable members of the mixed flow are the flows with maximal helicity. Mathematically, the

---

[1] Helicity is obviously a pseudoscalar.

conjecture would mean that most probably inside the helical fluctuations the relative helicity

$$H_{rel} = (\int \vec{v} \cdot \vec{\omega} d^3r)(\int \vec{v}^2 d^3r \int \vec{\omega}^2 d^3r)^{-1/2} = \pm 1. \quad (I.1)$$

Alternatively, we conjecture that the Euler flows with maximal helicity are less unstable, as compared with other helical flows. We observe that Euler flows yield zero for the non-linear terms of the NS equation. Thus, during the lifetime of helical fluctuation $\tau(L)$ the cascade of energy there is retarded. The basic assumption we have made is that simultaneously the energy dissipation is relatively small in helical fluctuations as well. Should this be true, then the Euler flows would have become quasi-steady flows for the NS equation as well. Thus, the regions of zero (or at least small) $[\vec{v} \times \vec{\omega}]$ are expected to be the regions of relatively small dissipation, i.e., ν curl $\vec{\omega} = \nu \vec{\nabla}^2 \vec{v}$ should be small almost everywhere in helical fluctuations.

To expand on the previous concept in order to make it self-similar and applicable to the inertial range of homogeneous turbulence, we assume that any volume $V(L_0)$ such that L_0 is a scale from the inertial range, may contain helical fluctuations of all scales $\ell_i < L_0$, belonging to the inertial range, and of different signs of helicity. Evidently, the helical fluctuations of all scales in the inertial range having the volume $V(\ell_{i+1}) < V(\ell_i)$ and the respective lifetime $\tau(\ell_{i+1}) < \tau(\ell_i)$, form a self-similar hierarchy of helical fluctuations. In all this hierarchy we expect the relative helicity to be close to unity, and the flows to be not far from the Euler flows with maximal helicity and minimal relative to each scale, dissipation. The helical fluctuations are separated from each other by sheets of instability, i.e., the vorticity sheets, the seats of large viscous dissipation. Since the fluctuations are of all scales the vorticity sheets are convoluted (curdled) at all scales. The implication is that the boundary between helical fluctuations of all scales is a fractal set with fractal dimensions D_F^i such that $2 < D_F^i < 3$. Thus, the concept of helical fluctuations plus the scaling assumption lead to a remarkable conclusion that the active part of turbulence, i.e., the regions with largest dissipation, are confined to fractal sets.

Clearly the above concept is a subject for comprehensive experimental scrutiny. Still, we have accumulated some data supporting at least some of the aspects of the helical picture. First is the analysis of data on coherent structures in laboratory and geophysical flows [4,10]. Also, numerical experiments conducted for different turbulent systems, the channel flow, decaying Taylor-Green vortex and homogeneous isotropic turbulence seem to indicate rather clearly the relevance of helicity to turbulence [11-14]. Finally, we mention that aesthetically the helical concept is attractive in that it embraces both the turbulent coherent structures and intermittence of fine scales of turbulence, the two fundamental features of turbulent flows.

II. FRACTAL CONCEPT AND INTERMITTENCE OF TURBULENCE

Historically there have been formulated several phenomenological or kinematic fractal models of turbulence. The simplest is the model of Novikov and Stewart [15]. This is usually called the model of fractally homogeneous turbulence, the FHT. Another model, probably the most relevant, is the multifractal hyperbolic model [16-20]. Though our real interest lies with hyperbolic models, for clarity we will discuss first a few properties of the FHT model. It is well known that the velocity structure functions in the FHT model are as follows [15]:

$$<\Delta v^n> \propto r^{n/3 - \mu/3(n-3)} \qquad (II.1)$$

where $\Delta \vec{v} = \vec{v}(\vec{x} + \vec{r}) - \vec{v}(\vec{x})$, Δv is the longitudinal projection of $\Delta \vec{v}$ on r-direction, and μ is the empirical exponent of intermittence. We note that if $\mu = 0$ we regain the Kolmogorov-like values of $<\Delta v^n>$. In particular, $<\Delta v^2> \propto r^{2/3}$ in physical space exactly corresponds to $E(k) \propto k^{-5/3}$ in k-space. We also remark that generation of a fractal set is always the result of a certain limiting procedure, in our case the $Re \to \infty$ or $\ell_d \to 0$. Thus, the translation of the results to the case of large but still finite Re as in the case of the Kolmogorov model implies the assumption of universal behavior in the above limit. It follows for the FHT model that the correlation in the rate of energy dissipation $\varepsilon(\vec{r},t) - <\varepsilon>$ is

$$<(\varepsilon - <\varepsilon>)(\varepsilon' - <\varepsilon>)> \propto \frac{<\Delta v^6>}{r^2} \propto r^{-\mu}. \qquad (II.2)$$

The phenomenological parameter μ unambiguously defines the fractal dimension $D_F = 3 - \mu$, the domain where a certain "active" part of turbulence, be it the molecular dissipation or the nonlinear transfer, is confined. It is also a matter of simple reasoning to conclude that the ratio of measures of fractal domain and the 3-D space is $\propto (r/L)^\mu$. This means that the "active" volume in the FHT model is $\propto r^\mu$.

Here we prove a slightly unusual and useful, for our further purposes, interpretation of the FHT model. The FHT model, like some other fractal models, violates the naïve scaling of Kolmogorov. By the naïve scaling we mean the following: It is usual to consider the velocity as a scaling field with the amplitude $v_{reg} \propto <\Delta v^2>^{1/2} \propto r^{1/3}$, so that $v_{reg}(ar) = a^{1/3} v_{reg}(r)$ or $[v_{reg}] = 1/3$. The exponent 1/3 gives full information about the velocity moments in the Kolmogorov picture. In the case of FHT we are compelled to introduce two scaling fields, corresponding to the averaging of velocity moments over the entire 3-D scale and that over the fractal domain D_F, only. Suppose formally that $\vec{v} = \vec{v}_{reg} + \vec{v}_{sing}$ where $\vec{v}_{reg}$ is nonzero everywhere in R^3 and $\vec{v}_{sing}$ is nonzero and homogeneous only in D_F, i.e., nonzero in a "volume" $\propto (r/L)^\mu$.[1] Then, it is not difficult to show that the following choice of exponents for the amplitudes v_{reg} and v_{sing}, $[v_{reg}] = 1/3 + \mu/6$ and $[v_{sing}] = 1/3 - \mu/3$, yields the structure functions (II.2). In particular,

$$<v_{sing}^2>_{3-D} \propto r^{2/3 - 2\mu/3} (\frac{r}{L})^\mu \propto r^{2/3 + \mu/3} \propto v_{reg}^2 . \qquad (II.3)$$

The above relation can actually serve as a definition of v_{reg}, as a homogeneous velocity component, obtained by averaging of the fractal singular velocity over the 3-D domain.

Next, we notice that space derivatives introduce no new anomalous dimensions. It is obvious that $[\omega] = [v] - 1$ and thereby for the nonlinear coupling term in the NS equation $[|\vec{\tau}|] = [|\vec{v} \times \vec{\omega}|] = 2[v] - 1 < 0$, so that $[<|\vec{\tau}^{sing}|>_{3-D}] = [|\vec{\tau}^{reg}|]$ and further $[<\vec{\tau}^{sing}>_{3-D}] = [\vec{\tau}^{reg}]$, where $\vec{\tau}^{reg} = \vec{\tau}\{\vec{v}_{reg}\}$ and $\vec{\tau}^{sing} = \vec{\tau}\{\vec{v}_{sing}, \vec{v}_{reg}\}$ [21].

A very remarkable class of fractal models is the hyperbolic models. These models, introduced in [16] as probably relevant to describe intermittent turbulent behavior, were later given significant support in the extremely valuable experimental observations of atmospheric turbulence and subsequent theoretical work [17-20]. The hyperbolic concept presumes the scaling of the basic amplitude field which in our notation is essentially $\vec{v}_{reg}$, but states that the probability of deviation from $\vec{v}_{reg}$ is a power law. In this sense, the hyperbolic models are superscaling, in that the statistical distributions of random fields of interest are scaling functions themselves. More formally, the probability distribution of a field $\Delta X = X(\vec{r} + \vec{x}) = X(\vec{x})$, or if X is a vector field of its longitudinal projection, is defined as follows:

$$Pr(\Delta X \gg \Delta X_{reg}) \propto (\frac{\Delta X_{reg}}{\Delta X})^\alpha \qquad (II.4)$$

---

[1]When we speak about velocity of a certain scale, what is really meant is the difference of velocities at this scale of separation.

with the meaning that all moments $\langle \Delta x^n \rangle = \infty$ for $n \geq \alpha$. Following [17-20], yet another interpretation of FHT model can be given. Consider a fluid domain in the shape of a cube of size L and split it into m cubes of size ℓ_1. Suppose that only in $n < m$ of these cubes ε or nonlinear transfer is not zero. If we iterate the splitting of cubes with the same requirements an infinite number of times we will recover the FHT model. This can be done formally as follows. Let us introduce a random function called a curdling operator W, such that $\langle W \rangle = 1$ and $\varepsilon_i = \varepsilon_{i-1} W$, $\Pr(W = \lambda^{-\mu}) = \lambda^{-\mu}$, $\Pr(W \neq \lambda^{-\mu}) = 0$, where ε_i is the dissipation in a cube of i-th generation and Pr means probability [17-20]. Evidently, such definition of W leads identically to the same values of structure functions as in the FHT model, i.e., it defines the same model. The situation is, however, quite unstable. Very innocuous at first sight, the change in definition of W leads to hyperbolic probability function and divergence of higher order moments. To demonstrate it, Lovejoy and Schertzer suggested the α-model. Let W be defined thus [17]: $\Pr(W = a_0 = \lambda^{1/h(D-D_\infty)} = \lambda^{-\mu_\infty/j}) = \lambda^{\mu_\infty}$, $h > 1$; $\Pr(W = a_1 \neq a_0) = 1 - \lambda^{\mu_\infty}$, where a_1 is determined from the condition $\langle W \rangle = 1$. The difference in definitions of W in the FHT and α models is clear. In the former, as we pass from one to another generation of cubes[1] the dissipation is always either zero or some definite value. In other words, the differential probability is a δ-function. In the α-model, with certain probability $W = a_1$. It is quite evident that after many steps of cascade the resulting dissipation may acquire any value. Since we preserve a self-similarity as we pass from one generation of cubes to another (self-similar cascade), the resulting probability function of ε should be self-similar as well. In other words, it should be a power law. Moreover, not only that, the moments $n \geq \alpha$ are divergent, when the averaging is done over the R^3 volume. Consider a smaller set $D_{A_i} < 3$, such that $D_{A_i} > D - \mu_\infty = 3 - \mu_\infty$. Then certain moments starting from $\langle \varepsilon^{m_i} \rangle_{D_{A_i}}$ where $m_i > n = \alpha$, are also divergent. As we decrease the dimension D_{A_i}, the value of m_i grows, until we reach $D_\infty = D - \mu_\infty$, where all the moments are convergent $\langle \varepsilon^n \rangle_{D_\infty} = \varepsilon^n|_{D_\infty}$. In other words, the hyperbolic probability introduces a new kind of fractal, defined generally by an infinite set of fractal dimensions, measures, converging in the limit to a certain D_∞. This is a set where the most singular part of dissipation (and related quantities) is seated. Less singular dissipation is seated at a larger set, etc. Generally, the structure of a fractal is very complicated, and in contrast to the FHT model, requires the knowledge of an infinite number of parameters (dimensions). It is remarkable that the only

---

[1]The geometry of cascade elements is quite irrelevant [5].

requirement needed, to lose the simple properties of the FHT model, is $\langle W^n \rangle > \lambda^{(1-n)\mu}A$ [17]. As is emphasized in [16], in practical terms the moments of hyperbolic random functions are not infinite. In particular in turbulence, the limitation is always present since Re is finite and this introduces a cutoff on the number of cascade steps. With respect to measurements, for any finite, though large value of Re, a natural cutoff is introduced by scales from the inertial range $\Delta r - r$. Indeed, the number of statistically independent measurements is proportional to (L/r). Following [17] we define the statistics of velocity field as follows:

$$\langle \Delta v^n \rangle = \frac{1}{P} \sum_{i=1}^{P} \Delta v_i^n = r^{\beta(n)} \qquad (II.5)$$

where P is the number of statistically independent measurements. Let us also introduce a dimensionless $X = (\Delta v_i)/(\Delta v_{i\,reg})$, where $[\Delta v_{reg}] = 1/3$, the Kolmogorov exponent. Hence

$$\langle \Delta v^n \rangle = r^{n/3} \cdot \frac{1}{P} \sum_{i=1}^{P} (X_i)^n . \qquad (II.6)$$

If $n < \alpha$, then $\langle \Delta v^n \rangle \propto r^{n/3}$, i.e., we have the Kolmogorov scaling law. If $n \geq \alpha$, then the situation is quite different, since the moments of X^n are divergent as $Re \to \infty$, and one applies a theory of Levy stable laws for divergent expectation values. Evidently we have the following relation:

$$\langle X^{n>\alpha} \rangle = \frac{1}{P} \sum_{i=1}^{P} X_i^n \simeq \frac{P(X_{max})}{P} X_{max}^n . \qquad (II.7)$$

It is useful to introduce a differential probability distribution

$$P(X)\,dX \propto \frac{dX}{X^{\alpha+1}} \qquad (II.8)$$

so that

$$\langle X^{n>\alpha} \rangle \propto X_{max}^{n-\alpha} . \qquad (II.9)$$

Comparison of (II.9) and (II.7) yields

$$\langle X^{n>\alpha} \rangle \propto \left\{ \frac{P(X_{max})}{P} \right\}^{\frac{n-\alpha}{\alpha}} . \qquad (II.10)$$

The hyperbolic law of probabilities is not sufficient to determine uniquely the corresponding multifractal structure of singular regions. Though for $n \geq \alpha$ and finite Re the structure functions are not universal, i.e., they are Re dependent, we can ask a question: what is their dependence on the separation scale r? The result is essentially dependent on the process of measurement. For isotropic turbulence in a vessel of size L and the separation scale r the number of statistically independent measurements $P \propto L/r$. The number of occurrences $P(X_{max}) \propto r^{-\eta}$, where η is generally an unknown parameter. Then formally

$$\frac{P(X_{max})}{P} \propto \left(\frac{1}{r}\right)^{1-\eta} . \qquad (II.11)$$

The factor $r^{1-\eta}$ can be associated with the fractal "volume," i.e., the relative volume occupied by a set of X_{max}, with $2 + \eta = D(X_{max})$ being the fractal dimension. We observe that $\eta > 0$. This is due to a very general observation by Mandelbrot that physically significant intermittence should be produced by a fractal set necessarily having the dimension more than two. Finally, we have the following estimate:

$$\langle \Delta X^n \rangle^\alpha \propto \left(\frac{1}{r}\right)^{(n-\alpha)(1-\eta)/\alpha} . \qquad (II.12)$$

Now we derive easily that $v_{sing}(X_{max}) \propto r^{1/3} X_{max} \propto r^{(1/3)-(1/\alpha(1-\eta))}$ and the respective time scale $t(X_{max}) \propto r/v_{sing} \propto r^{(2/3)+(1/\alpha(1-\eta))}$.

In our view, the hyperbolic law of probability is immensely relevant for turbulence, and for the above helical concept in particular. Indeed, as we perceive turbulence as a collection of helical fluctuations, where the flow is one of the unstable Beltrami flows, the $\vec{v} \times \vec{\omega}$ term and dissipation are nowhere zero, but take a spectrum of values. That is, the α-model or similar models are quite suitable to describe the hierarchy of helical fluctuations (structures). Moreover, it seems that some very special circumstances should take place, for any other than a hyperbolic probability law to realize. In more general terms, the hyperbolic model seems to reflect a fundamental *unpredictability* of turbulent flows. It would be interesting to look for an interaction with the recent ideas on this subject [22].

III. FRACTALS, PATH INTEGRALS AND TURBULENCE

One of the indispensable theoretical tools so familiar to physicists is the path integral. Though often quite ad hoc from the point of view of rigorous mathematics, the path integral formulation suits superbly many nonlinear problems of statistical mechanics and field theories. The path integral formulation of turbulence was developed quite independently [15]. Still, so far it has not found efficient practical applications. Here we will try to bring together the fractal concept and the path integral formulation to produce quantitative results with respect to the velocity structure functions in the limit of Re $\to \infty$. The work is tentative, and this allows a liberty of bold mathematical conjectures. The relevance of final results when they are set against experimental data should help to form a preliminary judgment with respect to the reason of the approach.

Let us consider the following functional:

$$Z = \lim_{\Delta \to 0} N(\Delta)^{-1} \int D\vec{v} \exp\{-\frac{F^{(2)}\{v\}}{\Delta^2}\} \qquad (III.1)$$

where

$$F^{(2)} = \int d^3r dt \{\frac{\partial \vec{v}}{\partial t} - \vec{\tau}\{v\} - \nu \vec{\nabla}^2 \vec{v} - \vec{\xi}_L\}^2 . \qquad (III.2)$$

The integration is done over all $\vec{v}$-field configurations in the infinitely dimensional space of functions, $N(\Delta)^{-1}$ is the normalizing factor, $\vec{\xi}_L$ is an arbitrary time dependent forcing at the large scale L only. It is not difficult to prove that the integrand in (III.1) is a functional δ-function [21] $\lim_{\Delta \to 0} N^{-1} \exp(-F^2/\Delta^2) = \delta\{\vec{v} - \vec{v}_{NS}\}$, where $\vec{v}_{NS} = \vec{v}_{NS}(\vec{\xi}_L)$ is a solution of the NS equation for a given forcing $\vec{\xi}_L$ and initial condition. As far as $\vec{\xi}_L$ is strictly deterministic the solution $\vec{v}_{NS}$ is also quite deterministic and (III.1) is just an identity. If a certain margin of inaccuracy is introduced in the definition of $\vec{\xi}_L$, the $\vec{v}_{NS}$ will presumably become chaotic, at least in the limit of Re $\to \infty$, and probably quite independently of a specific definition of $\vec{\xi}_L$. Let us choose the simplest $\vec{\xi}_L$ as a random function with a Gaussian probability distribution in functional space, e.g. (here and hereafter the nonessential normalizing factors are omitted),

$$W\{\vec{\tilde{\xi}}\}D\vec{\tilde{\xi}} = \{\exp - \int d^3k d\omega a_{ij} \tilde{\xi}_i \tilde{\xi}_j\}D\vec{\tilde{\xi}} \qquad (III.3)$$

where $\vec{\tilde{\xi}}(\vec{k},\omega) = \int \vec{\xi}_L(\vec{r},t) e^{i\vec{k}\vec{r}-i\omega t} d^3r dt$ and the choice of the matrix $a_{ij}(\vec{k},\omega)$ is determined by the choice of the forcing $\vec{\xi}_L$. In particular, if $\vec{\tilde{\xi}}(\vec{k},\omega) \to 0$ for $k > k_0$, then $|a_{ij}(k > k_0)|^{-1} \to 0$. Substituting (III.3) into (III.1) and doing the exact Gaussian integration over all configurations of $\vec{\tilde{\xi}}$ yields the generating functional [21]

$$W = \exp - \tilde{F}^{(2)} \qquad (III.4)$$

$$\tilde{F}^{(2)} = \int a_{ij} \vec{NS}_i \vec{NS}_j$$

where

$$\tilde{NS}_i = -i\omega v_i(\vec{k},w) - \frac{i}{2} P_{ijs}(\vec{k}) \int d^3q\, d\Lambda \cdot$$

$$\cdot v_j(\vec{k},\omega) v_s(\vec{k}-\vec{q},\omega-\Lambda) + \nu k^2 v_i(\vec{k},\omega). \quad \text{(III.5)}$$

The meaning of $W\{\vec{v}; a_{ij}\}$ as a probability distribution function in the infinitely dimensional space of functions is perfectly clear. The matrix a_{ij} plays formally a role of an inverse "temperature." Evidently, $W\{\vec{v}; a_{ij}\}$ allows one to calculate all velocity correlation functions as follows:

$$< \prod_{i=1}^n \vec{v}(\vec{r}_i t_i) > = \int D\vec{v} \prod_{i=1}^n \vec{v}(\vec{r}_i t_i) \exp - \tilde{F}^{(2)}. \quad \text{(III.6)}$$

It is needless to say that presently there are no general analytical methods to calculate complicated functional integrals, that are such that cannot be reduced to Gaussian integrals. Here we suggest a particular technique based on beforehand farfetched assumptions on the nature of turbulence and subsequent proof of self-consistency of the assumptions. Clearly the self-consistency is not, however, a proof of validity.

We assume that the nature of turbulence is adequately represented by the hyperbolic models as described previously. This allows us to introduce a sequence of most simplifying assumptions in what follows.

A mere consistency with the hyperbolic models tells us that there should be an exponentially large number of contributing $\vec{v}$-configurations in (III.6). Indeed, a naïve estimate of contributing $\vec{v}$-configurations (in 3 + 1 dimensional space) would be that the configurations are essential if $\tilde{F}^{(2)} \ll 1$. All other configurations give exponentially small contributions. But this estimate is good only provided that W is genuinely exponentially decaying Gaussian distribution. Should this be true, then all velocity moments would have been trivial and essentially determined by the inverse temperature $a_{ij}(\vec{k},\omega)$. In particular since $|a_{ij}|^{-1} \to 0$ for $k > k_0$, the only contributing $\vec{v}$-configuration would have been an almost purely decaying "low temperature" solution of the NS-equation,

$$|\tilde{NS}|^2 < \frac{|a_{ij}(k > k_0)|^{-1}}{k^3 \omega} \to 0, \text{ for each scale k and frequency } \omega.$$

Thus if we are to expect a nontrivial velocity moments behavior we must expect at the same time that the $\vec{v}$-configurations such that $\tilde{F}^{(2)} \gg 1$, are of dominant importance, though each of these configurations gives an exponentially small contribution to the path integral. This is possible only if the number of such configurations is exponentially large. It is also not difficult to realize that, strictly speaking, such nonanalytical behavior of (III.6) is possible if the number of configurations tends to infinity. But this can be only if Re $\to \infty$. This is reminiscent of the critical phenomenon, when a nonanalytical behavior of the statistical sum is expected at the critical temperature in the continuous limit, when the size of a system tends to infinity. What can we say about $\vec{v}$-configurations such that $\tilde{F}^{(2)} \gg 1$? Evidently they can be considered as solutions of the renormalized NS equation $\vec{\tilde{NS}} = \vec{\tilde{\xi}}'$, where $\vec{\tilde{\xi}}'$ is generally quite unknown random function. This function determines the extent of deviation from the purely decaying solution of the NS equation. On the other hand, it can be seen as a driving force in the NS equation. Let us decompose $\vec{\tilde{\xi}}'$ on a Gaussian and non-Gaussian part, $\vec{\tilde{\xi}}' = \vec{\tilde{\xi}}'_{reg} + \vec{\tilde{\xi}}'_{sing}$. Then we can formulate a new generating functional for the renormalized NS equation. Evidently this will be

$$W_{ren} = \exp - \tilde{F}^{(2)}_{ren} \qquad (III.7)$$

$$\tilde{F}^{(2)}_{ren} = \int a^{ren}_{ij} (\vec{\tilde{NS}} - \vec{\tilde{\xi}}'_{sing})_i (\vec{\tilde{NS}} - \vec{\tilde{\xi}}'_{sing})_j d^3k d\omega$$

where a^{ren}_{ij} is a renormalized temperature distribution determined by a "forcing" $\vec{\tilde{\xi}}'$. As $\vec{\tilde{\xi}}'$ corresponds to a higher, in comparison with $\vec{\tilde{\xi}}$, level of fluctuations at $k > k_0$, consistently $|a^{ren}_{ij}|^{-1}$ is a higher "temperature" at $k > k_0$. Now the velocity moments can be calculated as in (III.6) but instead of W the renormalized W_{ren} can be substituted. Now that we extended the number of $\vec{v}$-configurations such that $\tilde{F}^{(2)}_{ren} \gg 1$ should be accounted for, since many $\vec{v}$-configurations corresponding to $\tilde{F}^{(2)} \gg 1$ are among those corresponding to $\tilde{F}^{(2)}_{ren} \ll 1$. The hope is that there is such a fixed point $\vec{\tilde{\xi}}_f$ or, alternatively, such a "temperature" distribution $|a^f_{ij}|$, that a final number of terms would be sufficient to calculate at least some of the velocity moments. In other words, there is such a forcing that the corresponding generating functional W_f would be almost a Gaussian one. But what makes us believe that such renormalization does exist? The reason for this is the initial assumption of the relevance of the hyperbolic models, implying that a finite number of moments is almost trivial, $\langle \Delta v^n \rangle \propto \langle \Delta v^2 \rangle^{n/2}$, a relation typical for Gaussian or quasi-Gaussian probability distribution. On the other hand, it is quite clear that the renormalization is definitely not possible for the high order moments $n \geq \alpha$. If such renormalization were possible then these moments should have been almost trivial and could not have been divergent. On the other hand, the renormalization most probably should be possible for the moments $n < \alpha$. We note that the definition of $\tilde{\xi}_\zeta^f$ is very broad, so that it may include a dependence on the velocity field itself. This is an important concomitant of the renormalization procedure in general.

Thus we assume that the lower velocity moments are given by a solution of the following problem:

$$< \prod_{i=1}^{n<\alpha} \vec{v}(\vec{r}_i t_i) > = \int D\vec{v} \prod_{i=1}^{n<\alpha} \vec{v}(\vec{r}_i t_i) \cdot \exp - \tilde{F}_f^{(2)} . \quad (III.8)$$

Next, we suppose that contributing $\vec{v}$-configurations can be decomposed in the spirit of the above section, $\vec{v} = \vec{v}_{reg} + \vec{v}_{sing}$. The path integral in (III.8) can be understood as a limiting case of the summation $\sum_{\vec{v}_{reg}} \sum_{\vec{v}_{sing}} \prod_{i=1}^{n<\alpha} \vec{v}(\vec{r}_i t_i) W_f$, or, put differently, we can say:

$$\int D\vec{v} \prod_{i=1}^{n<\alpha} \vec{v}_i W_f = \int D\vec{v}_{reg} D\vec{v}_{sing}$$

$$\prod_{i=1}^{n<\alpha} \vec{v}_i W_f = \int D\vec{v}_{reg} < \prod_{i=1}^{n<\alpha} \vec{v}_i W_f >_{\vec{v}_{sing}} \quad (III.9)$$

and since $< \prod_{i=1}^{n<\alpha} \vec{v}_i > \simeq < \prod_{i=1}^{n<\alpha} \vec{v}_{i\,reg} >$ we obtain

$$< \prod_{i=1}^{n<\alpha} \vec{v}_i > = \int D\vec{v}_{reg} \prod_{i=1}^{n<\alpha} \vec{v}_{i\,reg} < W_f\{\vec{v}_{reg}, \vec{v}_{sing}\} >_{\vec{v}_{sing}} \quad (III.10)$$

where

$$<W_f>_{\vec{v}_{sing}} = \int D\vec{v}_{sing} W_f . \quad (III.10a)$$

The approximation we suggest is as follows. Instead of calculating faithfully the average over all possible v_{sing} we shall try to estimate the contribution of an averaged singular trajectory. Specifically, the following substitution is made:

$$\tilde{F}_f^{(2)}\{\vec{v}_{reg},\vec{v}_{sing}\} \to \langle \tilde{F}_f^{(2)}\{\vec{v}_{reg},\vec{v}_{sing}\}\rangle_{3-D} \qquad (III.11)$$

where averaging is performed as before over elements of 3-D space, or alternatively over a power law tail of the probability distribution function. Consequently,

$$\langle W_f\rangle_{\vec{v}_{sing}} \to \exp - \langle \tilde{F}_f^{(2)}\rangle_{3-D} \qquad (III.12)$$

which is of course a certain mean field approximation. Let us consider what such an approximation is able to achieve. As a result of decomposition of $\vec{v}$ on $\vec{v}_{reg}$ and $\vec{v}_{sing}$, the functional $\tilde{F}^{(2)}$ also decomposes on $\tilde{F}_{reg}^{(2)}\{\vec{v}_{reg}\}$ and $\tilde{F}_{sing}^{(2)}\{\vec{v}_{sing}\}$. The integrand in $\tilde{F}_{sing}^{(2)}$ is large, but in a small fraction of the volume tending to zero as $Re \to \infty$ and $r \to 0$. As a result, $\tilde{F}_{sing}^{(2)}$ should not necessarily be large as compared with $\tilde{F}_{reg}^{(2)}$. To be more precise, let $\tilde{F}_{reg}^{(2)}\{\vec{v}_{reg}\}$ be a scaling functional with a certain exponent $e_1(z)$. This means that a scaling transformation $\vec{r} \to a\vec{r}$, $t \to a^z t$, where z is an unknown dynamical exponent to be determined, yields the transformation $\tilde{F}_{reg}^{(2)} \to a^{e_1(z)}\tilde{F}_{reg}^{(2)}$. The statement is that simultaneously $\langle \tilde{F}_{sing}^{(2)}\rangle_{3-D} \to a^{e_1(z)}\langle \tilde{F}_{sing}^{(2)}\rangle_{3-D}$ with the accuracy of leading terms.

The damage we have caused in using the approximation (III.12) is that the resulting generating functional, as we shall see, is almost Gaussian. Hence no nontrivial correlation function can be calculated with its help. But within the framework of the hyperbolic models the moments are either "trivial" or do not exist, strictly speaking, at all. Still, quite nontrivial results will follow.

To start with, the convergence of the above average would require certain limitations with respect to the value of α in (II.4). To estimate it we suppose that ω^2 fluctuates the same way as ε does, a plausible assumption usually accepted in the phenomenological theories of turbulence and valid for the particular models considered here. Also, $\varepsilon \propto v^3/r = (v_{reg}^3/r)X^3$. Qualitatively, $\langle v^2\omega^2\rangle \propto \langle v^2\varepsilon\rangle \propto v_{reg}^5/r\langle X^5\rangle$ and exists provided $\alpha > 5$. Therefore, we shall accept this fairly safe inequality as an initial input.

The next important point is the choice of dynamical scaling exponent. The scaling demands that time or frequency do not appear directly in expressions for correlation functions, but only as $r^{-z}t$ or ω/k^z. In the Kolmogorov theory this time scaling can be easily determined. Indeed, the typical velocity of an element of scale r is $v \propto \bar{\varepsilon}^{1/3} r^{2/3}$. Thus the typical time is $t \propto r/\bar{\varepsilon}^{1/3} r^{1/3} = \bar{\varepsilon}^{-1/3} r^{2/3}$ or $\omega \propto \bar{\varepsilon}^{1/3} k^{1/3}$ and subsequently $z = 2/3$. In any fractal model the situation is much more complicated, since there is no one scaling velocity. For example, the FHT model is defined by $v_{reg} \propto r^{(1/3)-(\mu/6)}$ and $v_{sing} \propto r^{(1/3)-(\mu/3)}$. How to determine the time scaling in this situation? We are helped by the following simple reasoning. The smallest time is the time characterized by the largest velocity $t_F \propto r/r^{(1/3)-(\mu/3)} = r^{(2/3)+(\mu/3)}$. This time is instantaneous by comparison with the slow time $t_F \propto r/r^{(1/3)+(\mu/6)} = r^{(2/3)-(\mu/6)}$ in the limit $r \to 0$. However, any change in the fractal field component should be matched by a corresponding change of velocity field in structures. Analogously, an appropriate time for the hyperbolic model it is reasonable to suppose that $t_F = t(X_{max})$ $\propto r^{[(2/3)+(1/\alpha(1-\eta))]}$, hence $z = 2/3 + 1/\alpha(1 - \eta) = 2/3 + \mu/3$. Nevertheless, the characteristic time of viscous energy dissipation does not change in comparison with the Kolmogorov time, since the spectral function is the same as in Kolmogorov theory. Therefore, in the limit of small scales the typical time of velocity fluctuations is much faster than that corresponding to the physical time of energy dissipation. This, at first sight, "contradiction" in fact has quite important implications.

The advantage of $\exp - <\tilde{F}_f^{(2)}>_{3-D}$ functional is in that it does not have anomalously fluctuating terms. Thus an enormous power of scaling analysis can be used. In fact all correlation functions $n < \alpha$ can be calculated in the asymptotic limit of small scales and short time $(k,\omega) \to \infty$ [21]. It is more customary, however, to pass over again to a differential equation corresponding to this functional in the same spirit we have done it previously. It can be shown that the following equation is appropriate [21]:

$$\frac{\partial v_{i\,reg}}{\partial t} - \tau_i\{\vec{v}_{reg}\} - \nu(\vec{\nabla})^2 v_{i\,reg} \qquad (III.13)$$

$$= \xi_i - \phi v_{i\,reg} + a_{ijm} v_{j\,reg} v_{m\,reg} + \ldots \text{ higher order}$$

nonlinearities $+ \ldots$

The ξ_i term in (III.13) can be considered as a free term differently with our previous understanding of $\vec{\xi}_{reg}^f$. Eq. (III.13) can be understood as follows. The l.h.s. is the usual NS equation for the regular part of velocity and corresponding to the regular part of the $<\tilde{F}_f^{(2)}>$ functional. The r.h.s. is the averaged contribution of the singular fractal configurations. This contribution is subject to restrictions provided by the properties of the hyperbolic models and the assumption of scale invariance of the field $\vec{v}_{reg}$. These require that the nonlinear terms on the r.h.s. should be no more singular than that in the l.h.s. of (III.13). These serve as constraints on the scaling rules for the scaling *operators* ϕ, a_{ijm}, etc., as functions of position and time. In particular, $[\phi] = [t^{-1}] = -z = -2/3 - \mu/3$, $[a_{ijs}] = -1$, $[\phi \vec{v}_{reg}] = [\vec{\xi}] = [\frac{\partial \vec{v}_{reg}}{\partial t}] = +1/3 - z = -1/3 - \mu/3$. Otherwise, we cannot and should not try to constrain the action of the fractal velocity field component. If we believe in scaling on the one hand and the previous considerations with respect to the almost Gaussian nature of reduced generating functional then a solution of (III.13) should not depend on anything but the scaling properties of nonlinear couplings.

We would like to stress that the above scaling universal concepts are not new and found a tremendous success in various nonlinear problems, such as the second order phase transitions. Unfortunately, they cannot be applied to the NS equation directly, because of the highly irregular nature of the turbulent velocity field $\vec{v}$. But the scaling concepts can be successful again when we deal with the averaged $\vec{v}_{reg}$-field.

We note that the r.h.s. can be simply understood as a Reynolds stress caused by a fractal component of turbulence upon the homogeneous component. The difference, however, is that, in contrast to the usual reasoning leading to the concept of Reynolds stress, here the fractal and homogeneous components of turbulence do not have different scales in space and time, but rather they belong to different subspaces. On the other hand, we are able in effect to "close" the equation by virtue of fractal and scaling assumptions.

It is convenient to rewrite Eq. (III.13) in (k,ω) space notable for practical calculations. For simplicity we now define $\vec{v}_{reg}$ as a new $\vec{v}$. Then we have

$$v_i(\vec{k},\omega) = \xi_i^{ren}(\vec{k},\omega) G(\vec{k},\omega) - \frac{1}{2} i\lambda_0 G(\vec{k},\omega) \eta_{ijn}(\vec{k},\omega) \cdot$$

$$\cdot \int d^3q d\Lambda v_j(\vec{k},\omega) v_n(\vec{k}-\vec{q},\omega-\Lambda)$$

$$+ \text{higher order nonlinearities}) \qquad (III.14)$$

where the Green function $G = (-i\omega + Ck^z)^{-1}$, the scaling function $\phi(k,\omega)$ is substituted in the light of previous discussion by a simplest scaling power k^z, $C = \text{const}$, $\eta_{ijs}(\vec{k},\omega)$ is a vertex combined from the $\vec{\tau}^{reg}$ and "Reynolds stress" parts

$\tilde{a}_{ijs}(\vec{k},\omega)$, hence $[\eta_{ijs}(\vec{k},\omega)] = 1$. The scaling properties of $\xi^{ren}_{(\vec{k},\omega)}$, a homogeneous function by definition, can be determined from the knowledge of $[\vec{v}(r,t)] = 1/3$ and $[t] = 2/3 + \mu/3 = 2/3 + 1/\alpha$. Then it is easy to determine that in $(\vec{k},\vec{\omega})$ space

$$<\xi^{ren}_i(\vec{k},\omega) \xi^{ren}_j(\vec{k}',\omega')> = \frac{A}{k^y} \delta(\vec{k}+\vec{k}') \delta(\omega+\omega') (\delta_{ij} - \frac{k_i k_j}{k^2});$$

$y = 3 - \mu/3$, $A = $ const. (III.15)

Equation (III.14) together with (III.15) can be solved perturbatively with the help of modified renormalization group, RG methods. More precisely, the asymptotic solution $(\vec{k},\omega) \to \infty$ can be determined [21]. It should be stressed than the application of RG technique to Eq. (III.14) has little to do with recent attempts to apply the RG methods directly to the NS equation. The latter attempts, as was clearly demonstrated by Kraichnan [26], are bound to encounter exactly the same difficulties as any other perturbative scheme, in that they are unable to take account of the intermittent fractal nature of small scale turbulence. Failure to do so inevitably results in a wrong turbulent viscosity and forcing term.

In its turn the behavior of large-scale turbulence can be adversely affected by the inaccuracy of small scales description. Nevertheless, the RG technique becomes useful when applied to Eq. (III.14), where no anomalously fluctuating terms are left. The nontrivial asymptotically free solution of (III.14) yields a unique value of parameter α, μ, namely,

$$\mu = 0.5 \tag{III.16}$$

with logarithmic accuracy. At the same time we are able to pin down the value of α to a surprisingly close interval. Indeed, using the relation between μ and α and supposing that $\eta = 0$ we determine the maximal value of $\alpha_{max} < 6$. Recall that our previous estimate was $\alpha_{min} > 5$. Hence $5 < \alpha < 6$.

IV. IMPLICATIONS AND COMPARISON WITH EXPERIMENT

The knowledge of μ and α gives a certain information about the hyperbolic probability law and pseudomoments $<\Delta v^n>$. Indeed, our theoretical prediction is that all velocity moments starting at least from $<\Delta v^6>$ will, strictly speaking, diverge. The sixth moment divergence may be weak, however. As the fluctuations of the rate of energy dissipation ε coincide with the fluctuations of $\Delta v^6/r^2$ (at least we tacitly assume this throughout the paper), the implication is that, for example, the theoretical value of the correlation function at $\alpha = \alpha_{max} < 6$ is

$$\langle(\varepsilon - \langle\varepsilon\rangle)(\varepsilon' - \langle\varepsilon\rangle)\rangle \propto \ln\left|\left(\frac{L}{2}\right)\right|^4, \quad Re \gg 1 \qquad (IV.1)$$

and not the power law usually associated with this correlator.[1] The relation (IV.1) was derived with due respect to the logarithmic corrections for the value of α [21].

The subject of controversy may come from a nonuniversal form of $\langle\varepsilon\varepsilon'\rangle$ correlation function. Indirectly, there is reason to believe that the value of $\mu \sim 0.2$ is suitable in a rather extensive interval of Reynolds numbers. Whether this dependence can be reasonably extrapolated by a power of logarithm or this is ruled out is hard to say unambiguously on the basis of available data. It is plausible, however, that the measurement of μ can be subject to significant fluctuations, while the Reynolds number grows, and the energy dissipation fluctuations approach their asymptotic regime.

Also, logarithmic corrections to the value of μ may influence the values of lower velocity moments, in particular the skewness of the velocity derivatives, even though these moments are convergent [21].

A matter of significant interest is that the theory is consistent with the Kolmogorov spectrum with no power law corrections at all, but with a possibility of logarithmic corrections. In the framework of hyperbolic model the power law corrections to the Kolmogorov spectrum are ruled out. The result is [21]:

$$E(K) \propto \frac{1}{k^{5/3}(\ln kL)^{1/2}}. \qquad (IV.2)$$

The presence of logarithmic correction in (IV.2) is very hard to detect. Still, we point out a certain discrepancy in the value of the Kolmogorov constant as it follows from fairly low Reynolds number numerical simulations $C_k \approx 2.8$ [25], and the usually accepted value of $C_k \approx 1.8 \div 2$ as it is determined from higher Reynolds numbers laboratory experiments. This decrease of C_k may be consistent with logarithmic nonuniversal correction. Also, a general comment with respect to the spectrum (IV.2) is as follows. In the Kolmogorov theory there is only one time scale t_d and one corresponding length scale of viscous dissipation ℓ_d. We have mentioned that here we have two time scales, one for viscous dissipation coinciding with t_d and another much smaller time typical of the value of turbulent viscosity. As a consequence we shall have two length scales, ℓ_d and $\ell_d' \gg \ell_d$, the latter being dependent on the Reynolds number. We are not definite about the importance

---

[1] In fact at $\alpha = \alpha_{max}$ this correlator is a complicated function of $\ln\left|\frac{L}{r}\right|$ and behaves like (IV.1) only asymptotically for $\frac{L}{r} \to \infty$.

of this circumstance. Still the inclination is to suspect a nontrivial behavior of the spectrum at the edge of interval range $r \sim \ell_d$. We suspect that the inertial range does not merge with the viscous subrange smoothly.

The relevance of hyperbolic distribution was supported as a result of analysis of geophysical data [17-20]. On the other hand, in a variety of laboratory experiments, the probability function is a subject of concern as well. If, indeed, the hyperbolic distribution is the one realizing in turbulent flows, why did it go unnoticed in experiments? The answer may be as follows. On the one hand, the difficulties in measurement of probability distribution functions are notorious. Still, it may be that the hyperbolic distribution simply has not been searched for. Here we provide certain indications of the relevance of hyperbolic distribution on a basis of comprehensive experiment [24] and subsequent analysis of experimental data provided by this experiment. Specifically, we shall use the data for axisymmetric jet at $Re_\lambda = 536$. The figures from [24] provide details of the probability function $P(X)$ where $X = \Delta v / \langle \Delta v^2 \rangle^{1/2}$ and $\Delta \vec{v} = \vec{v}(\vec{x} + \vec{r}) - \vec{v}(\vec{x})$. In the notations of the present paper X defines the fluctuations from the $\Delta v_{reg} = \langle \Delta v^2 \rangle^{1/2}$, i.e., the fluctuations (relative) of $\Delta \vec{v}_{sing}$. The average is given by (II.9).

If Pr is hyperbolic with the exponent $\alpha = 6$, this means that $Pr = A/X^7$, where $A = A(X)$ can actually retain a logarithmic dependence on X. We note that this distribution is expected in the limit of $Re \to \infty$. As experimental values of Reynolds numbers in laboratory experiments are rather low, $Re \approx 20 \times 10^3$ (corresponding to $Re_\lambda = 536$), we cannot expect that for all values of X it will have the same form. It was emphasized in the original paper [16] that in practical terms one should expect to have a sufficiently long interval of values of X where a hyperbolic law holds. In particular, if $\alpha = 6$, there should be a sufficiently long interval of values of X, where $X^7 P(X) \approx$ const. This would mean that the value of $\langle X^6 \rangle$ is essentially determined by this interval. This, however, should not be true for the lower order moments. Close inspection reveals that this is indeed the case for $P_\lambda(X)$, as presented in [24]. A simple estimate shows that in the interval $3 < |X| < 6$, $X < 0$, the power law with $\alpha \approx 6$ is a good approximation. Indeed, in this interval, the value of A changes only in the interval $22 \leq A \leq 32$. We are deliberately crude in our estimate, since we believe that the real physical significance should be revealed rather obviously.

Also, it is a matter of simple estimations to show that the power law probability function with $\alpha \approx 6$ gives surprisingly good numerical values of the moments, starting from $\langle \Delta v^6 \rangle$ and up, with a quite small accumulation of mistakes up to $\langle\Delta v^{12}\rangle$. Thus we have come to a rather extraordinary conclusion, made on the basis of regular and reproducible experimental data, that indeed a power law with a theoretically calculated slope is a good approximation of the experimental curve of $P(X)$. A particular conclusion we draw is that in the range of Reynolds numbers usually employed in laboratory flows, a nonuniversal law for $\langle\Delta v^6\rangle$ can be easily confused with a genuine power law.

The comparison between the theoretical values of $\beta(n)$ for $n = 7,8,9,10,12,14,16,18$ and the experimental values from [24][1] can be seen in Table 1. For convenience we also placed the respective values of $\beta(n)$ as they would have followed from the lognormal model with $\mu = \mu_6 = 0.2$, $\beta(n) - (n\mu/18)(n - 3)$.

Table 1. Experimental data from [24].

	n=6	n=7	n=8	n=9	n=10	n=12	n=14	n=16	n=18
Exp $\beta(n)$	1.80	2.09	2.26	2.46	2.61	2.81	3.30	3.49	3.71
$\alpha = 6$	log corr.	2.16	2.34	2.50	2.67	3.00	3.33	3.66	3.99
Empirically corrected $\alpha = 5.8$	–	2.10	2.26	2.42	2.57	2.88	3.20	3.52	3.84
$\alpha = 5$	1.80	1.93	2.06	2.19	2.33	2.60	2.86	3.13	3.34
Lognormal at $\mu = 0.2$	1.80	2.02	2.21	2.40	2.55	2.80	2.95	3.02	3.00

It is emphasized in [24] that the lognormal model gives remarkably good agreement for this value of μ for the structure functions up to $n = 12$, but at $n = 14,16,18$ the deviation is too big to be explained by experimental inaccuracy. The results of [24] also clearly invalidate the FHT model. On the other hand, a look at Table 1 shows a remarkable agreement of our theoretical values of $\beta(n)*$ for $\alpha = 6$ and the experimental values for all n. Even more astonishing, this agreement can be made if we recall that logarithmic correction may slightly increase the effective value of μ, that is, to reduce a bit the value of α. The observation that the experimental data from [24] can possibly be explained by the two-segment linear $\beta(n)$, with a crossover at $n = \alpha$ and the value of α empirically chosen from the interval $5 < \alpha < 6$, was made in [17]. It was also pointed out that the value of α may follow from the theory. We can see from Table 1 that $\alpha = 6$, and especially

---

[1]The experimental values are taken as the average of three measurements in [24] for the circular jet at the local $Re_\lambda = 536$, circular jet at $Re_\lambda = 852$ and the duct flow at $Re_\lambda = 515$.

empirically corrected $\alpha = 5.8$ ($\mu \approx 0.52$), seems to be a consistently better fit for all $n \geq 7$, as compared with $\alpha = 5$. We point out also that at a rather large experimental jump of $\beta(n)$ between $n = 6$ and $n = 7$, $\beta(7) - \beta(6) = 0.29$, whereas for all other n, $\beta(n + 1) - \beta(n)$ is systematically smaller and on average consistent with the theoretical $\beta(n + 1) - \beta(n) = 0.16, 0.17$.

On the other hand, in [17] the value $\alpha \simeq 5$ was reported on the basis of geophysical data. It is noteworthy that the probability function $P_\lambda(X)$ measured in atmospheric turbulence [25], i.e., for much larger values of Re, indicates a clear power law slope but with $\alpha > 5$. It may be that the value of α decreases, i.e., the intermittence becomes stronger, with the growth of Re. To clarify this matter would require substantial experimental effort. More detailed comparison with experimental data is provided in [21].

A basic question is what impact the above theory of small scale turbulence may have upon the large scale, the subject of concern of LES methods. To see this we first observe that each individual term in the r.h.s. of (III.13) and (III.14) is much larger in the limit $(k\omega) \to \infty$ than the corresponding Kolmogorov like expressions. That is, if we consider the Reynolds term as a certain eddy viscosity, this eddy viscosity is larger than the Kolmogorov eddy viscosity. The Kolmogorov eddy viscosity is such that the corresponding average flux of energy is exactly the average viscous rate of dissipation $\bar{\varepsilon}$, and evidently this eddy viscosity is of one sign corresponding to a uniform drain of energy everywhere in the fluid. The Kolgomorov eddy viscosity term can be written as $\bar{\varepsilon}^{-1/3} k^{2/3} \vec{v}_{reg}$ in (k,ω) space. Evidently, locally for any term $\vec{Q}$ in the r.h.s. of (III.13), (III.14) $|\vec{v}_{reg} \vec{Q}| \gg \bar{\varepsilon}^{-1/3} k^{2/3} \vec{v}_{reg}^2$. If a steady state is to be preserved, then $\langle \frac{\partial \vec{v}^2}{\partial t} \rangle$ cannot exceed $\bar{\varepsilon}$. As a consequence it is quite clear that $\vec{v}_{reg} \vec{Q}$ cannot be of one sign everywhere in the fluid. The point is that if, as is usual, we would like to represent $\vec{v}_{reg} \vec{Q}$ as an eddy viscosity term, this is not of one sign everywhere. The structure of this eddy viscosity is such that in some regions it corresponds to a drain of energy from $\vec{v}_{reg}$ component, and at others to a generation of energy for this component. Evidently, the fractal component in this respect serves as a reservoir from which energy can be taken or given to. The average balance should not exceed $\bar{\varepsilon}$, i.e., by comparison with fluctuative exchange be almost a zero exchange. Intuitively the situation seems unstable. Furthermore, the large scales of turbulence seem to be a natural candidate for the development of an instability, should such exist at all. Indeed, at small scales Eq. [III.14] yields an asymptotically free solution. In other words, the nonlinearity turned out to be weak in this limit and everything is essentially governed by the linear terms. It is well known from the field theories, however, that if a theory is asymptotically free in the limit of small scales it is bound to develop complications at large scales. More formally, the effective coupling constant in (III.13) decreases with the decrease of scale. This is easy to check by making an iterative rescaling of (III.13). In the limit $(\vec{k},\omega) \to \infty$ the coupling term goes to zero. Revising our argument we find that in the limit of small scales the coupling term grows!

In the usual LES models the small scales are treated by means of closures or renormalization group *perturbative* methods, which is essentially the same. The result is that their action upon large scales is reduced to a certain eddy viscosity of Kolmogorov type, negative everywhere. Thus the small scales can serve only as a link of energy from large scales. In order to support the excitation of large scales one has to introduce so-called back scattering terms, etc. The point is that inaccurate consideration of small scales results in a lack of interaction between small and large scales of turbulence. We suspect that this may be the reason of, as a matter of fact, a failure of LES models to describe such fundamental features of turbulence as coherent structures.

A very long standing controversy is connected with the predominant role of local in $(\vec{k},\omega)$ space direct interactions versus nonlocal indirect interactions. The latter are difficult to associate with energy transfer in the space of scales. On the other hand, all troubles arising in the perturbation theories for the NS equation are connected with generation of singular in the limit of $(k,\omega) \to \infty$ terms produced by indirect interactions. In the theory advanced here the matter seems to be resolved in that the process of energy transfer is controlled by the direct interactions. The indirect interactions associated with the fast time processes are responsible for fluctuations in velocity field not connected with the energy transfer. In this respect it is instructive to relate the above reasoning with the results following from the helical concept. One of the conclusions of the helical concept is the existence of a fast topological time not directly connected with the characteristic time of energy dissipation or non-linear transfer. This has to do with the fact that for a given energy a flow is infinitely degenerate in terms of its topology. Thus a fast time necessity to describe topological transformations with no energy transfer [21]. This topological fast time is clearly consistent with the fast time derived above from different considerations. Another result of the helical concept is a large scale instability, an inverse cascade of some kind, due to the growth of helical correlation length [28,10]. This is also quite consistent with the above results. Recent direct numerical simulations of isotropic homogeneous turbulence seem to be supportive of both effects [29].

The model of small scales described here probably provides an intrinsic mechanism of strong interaction and backward influence of small scale fractal turbulence upon large scales of turbulence. Though at this stage the above reasoning may be seen as speculative, still we observe a clear compatibility with our earlier reasoning. This latter though can also be labeled as quite qualitative; nevertheless, it is based upon a very different kind of argument. That the possibility of large scale instabilities or inverse cascade emerges from considerations only remotely related to each other is encouraging. Clearly it is a long way from Eq. (III.14) to a model capable of describing inhomogeneous anysotropic atmospheric turbulence. Nevertheless, we believe that with due effort such a model can be formulated.

ACKNOWLEDGMENTS

The author is indebted to many colleagues, especially to A. Frenkel, B. Levich, D. Lilly, L. Shtilman, A. Tsinober, and E. Tzvetkov, for valuable comments and interest. The work was partially supported by the U.S. Department of Energy under contract NDE-AC02-80ER10559.

REFERENCES

[1] E. Levich, N.Y. Acad. Sci. 404, 73, 1982.

[2] E. Levich and A. Tsinober, Phys. Lett. 96A, 292, 1983.

[3] E. Levich, B. Levich and A. Tsinober, Proc. of IUTAM Symposium held in Kyoto, Sept. 1983; North-Holland, 1984.

[4] A. Tsinober and E. Levich, Phys. Lett. 99A, 99, 1983.

[5] H. K. Moffat, Prof. of IUTAM Symposium held in Kyoto, Sept. 1983; North-Holland, 1984.

[6] E. Levich and A. Tsinober, Phys. Lett. 101A, 265, 1984.

[7] H. K. Moffat, J. Fluid Mech. 159, 359, 1985.

[8] H. K. Moffat, J. Fluid Mech. 35, 117, 1969.

[9] V. Arnold, The asymptotic Hopf invariant and its application [in Russian], Proc. Summer School in Differential Equations, Erevan, Armenian S.S.R., Acad. Sci., 1974.

[10] E. Levich and E. Tzvetkov, Phys. Rep. 128, 1985.

[11] R. Pelz, V. Yakhot, S. Orszag, L. Shtilman and E. Levich, Phys. Rev. Lett. 54, 2505, 1985.

[12] L. Shtilman, E. Levich, S. Orszag, R. B. Pelz and A. Tsinober, On the role of helicity in turbulent flows, accepted for publication in Phys. Lett. A.

[13] L. Shtilman, R. B. Pelz, A. Tsinober, S. Orszag and E. Levich, Report at the APS meeting, 1985.

[14] R. M. Kerr, Report at the APS meeting, 1985.

[15] A. S. Monin and A. M. Yaglom, Statistical fluid mechanics of turbulence, Vols. 1 and 2; MIT Press, 1975.

[16] B. B. Mandelbrot, J. Fluid Mech. $\underline{62}$, 331, 1974.

[17] D. Schertzer and S. Lovejoy, Turbulence and chaotic phenomena in fluids, Proc. of IUTAM Symposium held in Kyoto, Sept. 1983; North-Holland, 1984.

[18] S. Lovejoy, A statistical analysis of rain areas in terms of fractals, 20th Conference on Radar Met., pp. 476-484; AMS, Boston, 1981.

[19] D. Schertzer and A. Lovejoy, The dimension and intermittency of atmospheric dynamics; in Turbulent shear flows, $\underline{4}$, ed. B. Launder; Springer, New York, in press.

[20] S. Lovejoy and B. Mandelbrot, Fractal properties of rain and a fractal model; Tellus, in press.

[21] E. Levich, Certain developments in the theory of turbulence, submitted for publication to Phys. Rep.

[22] S. Wolfram, Computer software in science and mathematics, Scientific American, 1985.

[23] C. DeDominicis and P. C. Martin, Phys. Rev. $\underline{A19}$, 419, 1979.

[24] F. Anselmet, Y. Gagne, E. J. Hopfinger and R. A. Antonia, J. Fluid Mech. $\underline{140}$, 63, 1984.

[25] R. M. Kerr, J. Fluid Mech. $\underline{153}$, 31, 1985.

[26] R. H. Kraichnan, Phys. Rev. $\underline{A25}$, 32, 1982.

[27] R. A. Antonia, B. R. Satyaprakash and A. J. Chambers, Phys. Fluids $\underline{25}$(1), 1982.

[28] E. Levich and A. Tsinober, Phys. Lett. $\underline{93A}$, 293, 1983.

[29] R. Pelz, L. Shtilman and A. Tsinober, Helicity fluctuations in isotropic homogeneous turbulence, submitted to Phys. Fluids.

LARGE-EDDY SIMULATION OF LOW REYNOLDS NUMBER CHANNEL FLOW
BY SPECTRAL AND FINITE DIFFERENCE METHODS

S. Gavrilakis, H.M. Tsai, P.R. Voke and D.C. Leslie
The Turbulence Unit, Queen Mary College, London University, U.K.

SUMMARY

We describe two simulations of the flow between parallel walls at a fixed, low turbulent Reynolds number. The simulations were carried out by two numerical codes, of which one is purely spectral while the other utilises finite differences. Both codes are described, and the relative merits of the two approaches are assessed. The near wall streak structures are well resolved by both the simulations.

INTRODUCTION

The value of numerical simulation as a complement to experiments on fluid flow is now widely recognised. For this purpose, Direct or Full Simulation is in principle the preferred technique, but for the foreseeable future it will be limited to the lowest turbulent Reynolds or Rayleigh numbers. The use of a subgrid model (SGM) eases this restriction, at the price of considerably increased uncertainty.

The aim of the Turbulence Unit at Queen Mary College is to improve the technique of LES, with particular emphasis on its application to engineering problems. We are therefore simulating flows driven by shear and buoyancy, at both the lowest and somewhat higher Reynolds and Rayleigh numbers. The understanding gained from this work now enables us to tackle certain geometrically simple engineering problems for which more established methods of calculation such as $k-\epsilon$ are not satisfactory. The study reported here is part of this overall programme.

STYLES OF SIMULATION

Having selected a particular flow for study, the simulator must make four further choices:

1. What should the Reynolds or Rayleigh number be.
2. Should natural or synthetic boundary conditions be used.
3. Should the spatial representation be by finite differences or by spectral expansions (or by some of combination of these).
4. Should the time advancement be explicit or implicit.

We shall discuss these four issues. Obviously they are interlinked; in particular, the choice of boundary conditions is determined by the Reynolds number being simulated and the grid being used. Natural boundary conditions are at present limited to low Re, while it is at the least difficult to combine synthetic boundary conditions with implicit time advancement.

Following the lead of Schumann [1] we are studying the use of synthetic boundary conditions at Reynolds numbers around 10^5 but we regard this work as derived rather than fundamental. These simulations are not reported here. It is simulations with natural boundary conditions which mimic and reveal the structure of the most crucial parts of the flow; LES is incomplete if it cannot do this adequately.

At the same time, the currently available SGMs [1], [2] are at least questionable near walls. We seek to limit the strain placed on these models by ensuring that the ratio of the maximum value of the subgrid eddy viscosity ν_e to the molecular viscosity ν is not much greater than 1. By assuming a balance between production and subgrid drain of kinetic energy, we can roughly estimate this ratio as a function of h^+, the channel half width on the wall (+) scale. The results are shown in table 1, for a mesh of 32 streamwise x 64 spanwise x 32 cross-stream mesh points.

TABLE 1

h^+	Re	ν_e/ν
194	6000	1
640	23000	4
5000	230000	43

Here Re is the channel Reynolds number, based on full width and mean velocity; the values are approximate. Our simulations show ratios smaller than these estimated values. The first two lines correspond to the experiments of Kreplin and Eckelmann [3] and of Hussain and Reynolds [4]. The third is typical of our simulations with synthetic boundary conditions. If the argument of the previous paragraph is accepted, it becomes obvious that we should simulate the Kreplin and Eckelmann experiment. In their pioneering work, Moin and Kim [2] made another choice; they were perhaps influenced by the fact that for them the Hussain-Reynolds experiment was "in-house".

Another line of argument points to the same choice. The greater part of our work is done on the University of London CRAY-1S which has 1 Mword of fast memory. With rather restricted disc facilities, the disc overheads are so great that we must work in core. With the CHANEL code (section 3) we are limited to 32000 degrees of freedom, and our standard calculation is 16 streamwise x 64 spanwise x 32 cross-stream. At h^+ = 194 the critical spanwise resolution is 9.5 in wall units, which is small enough to resolve at least the gross features of the streaks and rollers. At h^+ = 640, even with 128 Fourier modes, this resolution has deteriorated to 15.7; this may be why Moin and Kim [2] overestimate the streak spacing.

The main disadvantage of the low Reynolds number from the point of view of the simulator is that correlation lengths will be larger in physical units. The simulations reported here were carried out using a simulation box size of $2\pi h$ x πh x $2h$ in both cases, resulting in correlations at the half-length and half-span positions somewhat larger than are usually considered safe. However there is little evidence of the simulations failing to capture the main energy-containing eddies: it was more important to us in these simulations to be sure of resolving the turbulent streak structures adequately, and we believe we have done this.

In a purely numerical sense, spectral methods are more accurate in the upper part of the represented range of wave-numbers, and this must be an advantage in a Full Simulation. It is not so clear that this is the case in a LES; the upper part of the represented range will be so mangled by the effects of the SGS model that the filtering imposed by the finite differencing may even be advantageous. Finite differences avoid the need for incessant Fourier transforming, which typically consumes 50% of the running time of our spectral codes. Also finite differencing is appreciably simpler. Our FD codes are shorter, and were easier to write and to debug. This debate will no doubt continue for some time to come; there is certainly no agreement now as to which is the better method. We have chosen to write and use codes of both types, and to compare them.

Finally we come to the question of whether the time advancement should be implicit or explicit. The discussion is limited to the overriding of the viscous limit, by implicit (usually Crank-Nicolson) treatment of the viscous terms and at least part of the pressure term. There is general agreement that significant information will be lost if the material Courant limit is overridden. It is also conventional wisdom that, while explicit methods are satisfactory with synthetic boundary conditions, it is essential to use implicit methods with natural boundary conditions.

Our calculations suggest that this is not true for simulations of the Kreplin-Eckelmann experiment. In a simulation with 32 streamwise degrees of freedom, the viscous stability limit, which arises in the wall-adjacent cells, is hardly more pressing than the Courant limit, which comes from the central cells. On the other hand, with the 16 streamwise modes that our spectral code uses when run on the 1 Mword machine, the Courant limit is definitely less severe than the viscous limit, and the implicit treatment of the viscous terms by this code is therefore justified. The finite difference code ECCLES has more economical storage, and can be run with 32 streamwise degrees of freedom. Despite rather general scepticism, we believe that our decision to take advantage of the greater simplicity and speed of the explicit algorithms was justified for this case.

THE SIMULATION CODES

Two codes were used to produce the results reported here, CHANEL and ECCLES (Explicit Channel Code for Large Eddy Simulation). Both are written in Fortran and are normally run on a 1 Mword CRAY-1S. They will be described in turn.

CHANEL is a fully spectral channel code which uses Fourier expansions for the x (streamwise) and y (spanwise) directions, and Chebyshev expansions for the direction perpendicular to the two walls. The Fourier-Chebyshev expansion coefficients of the primitive variables u_1, u_2, u_3 and p are advanced in time by finite differencing, using the implicit second order accurate Crank-Nicolson scheme for the viscous terms and the pressure, and the explicit second order Adams-Bashforth scheme for the nonlinear and subgrid terms.

The solution method involves the elimination of the pressure term from the u_3 momentum equation using the u_1 and u_2 momentum equations and the continuity equation. We write the primitive equations in the following form, with H_i comprising the nonlinear, subgrid and viscous terms:

$$\frac{\partial u_i}{\partial t} = -\frac{\partial p}{\partial x_i} + H_i \qquad (1)$$

$$\frac{\partial u_i}{\partial x_i} = 0 \ . \qquad (2)$$

It is not difficult to show that these equations, with the boundary conditions appropriate to the channel flow, are equivalent to the following fourth order time advancement equation for u_3, and second order equation for the z component of vorticity. u_1 and u_2 are found from a pair of simple coupled equations for u_1 and u_2 that are solved trivially in wavenumber space for periodic boundary conditions in the x_1 and x_2 directions. (A special solution is required for zero Fourier wavenumbers.)

$$\frac{\partial \nabla^2 u_3}{\partial t} = \frac{\partial \mathrm{div}(H_i)}{\partial x_3} + \nabla^2 H_3 \qquad (3)$$

$$\frac{\partial \omega_3}{\partial t} = \frac{\partial H_2}{\partial x_1} - \frac{\partial H_1}{\partial x_2} \ . \qquad (4)$$

The boundary conditions used for these equations are that the value and the normal derivative of u_3 are zero at the walls, and ω_3 is zero at the walls. A solution for pressure is not required, since the solution method works entirely in terms of divergence free velocity fields. In this respect it is quite close to the projection method of Leonard and Wray [5].

The fourth-order equation is solved by introducing the variable $\Phi = \nabla^2 u_3$. Several solutions of the second order equation for Φ are found with different combinations of boundary conditions on the two walls, and the corresponding solutions of the Poisson equation for u_3 are then found, with u_3 zero on the walls; the solutions are finally combined to satisfy the wall continuity condition. This method is due to M. Antonopoulos-Domis [7] who designed the CHANEL code; it was implemented by B.A. Splawski. A comprehensive manual is available to purchasers of the code. The present CHANEL code incorporates the vorticity method, but the results reported here were obtained with an older version of the code which solved for the pressure algebraically from the u_3 momentum equation and then advanced the remaining two velocities using the primitive equations (1). In all cases the viscous and pressure terms (including part of the subgrid term involving the volume average of the eddy vicosity) are treated implicitly, and the remaining nonlinear terms explicitly.

It is important to note that the CHANEL code does not use prefiltering [6], nor does it employ dealiasing. Aliasing errors are important in full simulations and we should support their removal in this context; but in LES the high wavenumber Fourier and Chebyshev components that are affected by aliasing are subject to greater numerical errors arising from the subgrid model. In our view the subgrid drain of energy is sufficient to discount the adverse effects of aliasing errors.

The subgrid model used is close in all respects to that of Moin and Kim [2]. The values of c_1 and c_2 (the multiplying constants in the homogeneous and inhomogeneous parts) were 0.1 and 0.5 respectively, and the wall damping functions were based on those of Van Driest. However, since we do not use prefiltering, our length scale is based upon the collocation grid rather than on an explicit filter scale.

ECCLES is a finite difference code using grid-volume averaged velocities in a staggered grid, following Schumann [1]. The code has been written with a view to testing synthetic boundary conditions as well as the natural boundary conditions used for the simulations that are reported here, and both are incorporated.

The velocity field is advanced in time with the explicit Adams-Bashforth scheme, using a split-step method (also known as the projection method) which enforces continuity in the velocity field at each time level. If we define the vector field H_i as the sum of the nonlinear, subgrid, and viscous terms, the scheme [1] is expressed by:

$$\frac{u_i^* - u_i^n}{\Delta t} = \frac{3}{2} H_i^n - \frac{1}{2} (H_i^{n-1} - \frac{\partial p^{n-1}}{\partial x_i}) \qquad (5)$$

$$\nabla^2 p^n = \frac{2}{3\Delta t} \operatorname{div}(u_i^*) \qquad (6)$$

$$u_i^{n+1} = u_i^* - \frac{3}{2} \Delta t \frac{\partial p^n}{\partial x_i} . \qquad (7)$$

u_i^n is the i'th component of the velocity at time step n, and * denotes the intermediate or unprojected velocity. As in the spectral code, periodic boundary conditions are assumed for the streamwise and spanwise directions, for both velocity and pressure. No-slip or synthetic boundary conditions may be imposed on the velocity at the boundaries at the advanced time, the former being used for the simulation reported here. Because of the staggered grid only the normal component is defined at the boundary itself; for the other two components the "reflection" technique is applied to fictitious cells outside the computational domain.

The finite difference scheme employed leads to a set of equations that do not require a boundary condition on p. Superficially, we have:

$$\frac{\partial p}{\partial z}\bigg|_{z=0} = \frac{2}{3\Delta t} u_3^*(z=0) . \qquad (8)$$

However, the solution does not depend on the value assigned to u_3^* at the wall, which is chosen for convenience to be zero. It should be noted that this method is dependent on the staggered nature of the mesh and on the explicit time advancement.

The Poisson equation for pressure is solved by direct methods using fast Fourier transformation in the x and y directions and a tridiagonal solution in the inhomogeneous direction. The Gaussian elimination and back-substitution are carried out for every Fourier mode at a given z simultaneously, to allow vectorisation of this part of the code. Other parts of the calculation vectorise straightforwardly, with vector lengths typically of 1000 or 2000.

In channel flows most of the turbulent kinetic energy is generated near the walls. For a LES with natural boundary conditions it is necessary to have high resolution in the boundary layer in order adequately to reproduce the physical production of turbulence. This problem is avoided by the use of synthetic (or artificial or law-of-the-wall) boundary conditions [1]. All these types of boundary conditions have been used in the ECCLES code, but only those results obtained using natural boundary conditions at low

Reynolds numbers are reported here.

The Harlow and Welch finite difference scheme [8] is used for the nonlinear terms, which ensures that the volume integral of the convected square velocities is zero to within the accuracy of the velocity continuity residual. Since the projection method ensures that continuity is enforced to machine accuracy, no numerical instability is likely to arise from false energy production.

The subgrid-scale model used in the simulation reported here is, like that in the spectral simulation, derived from the model proposed by Moin and Kim [2]. Straightforward modifications are incorporated to generalise the model to the staggered finite-difference mesh employed by ECCLES. The same values of c_1 and c_2 were used (see above) and the same Van Driest wall damping factors.

CODE TESTING

It is always necessary to test large codes. This necessity is particularly acute for LES codes, since errors which would be obvious in a simple, symmetrical flow may be concealed by real fluctuations in a turbulent flow. In addition to numerical consistency checks, we have applied two types of fluid-mechanical test to both codes.

MK Test. This tests the ability of the code to predict the decay of certain specific disturbances of the Poiseuille profile at subcritical Reynolds numbers, following Moin and Kim [9]. The code must be able to reproduce the known forms and rates of the decay.

GHM Test. George, Hellums and Martin [10] have computed the critical amplitude and Reynolds number for a particular two-dimensional disturbance, and the decay or growth rate at other amplitudes and Reynolds numbers. This is a most sensitive test of the energy balance in a code. Because of differences in modal representation, it is not possible to obtain precise agreement with the results of George *et al*. We regard the agreement we were able to obtain as satisfactory.

STARTING THE SIMULATIONS

Any new simulation must be started from a synthesised initial field; CHANEL incorporates routines for generating such fields. The user specifies the mean velocity and the normal and shear stresses (or second order correlations) as functions of z. A library random number generator is then used to produce a Gaussian field with these means and variances. There is no inertial transfer at the start of such a simulation, since the triple correlations are all zero. The inertial transfer must build up as the simulation evolves; this happens quite naturally. The artificial initial field generation is only used when a suitable set of fields dumped from a previous simulation is not available.

In the work described here, the initial field generation routines were used only once, all later runs being started from previous dumps. The initial field for the finite difference simulation was created by a rather crude manipulation of a dump from the spectral code. It was very unphysical, containing discontinuous gradients and a high mean flow rate. It is reassuring that the simulation was able to recover from such an unpromising beginning.

The Reynolds number (based on channel half-width h and the wall shear needed to balance the imposed pressure gradient) was chosen to be $h^+=194$, and the simulation box size is $2\pi h \times \pi h \times 2h$. The spectral simulation was started with a very coarse modal representation: 8 (streamwise) x 16 (spanwise) x 17 (cross-stream). Realistic results were not expected from such a simulation, of course; but with such a low resolution it is possible to simulate cheaply for many non-dimensional time units (h/u_τ, known as a letot since it is a rough measure of the large eddy turn over time).

We have found that $\langle\langle w^2\rangle\rangle$, the volume average of the cross-stream velocity variance, is a sensitive measure of the state of the simulation, and we shall use this quantity to illustrate what is happening. The experimental value is 0.65 [3], and this was reproduced by the starting field. $\langle\langle w^2\rangle\rangle$ fell to about 0.15 during the first 1.5 letots of the simulation, and remained at this low level for a further 12 letots. There was nothing to suggest it would move away from this level. Clearly the combination of coarse resolution and inadequate subgrid model was injecting too little energy into the turbulence. The only surprise was that the turbulence appeared to be self-sustaining at this low level.

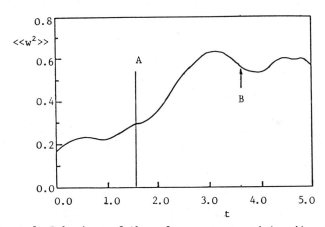

Figure 1. Behaviour of the volume average z-intensity with time

The resolution was then increased to 16 x 32 x 33 modes. The evolution of $\langle\langle w^2\rangle\rangle$ from this point onwards is shown in figure 1. The transfer of energy from the mean field to the turbulence is roughly correct, but required us in this case occasionally to restore the mean flow profile to the level found experimentally. Our simulations fix the mean pressure gradient, not the mass flow rate.

At the point marked A in figure 1 (t=1.5 approximately), there was a build-up of energy in the highest represented modes. The simulation was run for 6 timesteps with a resolution of 32 x 64 x 65. This is the finest resolution we have used, but since it requires the database to be stored on disc, it is very expensive to run at this resolution. (Each timestep took 900 seconds, mostly spent in disc transfers.) Surprisingly, this was sufficient to clean up the spectrum. Subsequently the simulation has been run in core using 16 x 64 x 33 modes, without any recurrence of the problem that occurred at t=1.5. The data presented here were taken at t=3.6, at the point marked B in figure 1.

RESULTS AND COMPARISONS

A comparison of the performance of the codes is shown in table 2. BUOYAN-77 is the commercial version of CHANEL, developed after the work reported here was completed. It is very highly vectorised and includes certain other features to reduce the running time. The figures given for BUOYAN-77 are for a simulation including a passive scalar field. ECCLES still gives 60% more letots per cpu hour, for a calculation of the same size as the spectral code. (Note that the finite difference calculation is twice as big as those done by the spectral codes.)

TABLE 2

Code	Mesh pts / modes	Degrees of freedom	Δt (letots)	letots / cpu-hour	ς (μs)
CHANEL	16x64x33	33,792	0.0015	0.44	360
BUOYAN-77	16x64x33	33,792	0.0015	4.2	38
ECCLES	32x64x32	65,536	0.0004	3.3	6.5

ς is the number of microseconds needed to advance one degree of freedom (all fields for one mesh point or one mode) one time step. ECCLES is very much faster *numerically* than the spectral code, owing to the inherent speed and simplicity of the explicit algorithms. Because of the viscous number limitation on the explicit code, the eleven-fold superiority in time stepping yields only a 60% advantage in letots per cpu-hour, at this Reynolds number. The ECCLES time step gives a maximum viscous number of 0.09. The theoretical limit is 1/6, but it is well known that one should not approach this limit too closely. The maximum Courant number was 0.11 and the fact that the two numbers are comparable is *a posteriori* justification for the explicit time stepping used by the ECCLES code.

Our comparison may be unrealistic, if we accept the common wisdom that a finite difference code needs twice as many degrees of freedom in each of the three dimensions (eight times as many overall) to produce statistics of a quality comparable to those from a spectral simulation. If this is the case, it is difficult to see how any finite difference code can compete with a spectral code when both are using natural boundary conditions. In fact we find the results from the finite difference code are in many cases not obviously worse than those from the spectral code. The reasons for this will be discussed below. Our conclusion regarding speeds of execution of the codes therefore agrees with the common view, namely that spectral codes may be more efficient for fundamental investigations using natural boundary conditions, but finite differences are superior for higher Reynolds numbers when synthetic boundary conditions are used. This conclusion depends on the relative stringency of the Courant and viscous stability limits that apply for the mesh spacings and Reynolds numbers being used.
The finite difference simulation reported here was run for much longer in

non-dimensional units (36 letots) than the spectral code (5.0 letots, 3.4 letots being on the finest mesh, 16 x 64 x 33). The statistics reproduced here were gathered between t=33 and t=36 for the finite difference simulation, and at t=3.6 for the spectral simulation.

Both codes were run with an imposed, constant pressure gradient, normalised to unity, rather than an imposed mean mass flow rate. Although we agree with the view that an imposed pressure gradient and an imposed mass flow rate are equivalent in the mean, once statistical equilibrium has been achieved, it is not obvious that they have precisely the same effect on a turbulent simulation that has not yet settled. The finite difference simulation reached a good statistical equilibrium at t=20 letots, with the maximum velocity at 17.8 u_τ (compared with 19.8 found in the experiment) and the average wall shear nicely balancing the imposed unit pressure gradient.

Because of the computational expense, it was not possible to run the spectral simulation until full statistical equilibrium was achieved. At t=3.6 letots, we believe all transients had died and realistic (numerical) turbulence was present; but after t=3.6 the total resolved turbulent kinetic energy started to decrease with time, in spite of the fact that the peak streamwise velocity was slowly decreasing and the momentum transfer through the walls was still a little higher than the momentum input from the imposed pressure gradient.

Since the flow was changing with time, it was not possible to obtain meaningful time-average statistics from this simulation run. The statistics presented here for the spectral code are instantaneous, planar average values taken at t=3.6. The statistics presented for the finite difference code are averaged over the planes parallel to the walls, and over 200 instantaneous times between t=33 and t=36.

Figures 2 to 4 show the variation of the three components of resolved turbulent kinetic energy. In spite of the instantaneous nature of the data from the spectral code (CHANEL), it would appear to be able to reproduce the peak of the $\langle u^2 \rangle$ variation a little better than the finite difference code. Note that no subgrid contribution has been included in these quantities. The usual estimate of $2/3(\epsilon\Delta)^{2/3}$ for each component, which is derived from assuming a $k^{-5/3}$ spectrum, will be much too big at these low Reynolds numbers, where the spectrum falls much faster. We plot the squared quantities rather than RMS values; the latter can give a misleading impression of the level of agreement between simulation and experiment.

The instantaneous plots of $\langle v^2 \rangle$ and $\langle w^2 \rangle$ from the spectral simulation suggest that time averaging is important, though some idea of the performance of the simulation can be gained from these plots. Comparison of instantaneous and time-averaged results from the finite difference simulation confirms this. The low predictions of $\langle w^2 \rangle$ and the overestimation of the distance from the wall to the peak are typical of our simulations of channel flow, as well as of those of some other workers [2]. The referee has pointed out that this suggests an incorrect dependence of the SGS model on the grid width, which could be corrected by using a transport equation for the subgrid energy. He has also pointed out that the use of anisotropy coefficients was helpful in correcting near-wall anisotropy with synthetic boundary conditions: the implication is that these coefficients might be equally helpful with natural boundary conditions.

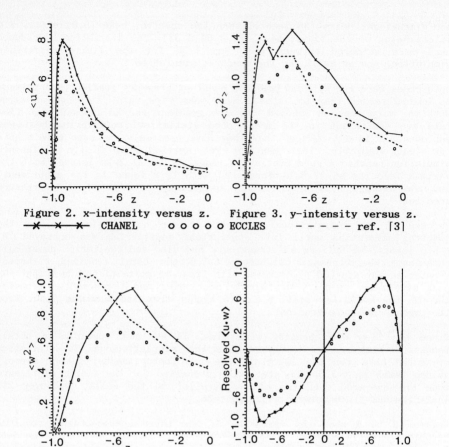

Figure 2. x-intensity versus z. Figure 3. y-intensity versus z.
—×—×—×— CHANEL   ο ο ο ο ο ECCLES   - - - - - ref. [3]

Figure 4. z-intensity versus z. Figure 5. Resolved Reynolds stress

Comparisons of our results are with the experimental data of Kreplin and Eckelmann [3] in all cases. These results demonstrate that most of the major energy-containing eddies have been captured by our simulations, in spite of the fact that the size of our periodic box was smaller than the known behaviour of the correlations would dictate.

Figure 5 shows predictions of the principal component of the resolved Reynolds stress across the channel. The time-averaged finite difference predictions are very satisfactory. The spectral predictions show instantaneous variations and a generally higher level of Reynolds stress than one would expect to see in a statistically stationary flow.

Predictions of skewness and flatness of u (figures 6, 7) also show some (presumably) instantaneous scatter in the spectral predictions. Without time-averaged spectral results it is difficult to say whether these graphs confirm or contradict the general belief that finite difference codes are less able to predict higher order statistics; this view would not appear to be borne out in our results.

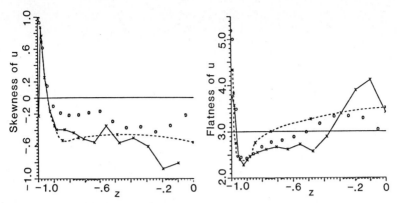

Figure 6. Skewness of u versus z.   Figure 7. Flatness of u.
—×—×— CHANEL   o o o o o ECCLES   – – – – – ref. [3]

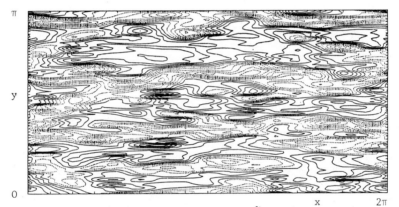

Figure 8. CHANEL. Contours of $\tilde{u}$ at $z^+=14.8$.

Figure 9. ECCLES. Contours of $\tilde{u}$ at $z^+=11.6$.

Figures 8 and 9 show plots of contours of $\tilde{u}$ close to one wall. Both plots are instantaneous. In estimating the streak spacing, it is helpful to bear in mind that the spanwise size of the box is 603 in wall (+) units. The streaks may appear clearer and the large structures smaller in the spectral predictions because finer contour intervals were employed. In fact the difference is marginal; in both cases the longest structures are about half the box length in size, and the streak spacing is about 100 in wall units. The spectral plot is nearer to the peak of the $\langle \tilde{u}^2 \rangle$ distribution, and this peak is higher than the finite difference result (see figure 2).

The estimation of the spanwise streak spacing λ^+ may be done in several ways. Simple inspection of the contour plots would suggest that λ^+ is about 100, which is satisfactory, and justifies our decision to opt for a low Reynolds number for this simulation and for good resolution of the important structures rather than a "safe" box size. Attempts to quantify the spacing more accurately need to be rather sophisticated. Simply looking at the spanwise correlation of the streamwise velocity is misleading, since although a negative crossover of the correlation is present, it suggests the spacing is greater than it clearly is; this is because the correlation is affected by parts of the flow where streaks are not the dominant feature.

Figure 10 also clearly reveals the streaks by means of contours of $\tilde{u}$ for a view along the flow direction. To appreciate the vertical extent of the streaks, it should be borne in mind that the vertical distance between the walls is 388 in wall units.

The correlations of most quantities fall to about 15 or 20% at the edges of the box, in the spanwise as well as the streamwise directions. This value is roughly in accordance with experiment (and with our expectation) for the streamwise direction. For the spanwise direction it is rather higher than we should like or expect. The correlation plots (not shown here) are all instantaneous, and possibly time-averaged correlations are needed to obtain a clear picture of the turbulence.

Figure 10. CHANEL. Contours of $\tilde{u}$ at fixed x.

CONCLUSIONS AND FURTHER WORK

From the results obtained from the codes ECCLES and CHANEL to date there is ample evidence that both are capable of reproducing many features of physical turbulence. The lower order statistics are very reasonable, though it is of some concern that the spectral code has not settled to a stationary state. With the faster version of the code, we may be able to obtain statistical stationarity after further simulation time has elapsed. The higher order statistics such as skewness and flatness are not as convincing, though of course time-average closure models cannot, by their nature, predict such quantities at all. We consider our agreement with experiment to be acceptable at this stage.

The streak plots obtained from CHANEL in 1983 and reproduced here represented the first evidence that the streak spacing could be predicted correctly by a simulation with adequate spanwise resolution. It is of considerable interest that a finite difference code can also simulate the streaks with roughly the correct spacing and features.

The comparison of the finite difference and spectral techniques shows that explicit (or nearly explicit) finite difference codes are very strong contenders for simulating wall-bounded flows at all Reynolds numbers, particularly since they lend themselves to the use of synthetic boundary conditions at higher Reynolds number than we have considered here. It is frequently assumed that finite difference methods are of no use for simulations at low Reynolds numbers; we have shown that is not the case.

Our results indicate that there is still much work of great interest that can be done on the channel flow, including the comparison of energy balances between the two types of code, the investigation of the origin and breakup of the streak structures, and the question of the effect of simulation box size on the other statistics.

ACKNOWLEDGEMENTS

We are particularly grateful to Professor M. Antonopoulos-Domis (Chair of Nuclear Technology, Aristotelian University of Thessaloniki) and to Dr A. Splawski (now of AERE Harwell): as Research Fellows in the QMC TU they played the leading roles in developing CHANEL and ECCLES respectively.

We have derived great benefit from our interaction with the LES effort at Stanford University and at NASA Ames, and particularly from the close cooperation we have over many years with Professor Joel Ferziger.

We should like to express our thanks to the referee for a number of helpful suggestions, and to the British Science and Engineering Research Council for continued financial support.

REFERENCES

[1] Schumann, U.: "Subgrid Scale Model for Finite Difference Simulations of Turbulent Flows in Plane Channels and Annuli", J. Comp. Phys. 18 (1975), 376 - 401.

[2] Moin, P. and Kim, J.: "Numerical Investigation of Turbulent Channel Flow", J. Fluid Mech. 118 (1982), 341.

[3] Kreplin, H. and Eckelmann, M.: "Behaviour of the Three Fluctuating Velocity Components in the Wall Region of a Turbulent Channel Flow", Phys. Fluids 22 7 (1979), 1233 - 1239.

[4] Hussain, A.K.M.F. and Reynolds, W.C.: "Measurements in Fully Developed Turbulent Channel Flow", J. Fluids Eng. 97 (1975), 568 - 578.

[5] Leonard, A. and Wray, A.: "A New Numerical Method for Simulation of Three-Dimensional Flow in a Pipe", *Proc. Int. Conf. Numer. Methods Fluid Dynamics; Lecture Notes in Physics,* 170, ed. Krause, E. (1982), Springer. 335 - 342.

[6] Moin, P., Reynolds, W.C. and Ferziger, J.H.: "Large-Eddy Simulation of Incompressible Turbulent Channel Flow", rep. TF-12, Dept. Mech. Eng., Stanford University, (1978).

[7] Antonopoulos-Domis, M.: "Numerical Simulation of Turbulent Flows in Plane Channels", *Proc. Third Int. Conf. Num. Methods in Laminar and Turbulent Flows* (1983), Pineridge Press Swansea U.K., 113 - 123.

[8] Harlow, F.H. and Welch, J.E.; "Numerical Calculation of Time-Dependent Viscous Incompressible Flow", Phys. Fluids 8 (1965), 2182.

[9] Moin, P. and Kim, J.: "On the Numerical Solution of Time-Dependent Viscous Incompressible Fluid Flows involving Solid Boundaries.", J. Comp. Phys. 35 (1980), 381.

[10] George, W.D., Hellums, J.D. and Martin, B.: "Finite-Amplitude Neutral Stability Disturbances in Plane Poiseuille Flow.", J. Fluid Mech. 63 (1974), 765.

LARGE EDDY SIMULATION OF TURBULENT CHANNEL FLOW BY 1- EQUATION MODEL

Kiyosi HORIUTI and Akira YOSHIZAWA

Institute of Industrial Science,
University of Tokyo,
7-22-1 Roppongi, Minato-ku, Tokyo 106, Japan.

Turbulent plane channel flow is numerically studied using Large Eddy Simulation(LES). Cyclic boundary conditions are imposed on velocity and pressure in the downstream and spanwise directions. The noslip boundary condition is imposed on the walls. Both Smagorinsky model and 1-equation model are applied, and the comparison is made. The importance of the diffusion term in subgrid scale (SGS) turbulent energy balance is pointed out.

INTRODUCTION

LES was first applied to turbulent channel flow by Deardorff,[1] who imposed the boundary condition on the wall according to the logarithmic law for the mean velocity profile. Later, Schumann[2] extended Deardorff[1] in a sophisticated manner, namely employing a transport equation for SGS turbulent energy, although the boundary condition obeying the logarithmic law was imposed on the wall. Recently, Moin and Kim[3] carried out an enormous computation which imposed noslip boundary condition on the wall. Horiuti[4] compared the two types of the boundary conditions imposed on the wall. As a turbulence model, the computations listed above basically employed the model proposed by Smagorinsky.[5] It should be noted here that, although Schumann[2] used a transport equation for SGS turbulent energy, no appreciable change was found in the computation.

In this paper, we focus on the improvement of the turbulence model employed in LES. Noslip boundary condition is imposed on the wall, and both Smagorinsky model and a transport equation for SGS turbulent energy, which was derived based on the two scale Direct Interaction (DI) theory by Yoshizawa[6] and by Yoshizawa and Horiuti,[7] are applied and the comparison is made.

GOVERNING EQUATIONS AND NUMERICAL METHOD

In LES, we define the grid scale or GS component of f by the convolution of f with a filter function $G(x)$[8] :

$$\overline{f}(x) = \int_{-\infty}^{\infty} G(x-x')\, f(x')\, dx' . \tag{1}$$

In the present study, a Gaussian filter is used as G in the streamwise and spanwise directions and a top-hat filter is used in the direction normal to the wall. Using this filtering procedure, the velocity field u_i and the pressure p are decomposed into grid scale (GS) and SGS components as

$$u_i = \overline{u_i} + u_i' \quad , \quad p = \overline{p} + p' \quad . \tag{2}$$

After applying the filtering to the Navier-Stokes and continuity equations, we get following filtered equations:

$$\frac{\partial \overline{u_i}}{\partial t} + \frac{\partial}{\partial x_j} \overline{u_i u_j} = -\frac{\partial \overline{p}}{\partial x_i} + \frac{1}{Re} \nabla^2 \overline{u_i} \quad , \tag{3}$$

$$\frac{\partial \overline{u_i}}{\partial x_i} = 0 \quad . \tag{4}$$

Here $i,j=1,2,3$ correspond to x,y,z respectively, where x is in the downstream direction, y is in the spanwise direction and z is in the direction normal to the walls. All velocity components and co-ordinates have been made dimensionless by means of the length scale h separating the parallel walls and the friction velocity.

In nonlinear terms of eq. (3), the Leonard terms in the x,y directions are explicitly computed and those in the z direction are represented by the truncation error inherent in the 2nd order central finite difference. Correlations between GS and SGS components are neglected. For SGS Reynolds stresses $\overline{u_i' u_j'}$, we introduce the SGS eddy coefficient,

$$\overline{u_i' u_j'} - \frac{2}{3} \delta_{ij} K_G = -K \left(\frac{\partial \overline{u_i}}{\partial x_j} + \frac{\partial \overline{u_j}}{\partial x_i} \right) \quad , \tag{5}$$

where K is the SGS eddy coefficient, and K_G is the SGS turbulent energy $\overline{u_l' u_l'}/2$. It should be noted here that in (5), the residual stress model employed in Moin and Kim [3] is not used here. [14] The equation for K_G can be written as follows:

$$\frac{\partial K_G}{\partial t} = -\overline{u_j} \frac{\partial K_G}{\partial x_j} + R_{ij} \frac{\partial \overline{u_i}}{\partial x_i} - \varepsilon + E_{DG} \quad , \tag{6}$$

where

$$R_{ij} = -\frac{2}{3} \delta_{ij} K_G + K \overline{e}_{ij} \quad , \quad K = C_\nu \Delta K_G^{1/2} \quad ,$$

$$\overline{e}_{ij} = \frac{\partial \overline{u_i}}{\partial x_j} + \frac{\partial \overline{u_j}}{\partial x_i} \quad , \quad E_{DG} = -\frac{\partial}{\partial x_i} \left(\frac{1}{2} \overline{u_i' u_j' u_j'} + \overline{p' u_i'} \right) .$$

Δ is a representative grid interval $\Delta = (\Delta_x \Delta_y \Delta_z)^{1/3}$ ($\Delta_x, \Delta_y, \Delta_z$ are grid intervals in the x,y,z directions, respectively), and C_ν is a numerical factor. Each term in the right hand side of eq.(6) is called convection, production, dissipation, and diffusion, respectively and have been evaluated based on the two-scale DI theory. [6],[7] Thus the model equation for K_G is given as

$$\frac{\partial K_G}{\partial t} = -\overline{u_j} \frac{\partial K_G}{\partial x_j} + \frac{1}{2} K (\overline{e}_{ij})^2 - C_\varepsilon K_G^{3/2}/\Delta + \frac{\partial}{\partial x_i} \left(\left(\frac{1}{Re} \right. \right.$$

$$\left. \left. + C_{kk} \Delta K_G^{1/2} \right) \frac{\partial K_G}{\partial x_i} \right) \quad , \tag{7}$$

where C_ε and C_{kk} are numerical factors.

Equations (3), (4), (5), and (7) form a closed system. This system can be called a 1-equation model. Three numerical factors $C_\nu, C_\varepsilon, C_{kk}$ appearing above may be estimated as 0.05,

1.0 and 0.1, respectively within the framework of the two-scale DI theory.[7] As stated in INTRODUCTION, a similar equation was also constructed by Schumann [2] using a dimensional analysis.

The Smagorinsky model (S-model) can be derived from the present system by assuming the balance of the SGS energy production and dissipation. Thus, the retention of only these two terms leads to S-model

$$K=(c\Delta)^2 \left[\frac{\partial \bar{u}_i}{\partial x_j}\left(\frac{\partial \bar{u}_i}{\partial x_j}+\frac{\partial \bar{u}_j}{\partial x_i}\right)\right]^{1/2}, \qquad (8)$$

where c is a dimensionless constant. In the two-scale DI theory, c is estimated as 0.081, which is very close to the value 0.1 obtained by a computer optimization. In this study, The damping of van Driest type [8] is applied to Δ near the walls.

In the actual computation, the SGS turbulent energy K_G is added to the pressure p and treated as a pressure head. Due to the strong anisotropy in a shear flow, however, there seems to be no systematic way to evaluate K_G. Therefore, it is difficult to evaluate the pressure separately. This point is considered to be one of defects of S-model [7], because occasionally, the evaluation of the turbulence energy drained in SGS is necessary and very important in turbulence modeling. On the other hand, in 1-equation model, K_G can be separately evaluated by eq. (7). Moreover, S-model assumes the balance of SGS production and dissipation. The validity of this assumption is still in a question, even in a plane channel flow, when noslip boundary condition is imposed on the wall. In this study, we employ both S-model and 1-eq. model, and make the comparison.

Similarly the energy budget equation for GS turbulent energy $\Gamma = <\bar{u}_l''\bar{u}_l''>/2$ can be derived as follows:

$$\frac{\partial \Gamma}{\partial t} = -\frac{\partial}{\partial z}<\bar{w}\ \Gamma> - <\bar{u}''\bar{w}''>\frac{\partial <\bar{u}>}{\partial z} - <(K+\frac{1}{Re})\frac{\partial \bar{u}_i''}{\partial x_j}\frac{\partial \bar{u}_i''}{\partial x_j}>$$

$$- <\bar{u}_i''\frac{\partial \bar{p}}{\partial x_i}> + \frac{\partial}{\partial z}<(K+\frac{1}{Re})\frac{\partial \Gamma}{\partial z}> . \qquad (9)$$

Each term in the right-hand side is called GS convection, production, dissipation, velocity pressure gradient and diffusion, respectively, $<\cdot>$ denotes the horizontal average, and $(\cdot)''$ denotes the deviation from the horizontal average.

In solving eq. (3) numerically, convective terms and pressure gradient terms on the right hand side are approximated by the Adams-Bashforth method, whereas for mean part of viscous terms, the implicit scheme of Crank-Nicolson type is used.[14] Convective terms are approximated in a momentum and energy conserving conservative form. Partial differential operators in the z direction are approximated with the second order central finite difference and those in the x and y directions are approximated by the pseudospectral method. The velocity and pressure are represented by the Discrete-Fourier expansion in the downstream and spanwise directions and a regular grid system is used. Consequently, equations (3) and (4) are solved as coupled equations.[9],[3] For details of the method employed in the present study, including

the comparison of the conservative form with the rotational form for convective terms, see Horiuti.[14]
   A system of linear block tri-diagonal difference equations is obtained as follows:

$$\alpha_k \tilde{u}_{k-1} + \beta'_k \tilde{u}_k + \gamma_k \tilde{u}_{k+1} + \sqrt{-1} \beta_{x,k} \tilde{p}_k = RHS_{k,1} , \qquad (10)$$

$$\alpha_k \tilde{v}_{k-1} + \beta''_k \tilde{v}_k + \gamma_k \tilde{v}_{k+1} + \sqrt{-1} \beta_{y,k} \tilde{p}_k = RHS_{k,2} ,$$

$$\alpha_k \tilde{w}_{k-1} + \beta'''_k \tilde{w}_k + \gamma_k \tilde{w}_{k+1} + d'_k \tilde{p}_{k-1} + d''_k \tilde{p}_{k+1} = RHS_{k,3} ,$$

$$\sqrt{-1} \beta_{x,k} \tilde{u}_k + \sqrt{-1} \beta_{y,k} \tilde{v}_k + e'_k \tilde{w}_{k-1} + e''_k \tilde{w}_{k+1} = 0 ,$$

where α_k s are resulting coefficients.
   The initial condition is given as the superposition of sinusoidal random fluctuations on the mean velocity profile measured by Laufer.[11] The region between boundaries to be treated has a downstream length L_x of 3.2 or 4.8 and a spanwise width L_y of 1.6 . The downstream and spanwise lengths are subdivided into NX and NY equal grid intervals, respectively and the vertical into NZ non-uniform intervals. Re is set at 1280 and computed cases are summerized in Table 1.

Table 1.   Computed cases

Case	NX	NY	NZ	L_x	L_y	Turb. model
1	32	32	32	3.2	1.6	S- model
2	32	32	32	3.2	1.6	1-eq. model
3	16	16	22	3.2	1.6	S- model
4	16	16	22	3.2	1.6	1-eq. model
5	64	64	62	3.2	1.6	S- model
6	64	64	62	3.2	1.6	1-eq. model
7	64	64	62	4.8	1.6	S- model

RESULTS AND DISCUSSION OF CASES 1 AND 2

   Figure 1 shows the mean velocity profile obtained in Case 1. This profile linearly stands near the wall and approximately fits the curve (plotted by a heavy line in the figure)

$$< \bar{u} > = \frac{1}{0.4} \log z_+ + 7.0 , \qquad (11)$$

in the logarithmic region. The von Karman constant obtained is in a good agreement with an experimentally measured value. The mean velocity profile from Case 2 is also shown in Fig.1, where no significant difference is found from Case 1.
   Figures 2 (a), (b) plot the profiles of the GS mean Reynolds stress $<\bar{u}''\bar{w}>$ and the total stress

$$<\bar{u}''\bar{w}> - < K (\frac{\partial \bar{u}}{\partial z} + \frac{\partial \bar{w}}{\partial x}) > - \frac{1}{Re} \frac{\partial <\bar{u}>}{\partial z}$$

which almost balances the mean downstream pressure gradient. However, a deviation from the mean gradient can be found particularly near the walls in both Cases 1 and 2. This deviation is considered to be due to the insufficiency of number of data averaged and probably inadequate resolution of grid points employed.

Figure 3 displays GS turblence intensities in Cases 1 and 2. In these figures, we find no significant difference in $<\overline{u'''^2}>^{1/2}$ and $<\overline{w'^2}>^{1/2}$, but we find an appreciable difference in $<\overline{v'^2}>^{1/2}$. That is, in Case 2 the peak position is closer to the wall than in Case 1. Overall agreement with Comte- Bellot [12] is better in Case 2 than in Case 1. Consequently, we find that the choice of turbulence model gives a minor effect on the mean velocity profile, whereas significantly affects the turbulence intensities.

To check this point further, we investigate the detailed flow structure. Fig.4 (a) plots the instantaneous contours of u'' in the $x-y$ plane located at $z_+ = 2.6$ at $t= 8.25$ from Case 1. Positive portions are contoured by solid lines and negative values by dashed lines. In this figure, an array of highly elongated regions of high-speed fluid u'' can be found, which corresponds to the experimentally observed streaks. However, the mean spacing of streaks in Case 1 is approximately 400 in wall units, which is considearbly larger than the generally accepted value of 100. Fig.4 (b) displays the corresponding contours from Case 2, where a streaky structure as in Fig.4 (a) can be found, but the elongated $u''>0$ portion seems to be less striking than in Case 1.

This difference can be clearly identified by the spanwise two point correlation function $R_{ii}(r_2)$ defined by

$$R_{ii}(r_2)=<\overline{u}_i''(x,y,z)\ \overline{u}_i''(x,y+r_2,z)>/<\overline{u}_i''^2(x,y,z)> . \qquad (12)$$

Figures 5 (a), (b) display $R_{ii}(r_2)$ at $z=0.014$ from Cases 1 and 2, respectively. As is recognized in Fig.4, Cases 1 and 2 show a marked difference. In Case 2, the most negative peak is located closer to the wall than in Case 1. Moreover, we can depict a long oscillatory tail in Case 1, which means that the streaky structure is more dominant in the spanwise direction than in Case 2. Although Case 2 exhibits a similar tail, the amplitude and the period are smaller than in Case 1. This period approximates the mean spacing of neighbouring streaks, which can be estimated to be 400 and 300 in Cases 1 and 2, respectively. From this point of view, 1-eq. model gives much improvements. Figs.6 (a),(b) plot $R_{ii}(r_2)$ at $z=0.022$ from Cases 3 and 4, respectively. The same tendency can be found in the figure and is pronounced when the number of employed grid points is small.

Contours of instantaneous distribution of SGS turbulent energy and each terms in eq. (7) in the $x-y$ plane at $z=0.0020$, $t= 8.25$ are plotted in Fig.7 (a). Root mean square values of production, convection, dissipation and diffusion terms in this $x-y$ plane are 222, 6, 283, and 70, respectively. Therefore, the diffusion term, neglected in deriving S-model, is not negligibly small compared with production or dissipation term. Fig.7 (b) displays the contours of the diffusion term in the $x-y$ plane at $z=0.0049$. In contrast to Fig.7 (a), portions of negative values are found in Fig.7 (b). That is, as we depart from the wall, the diffusion term changes a sign from positive to negative, which is considered to be related to experimentally observed ' burst ' and ' sweep '.[13)]

Figures 8 (a), (b) display the budget of the GS turbulent energy in Cases 1 and 2, respectively. However, Cases 1 and 2 show no significant difference, which reveals that the effect of an inclusion of convection and diffusion terms in the budget of SGS turbulent energy is localized in space. Both in Figs. 8 (a) and (b), although the production term compensates for the loss in the dissipation term in the central region, the diffusion term is found to be dominant near the wall, as in the budget of K_G, and changes a sign from positive to negative as we depart from the wall. Therefore, the characteristics of distributions of GS diffusion term is reflected on the budget of SGS turbulent energy.

Time lines computed using S-model are displayed in Fig.9 where a wire is set parallel to the y-axis at $x=0.05$, $z=0.01$. In spite of a relatively small number of grid points (32^3) used in Case 1, a qualitative agreement of Fig.9 with an experimental observation by Kline et al. [13)] is very good. Although S-model uses a very simple assumption that SGS production balances dissipation, reasonable results can be obtained by S-model, in respect to the global scale quantities. This proves that this simple assumption is valid as a first order approximation.

RESULTS AND DISCUSSION OF CASES 5 AND 6

Figures 10 and 11 plot the mean velocity profiles and the profiles of GS mean and total stresses, respectively from Cases 5 and 6. No significant difference is found in both cases, so the profiles of stress from Case 6 are omitted in Fig. 11. The velocity profile linearly stands near the wall and approximately fits the curve

$$< \bar{u} > = \frac{1}{0.4} \log z_+ + 6.2 \quad , \tag{13}$$

in the logarithmic region, and is appreciably improved from that in Cases 1 and 2. In addition, the total stress balances completely with the mean downstream pressure gradient.

Figure 12 display GS turbulence intensities from Cases 5 and 6, where we find that the peak value of $<u'^2>^{1/2}$ is larger than that in experimental measurements, [11)] and peak values of $<\overline{v^2}>^{1/2}$ and $<\overline{w^2}>^{1/2}$ are in a good agreement with experiments. Moreover, the peak position of $<\overline{v^2}>^{1/2}$ is closer to the wall than in Moin and Kim,[3)] and overall agreement with experimental measurement is very good. However no significant difference can be found between S-model and 1-eq. model.

The spanwise correlations $R_{ii}(r_2)$ at $z=0.013$ and 0.063 from Cases 5 and 6 are plotted in Figs.13 (a),(b), respectively. The position of negative peak locates at $y = 0.1$, and therefore the mean spacing of streaks is estimated as about 250 in wall units, which is almost same as that in Moin and Kim,[3)] but no appreciable difference can be found between S-model and 1-eq. model.

Figures 14 (a),(b) plot streamwise correlations $R_{ii}(r_1)$ at $z=0.013$ and 0.063 from Cases 5 and 6, respectively. It can be found in the figures that for larger values of r_1, $R_{11}(r_1)$ is significantly larger in Case 6 than in Case 5. In Moin and Kim,[3] it was pointed out that for large values of r_1, $R_{11}(r_1)$ is smaller than that in experiment by Comte-Bellot.[10] In the present results in both Cases 5 and 6, the streamwise correlation persists a longer distance in the downstream direction than in Moin and Kim.[3] Moreover, those in Case 6 show a better agreement with experiments. This improvement in Case 6 is considered to come from the inclusion of the diffusion term in SGS energy balance. To check the dependence of streamwise correlations on the size of computational box in the streamwise direction, L_x was elongated to 4.8 in Case 7. The results showed no essential difference from those in Case 5. For the detailed comparison with Moin and Kim,[3] see Horiuti.[14]

CONCLUSIONS

Turbulent plane channel flow is numerically studied using Large Eddy Simulation. The computation is carried out with both S-model and 1-eq. model. In the lower resolution cases, remarkable differences are found in the turbulence intensities and the spanwise correlations. This is considered to be due to the too dissipative nature of the Smagorinsky model.[7],[15] In the high resolution case, the difference is not apparent, but appreciable differences are shown in the streamwise two point correlations. It seems that the choice of SGS model is not crucial in the high resolution case, however we note that in 1-eq. model, SGS turbulent energy can be systematically evalated. Although the dependence on the parameters involved in 1-eq. model is still under an investigation, we have found that the diffusion term plays a dominant role in a budget of subgrid scale turbulent energy especially in the vicinity of the wall, and takes a positive value near the wall and changes a sign to negative as we depart from the wall. Correspondingly, in the budget of grid scale turbulent energy, the diffusion term is dominant in the vicinity of the wall and changes the sign from positive to negative as we depart from the wall.

REFERENCES

1) Deardorff, J.W.: J. Fluid Mech. 41 (1970) 453.
2) Schumann. U.: J. Comp. Phys. 18 (1975) 376.
3) Moin, P. and Kim, J.: J. Fluid Mech. 118 (1982) 341.
4) Horiuti, K.: Theor. Appl. Mech. 31 (1982) 407.
5) Smagorinsky, J., Manabe, S. and Holloway, J.L.: Mon. Weath. Rev. 93 (1965) 727.
6) Yoshizawa, A.: Phys. Fluids 25 (1982) 1532.
7) Yoshizawa, A. and Horiuti, K.: J. Phys. Soc. Jpn. 54 (1985) 2834.
8) Leonard, A.: Adv. Geophys. 18A (1974) 237.
9) Gottlieb, D., and Orszag, S.A.: Numerical Analysis of Spectral Method, CBMS Monograph no.26, SIAM (1977).
10) Van Driest, E.R.: J. Aero. Sci. 23 (1956) 1007.
11) Comte-Bellot, G.: Dr. Thesis, University of Grenoble (1963).
12) Laufer, J.: NACA Tech. Note, No.1053 (1950).
13) Kline, S.J., Reynolds, W.C., Schraub, F.A. and Runstadler,

R.W.: J. Fluid Mech. 30 (1967) 741.
14) Horiuti, K.: submitted to J. Comp. Phys. (1986)
15) Horiuti, K.: J. Phys. Soc. Jpn. 54 (1985) 2855

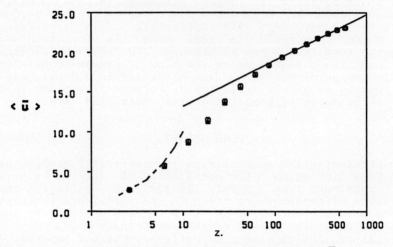

Fig. 1. Mean velocity profile $\langle \bar{u} \rangle$
Case 1 (◯), Case 2 (△)

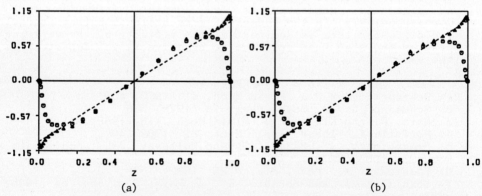

Fig. 2. Mean GS Reynolds stress. : ◯ : $\langle \bar{u}'' \bar{w} \rangle$; △ :
$\langle \bar{u}'' \bar{w} \rangle - \langle K (\frac{\partial \bar{u}}{\partial z} + \frac{\partial \bar{w}}{\partial x}) \rangle - \frac{1}{Re} \frac{\partial \langle \bar{u} \rangle}{\partial z}$
(a) Case 1, (b) Case 2

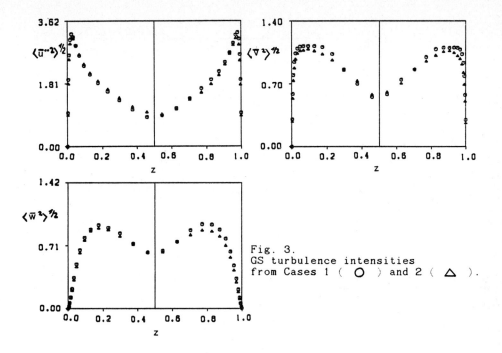

Fig. 3. GS turbulence intensities from Cases 1 (○) and 2 (△).

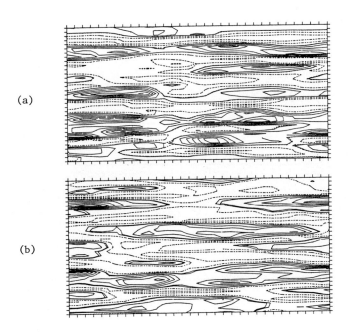

Fig. 4. Contours of $\bar{u}''$ in the x-y plane at $z_+=2.6$, $t=8.25$.
(a) Case 1, (b) Case 2

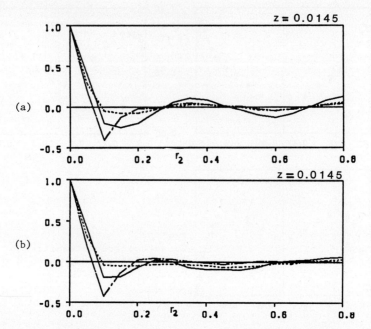

Fig. 5. Spanwise two point correlation function $R_{ii}(r_2)$;
——— : $R_{11}(r_2)$; - - - - - : $R_{22}(r_2)$; —·—·— : $R_{33}(r_2)$;
(a) Case 1, (b) Case 2

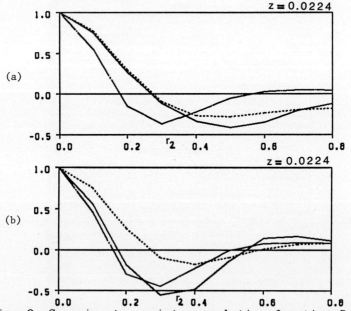

Fig. 6. Spanwise two point correlation function $R_{ii}(r_2)$;
see caption of Fig.5 for details. (a) Case 3, (b) Case 4

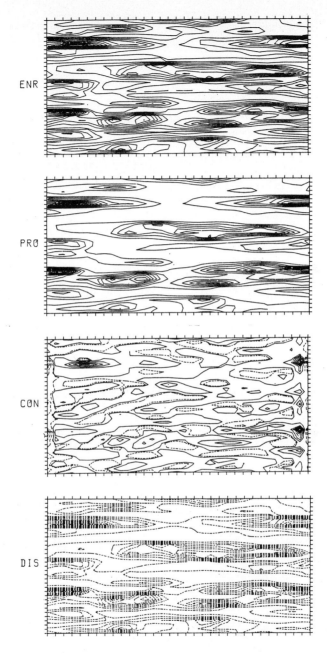

Fig. 7. Contours of instantaneous distribution of SGS turbulent energy and each terms of eq. (9), in the $x-y$ plane at $z_+=2.6$, $t=8.25$. ENR : SGS turbulent energy; PRO : production ; CON : convection; DIS: dissipation ; DIF : diffusion.
(a) $z=0.002$, (b) $z=0.0049$ (diffusion term only)

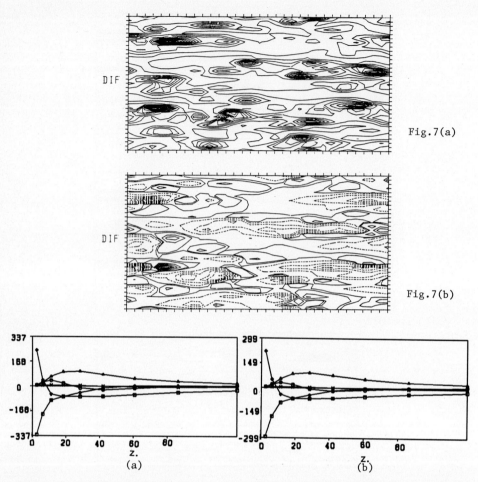

Fig.7(a)

Fig.7(b)

Fig. 8. Distribution of ensemble averaged GS turbulent energy.
△ : production; ○ : convection; × : velocity pressure gradient ; ◇ : diffusion; □ : dissipation.
(a) Case 1, (b) Case 2

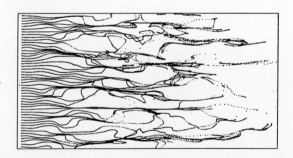

Fig. 9. Time line from Case 1; a wire is set at $z= 0.01$, $y=0.01 - 1.59$

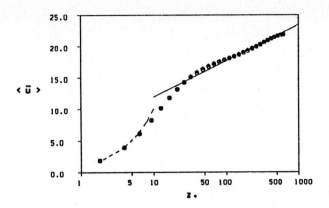

Fig. 10. Mean velocity profile $\langle \bar{u} \rangle$
Case 5 (○), Case 6 (△)

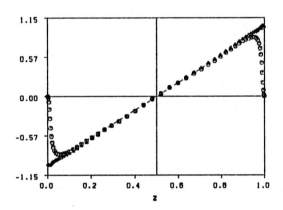

Fig. 11. Mean GS Reynolds stress and the total stress from Case 5.

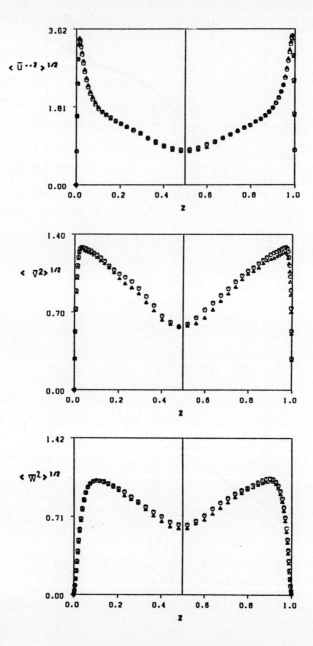

Fig. 12. GS turbulence intensities from Cases 5 (○) and 6 (△).

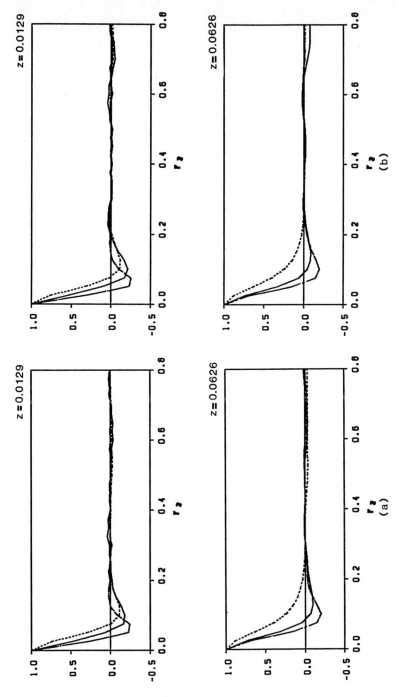

Fig. 13. Spanwise two point correlation function $R_{ii}(r_2)$; see caption of Fig.5 for details. (a) Case 5, (b) Case 6

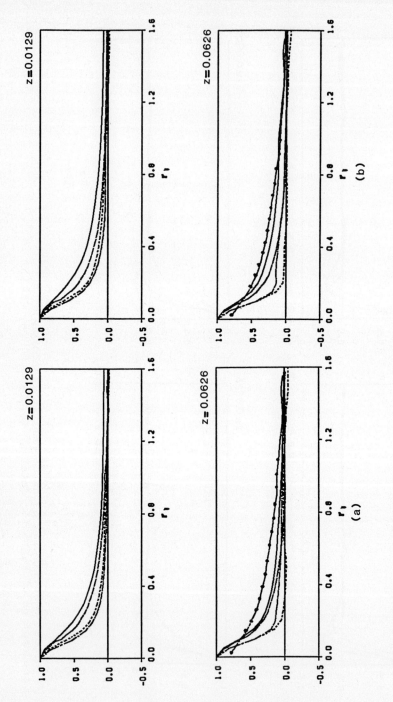

Fig. 14. Streamwise two point correlation function $R_{ii}(r_1)$; ———: $R_{11}(r_1)$; ---------: $R_{22}(r_1)$; —·—·—: $R_{33}(r_1)$; ——··——: Comte-Bellot [1] ; ———: Moin and Kim [3] ;
(a) Case 5, (b) Case 6

NUMERICAL PREDICTION OF TURBULENT PLANE COUETTE FLOW
BY LARGE EDDY SIMULATION

Toshio KOBAYASHI and Masanori KANO
Institute of Industrial Science, University of Tokyo
7-22-1 Roppongi, Minato-ku, Tokyo 106 JAPAN

1. INTRODUCTION

The recent advent of fast, large-core digital computers has enabled us to carry out complex turbulent flow calculations and Large Eddy Simulation which was developed by Deardorff[1], Schumann[2] and others is one of the most promising approaches to analyse the turbulence transport phenomena. Recently, Large Eddy Simulation(abbreviated as LES) succeeded in describing the detailed structure of wall turbulent flows[3]. Of course, the efforts have been continuously attempted towards the quantitative measurements of the near wall turbulence structures. For example, the spanwise spacings of the low speed streaks have been investigated in flat plate turbulent boundary layers experimentally[4]. Up to present, the suggestions have been made that the streaks are the inherent structures in the turbulent boundary layers and play a dominant role in turbulence production. However, there exist several ambiguities about streaks; (1) the spanwise spacing changes with the flow pattern, (2) the spatial positions occur at random, (3) it is quite difficult to measure the vorticity distribution, which plays an important role in the generation of streaks, and so on. The detailed LES calculation will be effective for these cases[5].

On the other hand, LES has been applied to more complicated turbulent flows. The authors have already tried to calculate the separated turbulent flow between two parallel flat plates, on which rectangular cylinders with square cross section are installed at equal intervals, and have compared the calculation results with results of a k-ε model and of experiment[6].

In the present paper, the LES code is applied to the turbulent plane Couette flow. Mean velocity profiles and turbulence statistics are calculated and compared with available experimental results. Furthermore, the effects of the Smagorinsky constant on the micro-structures of turbulence and the spanwise spacings of the streaks are discussed.

2. BASIC EQUATIONS FOR LES AND COMPUTATIONAL PROCEDURE

In LES, a physical quantity f is devided into the following two parts
$$f = \bar{f} + f' , \tag{1}$$
where $\bar{f}$ is the component of the large-scale field and f' is that of the small-scale flow field. The Gaussian function is used as filter function to generate the large-scale flow field:

$$G_i(x_i, x_i') = \left(\frac{6}{\pi \Delta_i^2}\right)^{1/2} \exp\left\{-6\frac{(x_i - x_i')^2}{\Delta_i^2}\right\} \qquad (i=1,2,3) . \tag{2}$$

Δ_i is the computational grid size in x_i direction. By operating the filter function to the whole flow field, the large-scale flow field is defined as

$$\bar{f}(x_1, x_2, x_3) = \int_D \prod_{i=1}^{3} G_i(x_i, x_i') f(x_1', x_2', x_3') \, dx_1' dx_2' dx_3' . \tag{3}$$

The computational flow field is composed of two parallel walls as shown in Fig.1. The upper wall moves in its plane with a uniform velocity U. The coordinate axis x_1 is defined as the direction in which the upper

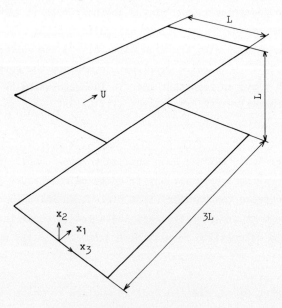

Fig.1 Model flow field

wall moves, x_2, as the direction normal to both walls and x_3, as the span-wise direction. The Navier-Stokes and continuity equations are filtered by Eq.(3), assuming an incompressible fluid. Velocity u_i, coordinates x_i, time t and pressure p are non-dimensionalized by U, the channel width L, L/U and ρU^2 (ρ: fluid density), respectively.

$$\frac{\partial \bar{u}_i}{\partial t} + \frac{\partial}{\partial x_j}(\bar{u}_i\bar{u}_j) = -\frac{\partial \bar{P}}{\partial x_i} - \frac{\partial L_{Eij}}{\partial x_j} - \frac{\partial C_{Rij}}{\partial x_j} - \frac{\partial R_{Eij}}{\partial x_j} + \frac{1}{4|R_e|}\frac{\partial^2 \bar{u}_i}{\partial x_j \partial x_j} \quad , \tag{4}$$

$$\frac{\partial \bar{u}_i}{\partial x_i} = 0 \quad , \tag{5}$$

where pressure $\bar{P}$ includes turbulence pressure and Re is the Reynolds number (= $UL/(4\nu)$, ν: kinematic viscosity). In Eq.(4) L_{Eij}, C_{Rij} and R_{Eij} are the Leonard, cross and Reynolds terms, respectively, and are defined as

$$\left. \begin{array}{l} L_{Eij} = \overline{\bar{u}_i \bar{u}_j} - \bar{u}_i \bar{u}_j \quad , \\ C_{Rij} = \overline{\bar{u}_i u'_j} + \overline{u'_i \bar{u}_j} \quad , \\ R_{Eij} = \overline{u'_i u'_j} - \delta_{ij} \overline{u'_i u'_i}/3 \quad . \end{array} \right\} \tag{6}$$

These terms are modelled in this paper as

$$L_{Eij} = \frac{\Delta_k^2}{24}\frac{\partial^2 (\bar{u}_i \bar{u}_j)}{\partial x_k \partial x_k} \quad , \tag{7}$$

$$C_{Rij} = -\frac{\Delta_k^2 \bar{u}_i}{24}\frac{\partial^2 \bar{u}_j}{\partial x_k \partial x_k} - \frac{\Delta_k^2 \bar{u}_j}{24}\frac{\partial^2 \bar{u}_i}{\partial x_k \partial x_k} \quad , \tag{8}$$

$$\left. \begin{array}{l} R_{Eij} = -2K\zeta_{ij} \quad , \\ K = (C\Delta)^2 [2\zeta_{ij}\zeta_{ij}]^{1/2} \quad , \\ \zeta_{ij} = \frac{1}{2}\left(\frac{\partial \bar{u}_j}{\partial x_i} + \frac{\partial \bar{u}_i}{\partial x_j}\right) \quad , \end{array} \right\} \tag{9}$$

where Δ, a representative length of the subgrid scale flow, can be expressed as

$$\Delta^2 = (\Delta_1^2 + \Delta_2^2 + \Delta_3^2)/3 \quad . \tag{10}$$

In this paper, L_{Eij} and C_{Rij} in Eqs.(7) and (8) are taken into consideration, though they are often neglected. R_{Eij} in Eq.(9) is the subgrid scale Reynolds stress and c is a Smagorinsky constant[7].

Equations (4),(5),(7),(8) and (9) are transformed into a series of difference equations and solved simultaneously using the SMAC method with central difference formulae in space and an Adams-Bashforth-type difference scheme in time.

According to the experimental data of the two-dimensional channel flow, the two-point correlation of x_1 direction near the wall is approximately three times as large as that of x_3 direction. Therefore, the geometrical dimensions of the computational region are determined to be $3L \times L \times L$ as shown in Fig.1. The computational region is subdivided into 30 mesh cells in x_1 direction, 30 in x_2 direction and 25 in x_3 direction, respectively. Considering the application of this code to three-dimensional flow problems of industrial concern, a pseudo-spectral method is not available in this paper.

The initial and boundary conditions are as follows: a measured mean velocity profile superimposed with white noise, namely uniform random numbers of 0.032 U, is used as initial condition. For velocity non-slip conditions are adopted at wall boundaries and cyclic conditions are incorporated in x_1 and x_3 directions. Convergence condition for the iteration procedure in SMAC method is

$$\text{Max}\left|\left[\text{dif}\left(\frac{\partial \bar{u}_i}{\partial x_i}\right)\right]_N\right| < 0.001 \quad , \tag{11}$$

where $\text{dif}(\partial \bar{u}_i/\partial x_i)$ denotes a difference formula for $\partial \bar{u}_i/\partial x_i$. The computational conditions are shown in Table 1, where the dimensionless constant c is estimated as 0.1, 0.05 and 0. Computations are carried out on the HITAC S-810 system computer of the University of Tokyo. The CPU time per timestep is about 1.4 s for cases I and II, and about 1.0 s for case III. The Reynolds number used for the presented computations is 3.75×10^3.

Table 1 Computational conditions

	The Number of Grids	δt	Grid Points	Method
Case I	30 × 30 × 25	0.002	Uniform	LES (c=0.1)
Case II	30 × 30 × 25	0.002	Uniform	LES (c=0.05)
Case III	30 × 30 × 25	0.002	Uniform	c = 0

3. COMPUTATIONAL RESULTS AND DISCUSSIONS

Starting from the prescribed initial conditions, calculations are continued until the spatially averaged velocity and turbulence intensity in $x_1 - x_3$ planes reach constant values. In each calculation the flow is considered to be steady at the dimensionless time $t \sim 5.0$ and the calculation in each case is terminated after 4000 steps.

In LES, physical quantities such as velocity, pressure and so on are calculated at each grid point and at each time step. In this paper, an average operation is utilized. $<f>$ means the spatial average of f in a $x_1 - x_3$ plane.

3.1 Mean Velocity Profiles

Figure 2 shows the mean velocity profiles $<u_1>$ for cases I, II and III. The experimental results of Robertson[8] and Burton[9] are also

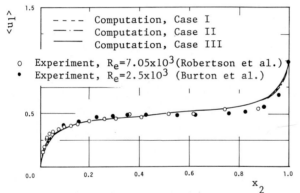

Fig.2 Mean velocity profiles

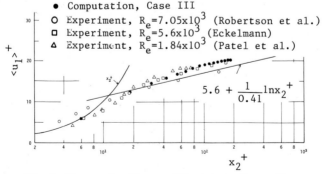

Fig.3 Mean velocity profiles near the wall

included in the figure. It is evident that the calculated mean velocity profiles are in good agreement with the experimental data and are independent of the Smagorinsky constant. Figure 3 shows the velocity profiles, non-dimensionalized by the friction velocity $u_\tau = \sqrt{\tau/\rho}$ (e.g. 2.37×10^{-2} for case III). The abscissa is the wall coordinate $x_2^+ = u_\tau x_2/\nu$. The figure demonstrates that the calculated results for case III agree well with the experimental results of Robertson[8]. For comparison, the experimental velocity profiles of Eckelmann[10] and Patel[11] for the two-dimensional channel flow are shown in the figure, and as it is anticipated, the logarithmic layer exists also in the case of Couette flow.

3.2 Turbulence Intensities

Velocity fluctuations are defined as follows:

$$u_i'' = u_i - \langle u_i \rangle .\qquad(12)$$

Figure 4(a) shows the profiles of RMS values of u_1'', and Fig.4(b) the profiles of the calculated $\langle u_i''^2 \rangle$, which includes a contribution from

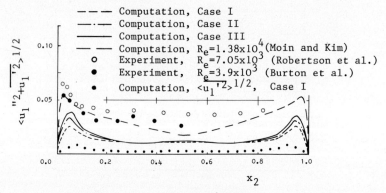

(a) Comparison with experiments

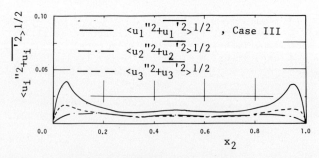

(b) Three component of $\langle u_i''^2 \rangle$

Fig.4 Turbulence intensities

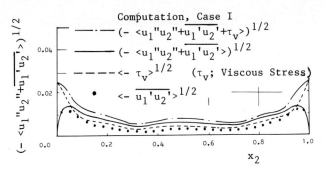

Fig.5 Shear stress profiles

the subgrid Reynolds stress in cases I and II. The computational results of Moin et al. for two-dimensional channel flow and the experimental data of Robertson and Burton for Couette flow are also included in Fig.4(a). From the figures it can be concluded that (1) all calculated profiles agree qualitatively well with the experimental data but are somewhat underestimated quantitatively, (2) the calculated $<u_1''^2>$ increases as the constant c decreases, (3) each profile of $<u_1''^2>$ has a constant value in the center region of the flow field, and (4) the turbulence intensities are anisotropic especially in the neighborhood of both walls.

For reference, the shear stress profile is shown in Fig.5.

3.3 Grid Scale Energy Balance

Grid scale energy equation is expressed as

$$\frac{\partial \langle k \rangle}{\partial t} = -\frac{\partial}{\partial x_2} \langle k \bar{u}_2'' \rangle - \langle \bar{u}_1'' \bar{u}_2'' \rangle \frac{\partial \langle \bar{u}_1 \rangle}{\partial x_2} - \left\langle \left(\frac{1}{4Re} + K \right) \frac{\partial \bar{u}_i''}{\partial x_j} \frac{\partial \bar{u}_i''}{\partial x_j} \right\rangle \\ - \left\langle \bar{u}_i'' \frac{\partial \bar{P}''}{\partial x_i} \right\rangle + \frac{\partial}{\partial x_2} \left\langle \left(K + \frac{1}{4Re} \right) \frac{\partial k}{\partial x_2} \right\rangle , \qquad (13)$$

where $k = \bar{u}_i'' \bar{u}_i''/2$ and contributions from the Leonard and cross terms are neglected. The right hand terms of Eq.(13) denote turbulence diffusion, production, dissipation, velocity-pressure gradient and viscous diffusion, respectively. Figure 6 shows the computational results of the grid scale energy budged. From the figure it can be concluded that (1) the production and dissipation are clearly the dominant terms in most of the region, (2) in the neighborhood of the wall, the balance of the governing equation for turbulent kinetic energy is lost in this paper and (3) the reduction of the Smagorinsky constant increases the magnitudes of all the terms of the energy equation.

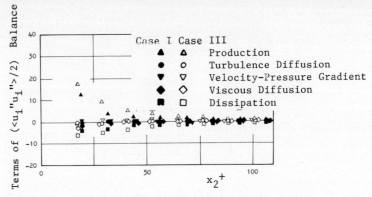

Fig.6 Balance of the resolvable turbulence kinetic energy

3.4 Detailed Structure of the Near Wall Flow

In the previous paragraph it is seen that the numerical constant c has an effect on the turbulence intensity, and now its effect on the spatial structures of the near wall flow is investigated in detail. Figure 7 shows the contours of the instantaneous distribution of u_1'' in the $x_1 - x_2$ plane at $x_3^+ = 0.08$, in which the contour interval is 0.01. In the figure corresponding to case III the higher density in the contour distribution is observed and this suggests that the RMS values of u_1'' in case III are bigger than those in case I.

It is seen in the turbulent channel flow that the organized structures, called streaks for example, appear near the wall. These streaks are also observed in Couette flow, as shown in Fig.8. The figures show that in the vicinity of the wall there exist low speed streaks and high speed streaks alternatively in the spanwise direction, and that the organized structure is weakened or vanishes as x_2 increases. The spanwise spacing of the streaks λ^+ can be obtained from the two-point correlation function in the $x_1 - x_3$ plane. This function is defined for u_1'' as

$$R_{11}(x_2, r_3) = \frac{\langle u_1''(x_1, x_2, x_3) \cdot u_1''(x_1, x_2, x_3 + r_3) \rangle}{\langle u_1''^2(x_1, x_2, x_3) \rangle} . \tag{14}$$

Figure 9 shows $R_{11}(x_2^+, r_3^+)$ for three horizontal planes ($x_1 - x_3$ planes). In the figure, for example, the minimum of $R_{11}(x_2^+ = 5.94, r_3^+)$ appears at $r_3^+ \sim 28.5$. This r_3^+ denotes the statistical width between two adjacent low and high speed streaks and leads to a λ^+ of 57.0. Figure 10 shows the spanwise spacings of the streaks obtained from the two-point correlation functions. Several experimental results of a flat plate turbulent boundary layer[12] are also shown in the figure, where Re_θ denotes the Reynoles number based on U/2 and the momentum thickness. The calculated

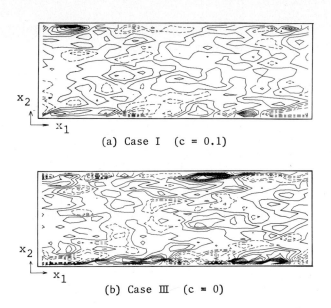

(a) Case I (c = 0.1)

(b) Case III (c = 0)

Fig.7 Contours of the streamwise component of velocity fluctuations
(solid lines : $u_1''>0$,   broken lines : $u_1''<0$)

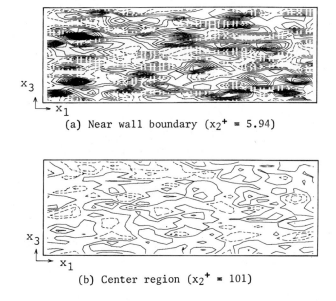

(a) Near wall boundary ($x_2^+ = 5.94$)

(b) Center region ($x_2^+ = 101$)

Fig.8 Contours of the streamwise component of velocity fluctuations

spacings are consistent with the experimental results, in view of the tendency that λ^+ increases with x^+.

Figure 11 shows the contours of the streamwise vorticity component in a vertical plane (x_1 - x_2 plane) at $x_3^+ \sim 29$. This component is defined as

$$\omega_{x1} = \frac{\partial u_3}{\partial x_2} - \frac{\partial u_2}{\partial x_3} \quad . \tag{15}$$

It seems that the axes of streamwise vorticities are inclined to the wall. The inclination is dominant in the neighborhood of both walls. In the range of x_2^+ = 5.94 ~ 41.6, the inclination angles determined from the figure lie between 10° and 30°. This suggests as in the turbulent channel flow[13] that the horse-shoe vortices exist also in this computed flow field.

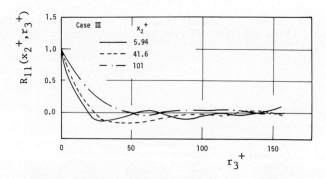

Fig.9 Two-point correlation functions of the streamwise component of velocity fluctuations

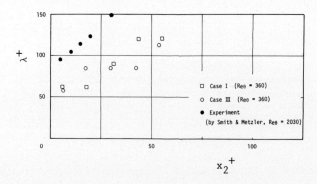

Fig.10 Spanwise spacings of streaks

Fig.11 An example of contours of the streamwise vorticity component (x_1 - x_2 plane, $x_3^+ = 29$)

4. CONCLUSIONS

The LES code developed for three-dimensional separated turbulent flow is applied to the plane turbulent Couette flow. The number of the computational grid points is $30 \times 30 \times 35$. The following results are obtained:

(1) In spite of the rather small grid number the mean velocity profiles are approximately independent of the Smagorinsky constant.

(2) However, turbulence intensities depend strongly on the Smagorinsky constant, and turbulence intensities increase with decreasing Smagorinsky constant.

(3) The turbulence is anisotropic especially in the neighborhood of the wall.

(4) The spanwise spacings of the streaks obtained from two-point correlation functions are in good agreement with observed results.

(5) The axes of the streamwise vorticities are inclined to the wall and the inclination angles lie between 10° ~ 30°.

REFERENCES

1) Deardorff, J.W., "A Numerical Study of Three-Dimensional Turbulent Channel Flow at Large Reynolds Numbers", J. Fluid Mech., Vol.41, No.2 (1970), p.453.

2) Schumann, U., "Subgrid Scale Model for Finite Difference Simulations of Turbulent Flows in Plane Channels and Annuli", J. Comput. Phys., Vol.18 (1975), p.376.

3) Moin, P. and Kim, J., "Numerical Investigation of Turbulent Channel Flow", J. Fluid Mech., Vol.118 (1982), p.341.

4) Kline, S.J. et al., "The Structure of Turbulent Boundary Layers", Vol.30, No.4 (1967), p.741.

5) Horiuchi, K., "Large Eddy Simulation of Turbulent Channel Flow by One-Equation Modeling", J. Phys. Soc. Jpn., Vol.54, No.8 (1985), p.2855.
6) Kobayashi, T., et al.,"Prediction of Turbulent Flow in Two-Dimensional Channel with Turbulence Promoters", Bull. JSME, Vol.27, No.231 (1984), p.1893 and Vol.28, No.246 (1985), p.2940.
7) Smagorinsky, J., "General Circulation Experiments with the Primitive Equations", Monthly Weather Review, Vol.93, No.3 (1963), p.99.
8) Robertson, J.M., and Johnson, H.F., "Turbulence Structure in Plane Couette Flow", J. Eng. Mech. Div., Proc. ASCE, Vol.96, EM6 (1970), p.1171.
9) Burton, R.A. and Carper, H.J., "An Experimental Study of Annular Flows with Applications in Turbulent Film Lubrication", J. Lubrication Technology, Trans. ASME, Vol.89, No.3 (1967), p.381.
10) Eckelmann, H., "Experimentelle Untersuchungen in einer Turbulenten Kanalströmung mit Starken Viskosen Wandschichten", Mitt. Max-Plank Inst. für Strümungsforshung, Göttingen No.48 (1970), 1.
11) Patel, V.C. and Head, M.R., "Some Observations of Skin Friction and Velocity Profiles in Fully Developed Pipe and Channel Flows, J. Fluid Mech., Vol.38, No.1 (1969), p.181.
12) Smith, C.R. and Metzler, S.P., "The Characteristics of Low-Speed Streaks in the Near-Wall Region of a Turbulent Boundary Layer", J. Fluid Mech., Vol.129, (1983), p.27.
13) Moin, P. and Kim, J., "The Structure of the Vorticity Field in Turbulent Channel Flow", J. Fluid Mech., Vol.155 (1985), p.441.

ADVECTIVE FORMULATION OF LARGE EDDY
SIMULATION FOR ENGINEERING TYPE FLOWS

D. Laurence
Electricité de France
Laboratoire National d'Hydraulique
6 quai Watier - 78400 Chatou, France

SUMMARY

Having in mind futur applications of large eddy simulation to engineering problems, a numerical code operating in physical space only is presented. Results are analysed in Fourier space and quality is shown to be comparable to that of pseudo-spectral codes. An advective formulation of subgrid scale effects is proposed and special care is devoted to the problems arising as a large mean field is present.

INTRODUCTION

Large Eddy Simulation, today, does not seem to be of such widespread practice as was forcasted a few years ago in connection with the rapid development of computer power. While researchers analysing fundamentals of turbulence make use of increasing computer capabilities to perform direct simulations, LES applications to engineering type flows are still scarce. This is due to the fact that eddy viscosity models are not fully satisfactory outside grid turbulence simulations and because of extra numerical difficulties arising as the homogeneity hypothesis is abandoned. These 2 points are analysed hereafter through a pseudo-Lagrangian approach.

I TOTAL FIELD DECOMPOSITION

Let $V(x,t)$ be the total velocity field in a fixed frame with respect to an experimental setup. Prior to performing a LES, this is decomposed as :

$$V(x,t) = \mathcal{V}(x) + \bar{v}(x,t) + v'(x,t). \tag{1}$$

In addition to the usual grid-scale subgrid-scale decomposition v/v', a supporting field $\mathcal{V}$ (i.e. average over an infinite time scale) is always present. It is easily accounted for in homogeneous flows by choice of the computational frame or by a simple function of the space variable $\bar{X}$ only. The computed variable is then $u = \bar{v}$.

In engineering applications, with which we are concerned, $\mathcal{V}$ is usually complex and may be even unknown initially, so the computed variable must be $u = \mathcal{V} + \bar{v}$. The non-linear term of the u equation, $u_j \frac{\partial}{\partial x_j} u_i$, then contains not only pure non-linear turbulent interactions and production effects (with time scales $T_t = q^2/\mathcal{E}$, $T_p = (d\mathcal{V}_i/dx_j)^{-1}$ respectively), but also transport for which Eulerian numerics introduce an additional time scale $T_{num} = h/max.|\mathcal{V}|$ (h : mesh step, $2q^2$: kinetic energy, $\mathcal{E}$: dissipation). Thus, if $\mathcal{V}$ is large with small gradients, T_{num} may be smaller than the "more physical" scales T_t and T_p. To obey this CFL criterion may be unnecessarily expensive since transport effects can be taken care of by introducing some Lagrangian approach.

II   NUMERICAL SCHEME AND FILTERING

The standard scheme [1] uses Smagorinsky's model for the subgrid stresses and the resolution is based on fractional time steps. It's originality lies in the use of a characteristics method for the non linear term (advection of momentum) : A 3D particle trajectory $\vec{X}_p(t)$ starting from grid point $\vec{X}_M$ at time $t^{n+1} = t^n + \Delta t$ is computed backwards in time. The advected velocity is then $\tilde{u}^{n+1}(X_M, t^{n+1}) = U(X_N, t^n)$ with $\vec{X}_N = \vec{X}_p(t^n)$. $\vec{X}_N$ not being a node point, high order interpolation must be introduced. This is where most of the numerical filtering occurs.
In order to retain as much information as possible, this implicit filtering is used instead of the usual Gaussian filter :

$$\bar{v}(x,t) = \int_\Omega G(x-y) \cdot v(y,t) \cdot dy \quad ; \quad (v = V - \bar{V})$$

or $\bar{v}(x,t) = G(k) \cdot v(k,t) \quad ; \quad G(k) = \exp\left[-\Delta_G^2 K^2/24\right]$ in Fourier space.

The implicit filtering, g_I, operates at each time step so the total filtering of a computation performed through N time steps is $G_I(k) = (g_I(k))^N$. g_I is gaussian-like and thus it's width can be estimated by comparison with grid turbulence experiment [14] . (Initial turbulent field matches exactly the spectrum at station t $V°/M = 42$ and has non-zero skewness, cut-off wave number is well located in the inertial range. Spectra are compared at station t $V°/M = 98$, on fig. 1).

This implicit filtering can be checked by performing again the computation, this time adding a (spectral) defiltering after every time step, using the estimated width ΔG_I (fig. 2).
The non dimensional width δ ($\Delta G_I = \delta \cdot h$) ranges from 1.6 to 3.3 for the total run, depending on the numerical scheme [2]. This "empirical numerical analysis" provides guidance for comparison between alternative schemes.

III   ADVECTIVE FORMULATION OF SGS MODELS

The characteristics scheme brings natural separation between advected (U^{ed}) and advecting (U^{ing}) velocities. The advection step is :

$$\int_{t^n}^{t^{n+1}} \left[\frac{\partial u^{ed}}{\partial t} + u_i^{ing} \frac{\partial}{\partial x_i} u^{ed}\right] dt = 0 \rightarrow \tilde{u}^{n+1}$$

(higher order in time is actually achieved with a non-zero R.H.S.).

The standard scheme is $U^{ing} = U^{ed} = U$. Subgrid scale eddies can be represented by perturbating the particle trajectory by a random flight model. One sets $U^{ing} = U + a$, then if a is purely random (not correlated with U ; $\langle a \cdot U \rangle = 0$), and its it's variance is scaled by $\langle a^2 \rangle = 2\nu_T/\Delta t$, the random flight is equivalent to diffusion and the advection step inclueds the effect of an eddy viscosity ν_T. This results in resetting the sub-grid stresses into an advection form from which they actually originate and shows that modelling through perturbated advection could be a more promising approach since it at least contains the classical eddy viscosity models.

New models the effects of which are not restricted to energy drain can be found by imposing a non-zero $<a.u>$ correlation. The underlying idea is Bardina Fertziger & Reynolds's scale similarity model [3] : the subgrid scales are modeled using the smallest computed eddies "trapped" between single and double filtering. The force we actualy add to the U_i equation is :

$$F_i = (\bar{v}_j - \bar{\bar{v}}_j) \cdot \frac{\partial}{\partial x_j} \bar{v}_i \tag{2}$$

which is equivalent to the BFR model :
as it enters the U_i equation through it's divergence :

$$\frac{\partial}{\partial x_j} M_{ij} = \frac{\partial}{\partial x_j} \left[\bar{v}_j \bar{v}_i - \bar{\bar{v}}_j \bar{\bar{v}}_i \right] = \bar{v}_j \frac{\partial}{\partial x_j} \bar{v}_i - \bar{\bar{v}}_j \frac{\partial}{\partial x_j} \bar{\bar{v}}_i \tag{3}$$

and after the term responsible for non Galilean Invariance is extracted :
(see Speziale [4] and [2]). $\frac{\partial}{\partial x_j} M_{ij} = \frac{\partial}{\partial x_j} M_{ij} - \bar{\bar{v}}_j \frac{\partial}{\partial x_j} \left[\bar{v}_i - \bar{\bar{v}}_i \right]$.

This model, plus the random flight model, in which ν_T has the same value as in Smagorinsky's model, give satisfying results for grid turbulence decay (fig. 3), although a very large time step is taken. The effect of the scale similarity model is to enhance transfer of energy between larger and smaller computed scales. Thus improving the shape of the spectra improves in turn the Reynolds stress anisotropy in the homogeneous shear experiment (fig. 7-10). The excessive value of b_{11} ($b_{11} \simeq 2.6$) on fig. 9, while experiment shows a final value $b_{11} \simeq 1.5$) is decreased as well as the discreptancy of b_{22}. Excess of anisotropy may be related to excess of energy on small wave numbers.

IV SKETCH OF SCALE SIMILARITY MODELS IN SPECTRAL SPACE

Effects of scale similarity models cannot be investigated on the integral variables. Since they imply a decomposition along the wave number axis, spectral analysis is required. The EDQNM approach can be very helpful here (eg : Aupoix, Bertoglio, Chollet, Lesieur, etc ..., present conference) but we will limit ourselves to a rough sketch.

Consider the exact equation of the filtered field ($\mathcal{V} = 0$) :

$$\frac{\partial \bar{v}_i}{\partial t} + \overline{(\bar{v}_j + v_j')\frac{\partial}{\partial x_j}(\bar{v}_i + v_i')} = -\frac{1}{\rho}\frac{\partial}{\partial x_i}\bar{P} + \nu \Delta \bar{v}_i. \tag{4}$$

The advection term in spectral space is :

$$\overline{(\bar{v}_j + v_j')\frac{\partial}{\partial x_j}(\bar{v}_i + v_i')} \rightarrow (-i) \cdot \iiint \Big\{ \underbrace{G(\vec{K}) G(\vec{K}') G(\vec{K}-\vec{K}')}_{I}$$

$$+ \underbrace{G(\vec{K}) G(\vec{K}') \left[1 - G(\vec{K}-\vec{K}')\right]}_{II} + \underbrace{G(\vec{K}) G(\vec{K}-\vec{K}') \left[1 - G(\vec{K}')\right]}_{II'} \tag{5}$$

$$+ \underbrace{G(\vec{K}) \left[1 - G(\vec{K}')\right]\left[1 - G(\vec{K}-\vec{K}')\right]}_{III} \Big\} v_i(\vec{K}') K_j v_j(\vec{K}-\vec{K}') d\vec{K}'$$

I	−	$\overline{v}.\overline{v}$	is the computed term
II + II'	−	$\overline{v}\ v' + \overline{v'\ v}$	are the cross terms
III	−	$\overline{v'\ v'}$	is the true subgrid term.

The usual procedure here (Kraichnan (76), Leslie and Quarini [5], Bertoglio [6]) is to introduce the Fourier transforms of the 2-point-correlations $\Phi_{ij}(\vec{K})$, integrate over all angles and introduce a closure for the triple correlations. This leads to eq. 6 introduced further on, but masks the local anisotropy of the combined effects of the filters and the "triangle" requirement. Also, some effects of the models might not show directly on $\varphi_{ij}(K)$ although they can be strong on higher moments (eg : see velocity derivative skewness, B.F.R. [3]). We will thus limit ourselves to the primitive equation (5) and to a rough sketch of the regions in which the choice of $\vec{K}'$, for a given $\vec{K}$ will not lead to a cancellation through filtering and incompressibility. For instance, term I is illustrated by fig. 19. Action of the filter G is represented by a disk. Since Gaussian filters are considered, the boundaries are actually not sharp, but decrease rapidly outside the limit circles. Regions having such a line in common actually overlap. "Scale similarity" assumption is that triangles having similar shapes and scales have a similar contribution. Terms II, II' and III are illustrated on figs. 20-21. The cone generated at the tip of K represents incompresibility : $K_j V_j (\vec{K} - \vec{K}')$ tends to zero as $(\vec{K} - \vec{K}')$ tends to be aligned with K.

Scale similarity models involve products as $G(P)\left[1 - G(P)\right]$. This is maximum for $G(P) = 1/2$, $P = K_s = \sqrt{24 \log 2/\Delta^2} \sim 4/\Delta$, with $G(K) = \exp(-\Delta^2 K^2/24)$. K_s is close to the maximum resolvable scale $K_m = 2\pi/\Delta$, if Δ is twice the mesh step. Terms as $\overline{v}i - \overline{\overline{v}}i$ can thus be represented by a ring near the K_m circle.

The B.F.R. model (fig. 22) does seem to be a good representation of the terms II and II' and the figures clearly show that III remains to be modeled. This is well done by an eddy viscosity model, due to the larger separation of scales involved in III.

Another model proposed by B.F.R., $\partial/\partial x_j \left[(\overline{v}_j - \overline{\overline{v}}_j)(\overline{v}_i - \overline{\overline{v}}_i)\right]$, is sketched on fig. 23. It was felt that it did not account for terms II and II', but they still seem to be partly represented especially when incompressibility is considered. Also, this model is G.I.

Finally, our model $(\overline{v}_j - \overline{\overline{v}}_j)\dfrac{\partial}{\partial x_j} \overline{v}_i$, seems to cover the same area (fig. 24).

Effects of scale similarity models on the energy spectra (isotropic case) will depend on the respective positions of K_s where $G^2(K)\left[1 - G^2(K)\right]$ is maximum and Ko wave number where the energy transfer rate T(K) changes sign (T(Ko) = 0), fig. 25. The global effect is then extra transfer of energy between computed scales when $K_s \sim K_o$. This absence of global drain of the model tends to be related to equilibrium between true drain and backscatter. Using assumptions of [5] for an inertial range spectrum. One can write :

$$\text{true drain} = \sum^{\Delta} A(K,P,R) \cdot R^{-\frac{2}{3}} \quad K^{-\frac{2}{3}} \quad P^{-1} \ . \tag{6}$$

$$\text{backscatter} = \sum^{\Delta} A(K,P,R) \cdot K^3 \quad P^{-\frac{8}{3}} \quad R^{-\frac{8}{3}} \quad G^2(K).$$

$$(K = |\vec{K}|, \ P = |\vec{K}'|, \ R = |\vec{K} - \vec{K}'|),$$

$$A = A'\left[1 - G(P)G(R)\right] \text{ is common to both terms.}$$

Because of $K^3 G^2(K)$, maximum backscatter is for K in a ring close to the Km circle. Observing powers of P and R, triangles leading to a larger contribution to backscatter with respect to true drain should have P and R both small (but not too small because of the $[1 - G(P)G(R)]$ term. This kind of triangles can be found in the cross terms, particularly in the sub-regions modelled by the $(\bar{v}i - \bar{\bar{v}}i)(\bar{v}j - \bar{\bar{v}}j)$ model.

V GALILEAN INVARIANCE

As was pointed out by Speziale [4] the Leonard term $L_{ij} = \overline{\bar{v}i \, \bar{v}j} - \bar{v}i \, \bar{v}j$, and the cross terms $L_{ij} = \overline{v_i' \bar{v}_j} + \overline{\bar{v}_i v_j'}$, taken separately are not Galilean invariant (G.I.). The same can be said of certain proposed models. This was of little consequence for previous homogeneous turbulence simulations where $\mathscr{V}$ (the supporting field) has been excluded. In engineering applications, the problem is more crucial but can be solved by using the three-level decomposition of eq (1) and writing models in terms of $\bar{v}$ and space derivatives of $\mathscr{V}$ only.

However, the numerical schemes themselves are not G.I. as soon as some linearisation has to be introduced.

This is true for the characteristics scheme by which the following equation is solved and where U^f is a "frozen" velocity field in some reference frame $\mathscr{O} \to [x,t]$.

$$\int_{t^n}^{t^{n+1}} \left[\frac{\partial u}{\partial t} + u_j^f \frac{\partial u}{\partial x_j} = 0 \right] \quad ; \quad \frac{\partial u^f}{\partial t} = 0 \text{ in } \mathscr{O}. \tag{6}$$

Observing this equation from a different reference frame $\mathscr{O}^* \to [x^*, t^*]$, we have :

$$\begin{cases} t^* = t \\ X_k^* = X_k + W_k \, t + b \\ U_k^* = U_k - W_k \end{cases} \tag{7}$$

W translational velocity

$$\int_{t^n}^{t^{n+1}} \left[\left(\frac{\partial}{\partial t^*} + W_k \frac{\partial}{\partial X_k^*} \right) U^* + U_k^f \frac{\partial U^*}{\partial X_k} = 0 \right] \text{ in } \mathscr{O}^*.$$

Introducing $(U_k^f)^* = U_k^f - W_k$, cancels the term $W_k \frac{\partial}{\partial X_k^*} U^*$ but eq 7 cannot be solved by the standard characteristics scheme because $(U_k^f)^*$ is not "frozen" in $\mathscr{O}^*$:

$$\frac{\partial (U_k^f)^*}{\partial t} = \frac{\partial (U_k^f)^*}{\partial t} - W_j \frac{\partial (U_k^f)^*}{\partial X_j} = - W_j \frac{\partial (U_k^f)^*}{\partial X_j} \neq 0$$

$(U_k^f)^*$ in $\mathscr{O}^*$ is evolving in time through the translation by W.

As previously for the non G.I. models, discrepancies are small if the variable U is $U = \bar{v}$ only, instead of $U = \mathscr{V} + \bar{v}$, but for channel flow, for instance, influence of the computational frame, or (equivalently) treatment of mean advection, has been noted (e.g. see [7]). Also, when computing homogeneous shear with Baron's shear-periodic boundary conditions [8] (see also [9]) partial statistics in the $\bar{X}_1 - \bar{X}_2$ planes are not exactly homogeneous in the $\bar{X}_3$ direction (turbulent kinetic energy decreases from its value in the centre plane, as $V_1 = S \, X_3$ increases).

Indeed, if $\mathscr{V}$ is large, the approximation :

$$U_k(t) \frac{\partial}{\partial X_k} U_i(t) \simeq U_k^f(t^n) \frac{\partial}{\partial X_k} U_i(t)$$

as $t \to t^{n+1}$ includes large phase errors for high wave numbers. Interaction is forced between uncorrelated eddies since only one is being advected by $\vec{V}$ while the other is "frozen". Numerical diffusion then results as can be seen on figs. 4-6.

Fortunatly, advection is local so an optimal computational frame, Op*, can be found for each node point $\vec{X}p$. The translational velocity is chosen to be the mean velocity of the dependancy domain of $\vec{X}p$ (all points involved for integrating the advection step between t^n and t^{n+1}, (U*) is then frozen in this frame G*p and the usual scheme is applied. Fig. 6 shows that the scheme is no longer sensitive to $\vec{V}$.

In this context, Schumann's improved advection scheme, using the three level decomposition (eq. 1) can be considered as the most accurate characteristics scheme for $\vec{V}$ advection since Fourier interpolation is used (but feasable only because the frame shift $0 \to 0*p$ is constant in each $\vec{X}_1 - \vec{X}_2$ plane).

VI WEAK FORMULATION OF THE CHARACTERISTICS SCHEME

The characteristics scheme is used in nearly all codes at L.N.H. because of it's unconditional stability and good accuracy in engineering problems [10]. Applied to LES it can lead to a new field of S.G.S. modeling through it's interpretation in terms of particle trajectory. But it's major drawback lies in the need of interpolation. This is now improved through the weak formulation : the transport equation of a variable f, in a domain Ω bounded by Γ is projected onto time dependent test functions ψ :

$$\int_{t^n}^{t^{n+1}} dt \times \int_\Omega (\frac{\partial}{\partial t} f + \vec{u} \cdot \vec{\nabla} f) \cdot \psi \, d\Omega = 0 . \tag{8}$$

After integrating by parts :

$$\int_\Omega f^{n+1} \cdot \psi^{n+1} d\Omega - \int_\Omega f^n \cdot \psi^n \, d\Omega - \int_{t^n}^{t^{n+1}} dt \int_\Omega f \cdot (\frac{\partial \psi}{\partial t} + \mathrm{div}\, \psi \vec{V}) \, d\Omega \tag{9}$$

$$= \int_{t^n}^{t^{n+1}} dt \int \psi \cdot f \cdot \vec{u} \, d\Gamma .$$

The R.H.S. is zero for solid or periodic boundaries, the last term on the L.H.S. vanishes as the ψ functions obey :

$$\frac{\partial \psi}{\partial t} + \mathrm{div}\, \psi \vec{u} = 0 . \tag{10}$$

The standard scheme is used for solving (10), and (9) simplifies to :

$$\int_\Omega f^{n+1} \psi^{n+1} d\Omega = \int_\Omega f^n \psi^n \, d\Omega \tag{11}$$

+ (eventually boundary terms).

Proper integration of (11) enables the scheme to be very conservative, even for energy $<f^2>$ although it is not introduced explicitly. It is now commonly used in 2D [11] and a highly vectorized 3D version is presently tested for LES. The 1D version has been applied to grid turbulence (a low CFL allowing splitting in directions) and has remarkably improved the results (fig. 13).

Posterior to the Colloquium, results with the 3D version where obtained. The higher accuracy of the present scheme made it necessary to incorporate the Leonard term. Constants of the smagorinsky and B.F.R. like models where then fitted to match the spectrum at t V°/M = 98. Then continuing the computation shows that very good agreement is found again further downstream at station t V°/M = 172 (fig. 14).

It must be recalled here that the code operates solely in physical space. Although in spectral space a turbulent viscosity can be written as a function of wave-number, ν_T = f(k), so as to "mould" a spectrum in just about any shape, to obtain a sharp cut-off behaviour in Fourier space of the implicit filtering (as exhibits fig. 14) is not straightforward in physical space. This is a crucial improvement since the total filtering is no longer dependant on the number of time steps (see § II).

VII LAGRANGIAN "Λ VORTEX" MODEL FOR WALL TURBULENCE

LES simulations of wall turbulence require high mesh resolution for even moderate Reynolds number and also a very lengthy computation for the establishment of a proper initial field during which maintenance of the turbulent energy level is difficult. This level is self-sustained only once energy producing structures (as shown by Kim and Moin [12]) stand out "by chance" of the random field while on the other hand, the S.G.S. model drains energy right from the first time step.

In an engineering application where wall turbulence might not be the only item of interest, the expensive transitional stage can be avoided and coarseness of the grid near the wall can be compensated through the following steps :

. prescribe shear velocity u* .

. choose a rough model of energy producing eddy and scale it in wall units (e.g. the Λ vortex of Perry and Chong [13]).

. perform a Lagrangian computation of the evolution of the structure (this enables the initially randomly defined eddy to bear some correlation with the Eulerian field : e.g. growth will be associated with bursting events).

. restore the structure to the filtered field if and when scales are compatible .

. control the energy input rate (number of Λ vortices) through a balance equation (input proportional to the defect of the observed head loss).

This enables the computation to be almost immediatly on an energy producing stage (shear stress <U.W> $\neq$ 0 after 15 time steps) (fig. 15-17). The procedure can be used to rapidly generate an initial field, gives suitable statistics, but will require some theoretical justification in terms of subgrid scale modeling before it can be maintained, after the initial stage, in a LES computation.

CONCLUSION

Direct simulation is now preferred to LES for fundamental research but should soon regain interest in the engineering field. This is possible because robust physical space codes required for non-homogeneous computations can feature correct behavior of results in spectral space close to that of pseudo spectral codes, even for high wave numbers (yet not competitive with respect to c.p.u. time requirements). This has been achieved by omitting the usual explicit filtering in order to directly mesure the implicit numerical filter, and subsequently reducing it by improving the numerical scheme.

Interaction through cut-off wave number is complex and does not reduce to pure dissipation effects. Use of aditional S.G.S. models significantly improve results. Still, fundamental analysis (in spectral space) is still required for these new models. A profitable approach here is to first write whatever models can be introduced in physical space and subsequently analyse them in spectral space (instead of the reverse).

Special care must be devoted to pure transport effects of turbulence which arise as a large mean flow results from the non-homogeneity. Galilean invariance of models and numerical schemes is essential here. In this non-homogeneous context, transport effects on the SGS level will probably have to be included through "Schumann-like" models.

LES for high Reynolds channel flows can be attempted on a coarse grid by introduction of more empirical data and disaveraging techniques. Structural "ad hoc" models can be introduced through discrete vortex methods, at least in the initial stage, in order to trigger energy producing mecanisms.
The author is thankful to Dr. BERTOGLIO and BARON for helpful discussions, and to an unknown reviewer for his remarks in turning the manuscript into an acceptable paper.

REFERENCES

[1] F. BARON, D. LAURENCE : "Large Eddy Simulation of a Confined Turbulent Jet Flow". Turb. Shear Flow IV (1983).

[2] J.P. BENQUE, A. HAUGUEL, D. LAURENCE : "Large Eddy Simulation of Turbulence in Physical Space". Macroscopic Mod. of Turb. Flows, Sophia Antipolis (1984), lecture notes in physics, Springer.

[3] BARDINA, FERZIGER, REYNOLDS : "Improved Turbulence Models based on Large Eddy Simulation". Rep. TF-19, Stanford U. Cal. (1983).

[4] C.G. SPEZIALE : "Galilean Invariance of Subgrid-Scale Stress models". J. Fluid Mech. Vol. 156 (1985).

[5] D.C. LESLIE, G.L. QUARINI : "Application of Turbulence Theory to the Formulation of Subgrid Modelling". J. Fluid Mech. Vol. 91 (1979).

[6] J.P. BERTOGLIO, J. MATHIEU : "Study of Subgrid Models for Sheared Turbulence". Fourth Turb. Shear Flow Symp., Karlsruhe (1983).

[7] L. SCHMITT, K. RICHTER, R. FRIEDRICH : present conference.

[8] F. BARON : "Macrosimulation Tridimensionnelle d'Ecoulements Turbulents Cisaillés". Thesis and E.D.F. report HE41/82.21 (1982).

[9] S. ELGHOBASHI, T. GERZ, U. SCHUMANN : Present Conference.

[10] A. HAUGUEL : "Numerical Modelling of Complex Industrial and Environmental Flows". Int. Symp. on Ref. Flow Modelling, Iowa (1985).

[11] J.M. HERVOUET : "Application of the Method of Characteristics in their Weak Formulation to Solving Two-dimensional Advection Eq.". (E.D.F. report E41/85.22).

[12] J. KIM, P. MOIN : Present Conference.

[13] A.E. PERRY, M.S. CHONG : "On the Mecanism of Wall Turbulence". J. Fluid Mech. vol. 119 (1982).

[14] COMPTE-BELLOT, CORRSIN : "Simple Eulerian Time Correlation of Full and Narrow-band Velocity Signals in Grid Generated, Isotropic, Turbulence". J. Fluid Mech. (1971), vol. 48.

[15] CHAMPAGNE, HARRIS, CORRSIN : "Experiments on Nearly Homogeneous Turbulent Shear Flow". J. Fluid Mech. (1970).

DECREASING GRID TURBULENCE [14]

Computed Energy Spectra at station $tU°/M=98$ (Initial:$tU°/M:42$)

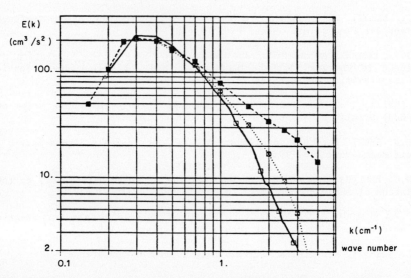

Fig. 1 : Standard scheme, $CFL_{rms} = 0.5$ ($CFL_{max} = 2.5$).
Solid lines are LES results, dashed lines are experimental results, dotted lines are filtered exp. results (Gaussian, $\Delta = 2h$ for comparison).

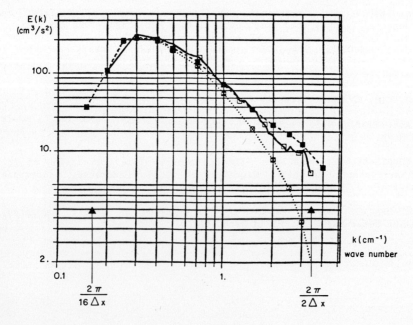

Fig. 2 : same as fig. 1, + spectral defiltering at each time step.

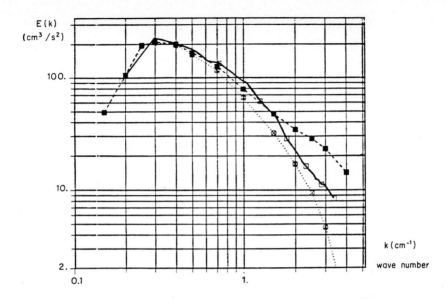

Fig. 3 : same as fig. 1, with random flight + scale similarity model.

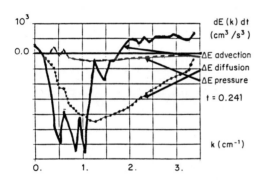

Fig. 4

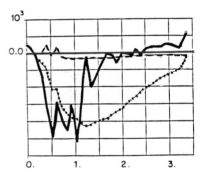

Fig. 5

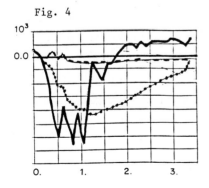

Fig. 6

Energy Variation through the Advection step

Fig. 4 : Standard run, $\mathcal{V} = 0$
Fig. 5 : same as 4, with $\mathcal{V} \neq 0$
Fig. 6 : same as 5, with code computed optimum local frame.

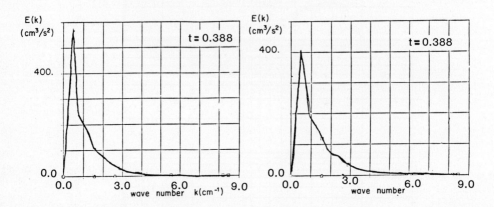

Fig. 7 : Final energy spectrum, standard scheme.
Fig. 8 : same as 7, +scale similarity model.

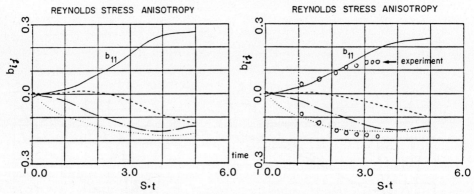

Fig. 9 : standard scheme.

Fig. 10 : same as 8, +scale similarity model.

Fig. 11 : same as 10, + advection in optimum local frame.

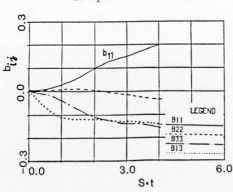

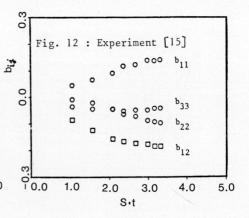

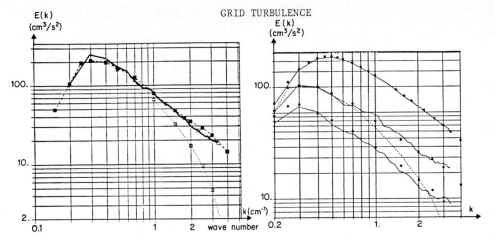

Fig. 13 : Weak formulation of advection, CFL_{rms} = 0.1 (see fig. 1).

Fig. 14 : Results of the final code (3D weak formulation, Smagorinsky, B.F.R. and Leonard terms), comparison at the 3 exp. stations.

TURBULENT CHANNEL FLOW

(LES results after 15 time steps only, turbulence is initially zero and is generated by Λ vortices).

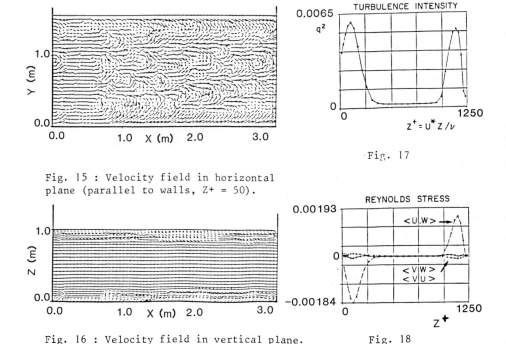

Fig. 15 : Velocity field in horizontal plane (parallel to walls, $Z+$ = 50).

Fig. 16 : Velocity field in vertical plane.

Fig. 17

Fig. 18

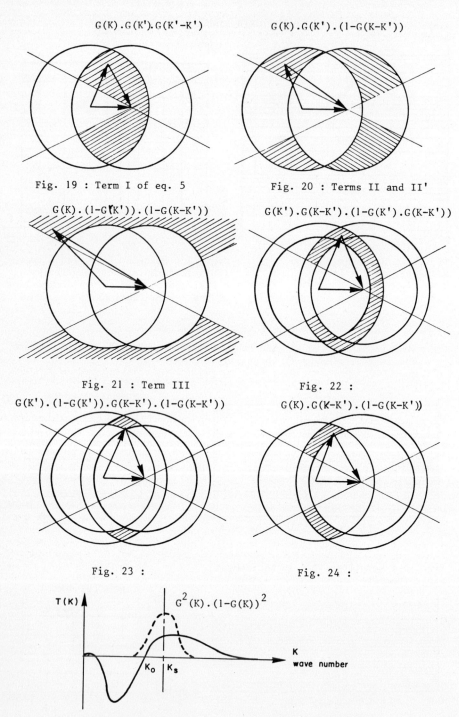

Fig. 19 : Term I of eq. 5

Fig. 20 : Terms II and II'

Fig. 21 : Term III

Fig. 22 :

Fig. 23 :

Fig. 24 :

Fig. 25 : Energy transfer rate and scale similarity model filter.

LARGE-EDDY SIMULATION OF TURBULENT BOUNDARY LAYER AND CHANNEL FLOW AT HIGH REYNOLDS NUMBER

L.Schmitt, K.Richter, R.Friedrich

Lehrstuhl für Strömungsmechanik
Technische Universität München
Arcisstraße 21, 8000 München 2
Federal Republic of Germany

SUMMARY

Statistically steady and two-dimensional high Reynolds number turbulent flat plate channel and boundary layer flow is simulated by numerical integration of the three-dimensional, time-dependent transport equations for grid scale (GS) quantities. Detailed comparisons of the numerical results are made with measurements of statistical mean values. The structure of the fluctuating and instantaneous fields is displayed. Especially the influence of the treatment of the mean convection terms on the results is discussed.

INTRODUCTION

In the past the large-eddy simulation (LES) technique has been applied successfully to such wall bounded shear flows where periodicity boundary conditions at the free edges of the computational domain can be taken [5,17,6,11]. These simulations showed good general agreement with experiments and provided reliable information which cannot be obtained from experiment.
On the long range our aim is to simulate developing and recirculating turbulent flows. In these cases periodical boundary conditions are unsuitable in general. Therefore, we have developed a computer code allowing us to apply other kinds of boundary conditions, too. This code is based on an extended version of Schumann's 'volume balance procedure' [17]. Together with the corresponding subgrid scale (SGS) model this procedure suits our intentions very well, because it is applicable to complex geometries, can account for non-isotropic mesh cells and flow inhomogeneities and gives a clear definition of the needed GS boundary values.
In this paper we describe the application of this code to turbulent flat plate boundary layer flow with zero pressure gradient and to turbulent channel flow with heat transport. Compared to [15] the present boundary layer simulations are strongly improved. For this flow case non-periodical boundary conditions in mean flow direction are used. At the crucial inflow boundary a method which self-generates GS fluctuations with controlled statistics is applied. Besides this, simulations of turbulent channel flow are performed with periodic boundary conditions with the aim to control the numerical method in a situation where there is no effect of the boundary conditions at the free edges.

GOVERNING EQUATIONS FOR THE GS-QUANTITIES

Our generalized 'volume balance procedure' means nothing but applying the integral conservation equations for mass, momentum and enthalpy to finite volumes ΔV of an orthogonal grid [15]. In a Cartesian coordinate system the set of equations reads:

$$\sum_j (\Delta A_j^+ \; {}^j\overline{v_j}|^+ - \Delta A_j^- \; {}^j\overline{v_j}|^-) = 0 , \qquad (1)$$

$$\Delta V \, ({}^V\overline{v_i}(t^+) - {}^V\overline{v_i}(t^-))$$
$$+ \Delta t \sum_j [\quad \Delta A_j^+ \, ({}^{t,j}\overline{v_j} \; {}^{t,j}\overline{v_i} + {}^{t,j}\overline{v_j'v_i'} + {}^{t,j}\overline{p} \, \delta_{ji} - {}^{t,j}\overline{\tau_{ji}})|^+$$
$$\qquad - \Delta A_j^- \, ({}^{t,j}\overline{v_j} \; {}^{t,j}\overline{v_i} + {}^{t,j}\overline{v_j'v_i'} + {}^{t,j}\overline{p} \, \delta_{ji} - {}^{t,j}\overline{\tau_{ji}})|^- \;]$$
$$- \Delta t \, \Delta V \; {}^{t,V}\overline{F_i} = 0 , \qquad (2)$$

$$\Delta V \, ({}^V\overline{T}(t^+) - {}^V\overline{T}(t^-))$$
$$+ \Delta t \sum_j [\quad \Delta A_j^+ \, ({}^{t,j}\overline{v_j} \; {}^{t,j}\overline{T} + {}^{t,j}\overline{v_j'T'} + {}^{t,j}\overline{q_j})|^+$$
$$\qquad - \Delta A_j^- \, ({}^{t,j}\overline{v_j} \; {}^{t,j}\overline{T} + {}^{t,j}\overline{v_j'T'} + {}^{t,j}\overline{q_j})|^- \;]$$
$$- \Delta t \, \Delta V \; {}^{t,V}\overline{Q} = 0 . \qquad (3)$$

The velocity v_i, pressure p, shear stress τ_{ij}, volume force F_i, temperature T, heat flux q_i and the volumetric heat source Q are properly normalized. The GS mean values are defined as

$$^{(.)}\overline{\phi} = 1/\Delta(.) \int_{\Delta(.)} \phi \, d(.) \qquad (4)$$

and the SGS fluxes ${}^{t,i}\overline{v_i'v_j'}$ and ${}^{t,i}\overline{v_i'T'}$ arise from the nonlinear convection terms by introducing the decomposition

$$\phi = {}^{(.)}\overline{\phi} + \phi' . \qquad (5)$$

The brackets (.) denote either averaging over the volume ΔV or the surface ΔA_i of a mesh cell or an additional average over the time interval $\Delta t = t^+ - t^-$.
In equations (1) – (3) models are needed to determine the SGS fluxes and assumptions for the relations between GS quantities at different surfaces since their number is larger than the number of equations.

MODEL FOR THE SGS FLUXES

For the SGS fluxes we adopt eddy viscosity concepts, which account for possible bad resolution, flow inhomogeneities and grid volume anisotropies:

$$^{t,i}\overline{v_i'v_j'} = -\,^{ij}\mu_{iso}\,(\,^{i}\overline{D_{ij}} - \langle\,^{i}\overline{D_{ij}}\rangle\,) - \,^{ij}\mu_{inh}\langle\,^{i}\overline{D_{ij}}\rangle \quad , \tag{6}$$

$$^{t,i}\overline{v_i'T'} = -\,^{i}a_{iso}\,(\,^{i}\overline{\partial T/\partial x_i} - \langle\,^{i}\overline{\partial T/\partial x_i}\rangle\,) - \,^{i}a_{inh}\langle\,^{i}\overline{\partial T/\partial x_i}\rangle \quad . \tag{7}$$

$\langle\ \rangle$ indicates statistical averages. For more details we refer to [17], [6] and [15]. It should be noted, that up to now we are not solving a transport equation for the SGS energy $^{V}\overline{E'}_{iso}$, but determine it by equating the isotropic production term and high Reynolds number dissipation term in Grötzbach's transport equation [6].

STANDARD NUMERICAL SOLUTION METHOD

We assume that the computable GS quantities are the grid surface mean values of the velocity components $^{i}\overline{v_i}$ appearing in the equation for the mass transport (1), the pressure mean value $^{t,V}\overline{p}$ and the temperature mean value $^{V}\overline{T}$ which are defined for the same grid volume. The unknown GS quantities are related to the computable ones by *proper* selection of interpolation functions. Standard is linear interpolation which leads for the GS quantities to an energy-conserving 'finite-difference' method on a staggered grid.

The integration of velocity from time level n to n+1 proceeds as follows (averaging bars have been omitted, C_i denotes convection, D_i diffusion and F_i source terms):

$$v_i^\circ = v_i^* = f_1 v_i^{n-1} + f_2 v_i^n + f_3(C_i^n + D_i^{n-1} + F_i) \quad , \tag{8.1}$$

$$v_i^{n+1} = v_i^\circ - f_3\,\delta_i p^{n+1} \quad . \tag{8.2}$$

The factors f_i in (8) are determined such that the time integration is done in cycles, starting with an Euler step, followed by a number of leapfrog steps and closed by an averaging step. The temperature is integrated likewise with $T^{n+1} = T^\circ$. Except for the Euler and averaging steps this scheme is of second-order accuracy with respect to the dominating convection terms. The pressure p^{n+1} follows from the solution of a Poisson-like equation which results from inserting (8.2) into the continuity equation (1). This equation is solved by combinations of Fast Fourier Transformation, Cyclic Reduction and Tridiagonal Matrix Algorithm depending on the selected boundary conditions for the velocity components.

COMPUTATIONAL DOMAIN AND REFERENCE VALUES

The turbulent flow is studied in a fixed space section extending 8 units in the mean flow (longitudinal) direction x, 4 in the homogeneous (spanwise) direction y and 2 in the direction normal to the wall, z. This domain is large enough to allow for zero two-point correlations of all flow variables in x- and y-directions.
For most of the discussed results an equally spaced grid of 64x32x32 grid volumes has been used.
All flow quantities are made dimensionless by means of the boundary layer thickness δ (or channel half width), the friction velocity u_τ and the wall heat flux at the inflow edge (time mean values averaged over both walls of the channel). In terms of these quantities the Reynolds number is 3240 and the applied time step

0.001. The Prandtl number is 0.7. Scaled with the free stream velocity the Reynolds number corresponds approximately to the one of Klebanoff [8].

BOUNDARY CONDITIONS

For the channel flow simulation periodic boundary conditions are used in x- and y-directions.
This kind of boundary condition is also applied in the spanwise direction of the boundary layer. But there the development of the mean flow in streamwise direction prohibits periodic boundary conditions for the GS quantities. Instead, we prescribe inflow/outflow conditions. At the inflow edge realistic fluctuating GS quantities are needed for every mesh cell and time step. Since no experimental data are available the GS quantities are generated by taking computed values from a location near the outflow edge and by adjusting these values such that they yield prescribed rms and mean values (rescaling of geometry is till now avoided because of limited vertical resolution and the difficulty of interpolating GS quantities properly). The imposed ensemble mean values have been derived from the 1/7-power law [14] and the rms values correspond to those of Klebanoff [8]. This procedure gives rise to a physically reasonable turbulence structure at the inflow edge and should be permitted because of the weak inhomogeneity of the mean flow in x-direction and the fact that the turbulent structures arise periodically and have a limited life-time [3]. At the outflow edge the GS quantities are extrapolated.
At the free stream edge we assume $^x\overline{u} = u_\infty$, $^y\overline{v} = 0$, $^V\overline{E'}_{iso} = 0$. The normal velocity component $^z\overline{w}$ is extrapolated allowing for a vertical mass flux.
Note that for velocity components normal to bounding surfaces the solution procedure requires values on the new time level. This results in von Neumann boundary conditions for the pressure.
Since the mesh cell nearest to the wall covers the viscous and buffer layers and extends well into the logarithmic regime, we use in both flow cases wall boundary conditions as suggested by Schumann [17] and Grötzbach [6] which make the wall fluxes consistent with universal wall functions.
A more detailed discussion of the boundary conditions is given in [16].

INITIAL CONDITIONS

As initial conditions we prescribe the statistical mean values and superimpose random Gaussian fluctuations fulfilling the rms values from experiment and zero velocity divergence. We assume, that turbulence develops and reaches a statistically steady state by repeated application of the solution procedure.

DISCUSSION OF RESULTS

If ⟨ ⟩ denotes a statistical average Reynolds' fluctuations are defined through $\phi'' = \phi - \langle \phi \rangle$. Averages are taken at fixed times over planes which are parallel to the walls. In addition, these values are time-averaged during the course of the integration. In order to obtain better statistical samples this procedure has been applied also in the boundary layer case. Because of the slow variation of mean values in x-direction (high Reynolds number) this seemed us to be a permissible assumption which however prevents us to show the streamwise flow development. In studies to follow we intend to drop this restriction. Note that all figures are unsmoothed reproductions of computer graphics in which the marked points represent the actually calculated values. Connecting lines between points serve only for better readability.

Boundary layer results

We discuss results which have been produced with the standard solution method (nonstandard is an improvement in the calculation of convection terms; see below). The profile of the mean turbulent shear stress (fig.1d) demonstrates that the implemented method for the generation of inflow boundary values works well. Starting with zero initial values a physically reasonable Reynolds shear stress profile (except for a slightly too high value close to the wall) has developed. The major part of the momentum transport is accomplished by the resolved fluctuations.

Figure 1a shows the mean velocity profiles. The shape factor H (= displacement thickness/momentum thickness) has the value 1.27 typical for the present Reynolds number. While the $\langle v \rangle$-component is zero, the vertical velocity component is so small ($\approx 0.04\ u_\tau$) that its value cannot be taken from the graph.

In figures 1b and 1c the resolvable rms velocity fluctuations and the total kinetic energy are compared with Klebanoff's [8] data. The overall agreement is good. But striking are too high values of the longitudinal velocity component and consequently of the energy.

The skewness and flatness factors of the velocity fluctuations are shown in fig.1e and fig.1f. Their rapid variation near the edge of the boundary layer suggests a region of intermittent turbulence. This and the opposite trend of the skewness profiles of u- and w- fluctuations are in agreement with experiments [8,1].

The resolvable rms pressure fluctuations are plotted in fig.1g. The rms value in the wall adjacent mesh cell is 2.94 which corresponds to experimental values at the wall (see [18]). No measurements are known for other z-positions.

Figure 1h shows the profiles of the resolvable rms vorticity fluctuations. Spanwise and streamwise components attain their maximum 'at the wall' and decrease towards the boundary layer edge. Away from the wall the profiles of all components are nearly identical.

The figures 2a – f shall give some impression of the computed flow fields. The contour line plots of the fluctuating velocity components and the SGS energy in (x,z)-planes show the typical resulting flow patterns. As one expects for the boundary layer, the fluctuations are confined to the region near the wall. In fig.2e the fluctuating velocity vectors in the (x,y)-plane nearest to the wall are displayed. The regions with high positive values of the u-fluctuations find their equivalent in the contours of the instantaneous resolvable Reynolds shear stress (fig.2f). The spotty occurence of the shear stress indicates high intermittency of this quantity.

In this section Lumley's 'orthogonal decomposition method' [9] is applied in the inhomogeneous z-direction in order to extract deterministic structures from the simulated flow field. An important feature of this approach is the unambiguous determination of the contribution of the individual structures to turbulence kinetic energy and shear stress.

The procedure decomposes the instantaneous velocity field v_i into a series of deterministic functions $\Phi_i^{(n)}(z)$ with random coefficients

$$v_i = \sum_n a_n\ \Phi_i^{(n)}(z) \tag{9}$$

using the Karhunen-Loeve expansion [10]. The resulting eigenvalue problem with the correlation tensor as the kernel is solved numerically.

Figures 3a – c show the three dominating eigenfunctions for the domain $0 \leq z \leq 1$. Note that the x- and z-components of these eigenfunctions have opposite sign, and hence, make a negative contribution to the shear stress. In table 1 the contribution of the dominant eigenfunction to the total resolved turbulence kinetic energy and the Reynolds shear stress is shown. In addition, the ratio of the first and second eigenvalues is given. Figures 3d – g show the convergence of the Karhunen-Loeve expansion for the turbulence stresses as the number of terms is increased. The convergence behavior including the fact, that some of the higher-order eigenfunctions make a positive contribution to the Reynolds shear stress shows similarity to what Moin [10] analysed in the case of channel flow.

In order to display the three-dimensional structure of our simulated flow field we calculated instantaneous vortex lines. Starting from an initial location in the three-dimensional vorticity field the defining equation

$$d\mathbf{x}/dh = {}^V\overline{\boldsymbol{\omega}} / |{}^V\overline{\boldsymbol{\omega}}| \tag{10}$$

was integrated numerically (**x** is the location of the vortex line in space and h is the distance along the vortex line).

It turned out, that if we start from arbitrarily chosen points no definite or coherent behavior of the lines can be observed in general. Sometimes the vortex lines meander over the whole field. This is due to the rapid variation of the vorticity fluctuations and the resulting sensitivity of vortex line locations to small perturbation of $\overline{\boldsymbol{\omega}}$. Common to all lines is that they are aligned more or less into the positive y-direction (at the upper wall of the channel into the negative y-direction) and that they contain very often loops which are inclined with respect to the wall.

That a part of a vortex line may represent actually a vortical structure can be shown by starting a set of vortex lines close to that vortex line and within the expected structure [12]. Then the coherence of the structure will have a self-correcting effect in realigning the vortex lines and they will probably be confined to the core of the structure for large values of h. Figures 4b–d show different views of such an agglomeration of vortex lines with a characteristic inclination to the wall and demonstrate that the observed vortex line loops may be definite features of the flow field. But this has also to be confirmed by the calculation of two-point correlations [12].

That in large portions of the flow the inclination angle of vorticity vectors attains characteristic values is shown in figures 4e–h with the help of histograms describing the distribution of this angle at various z-positions. Here, at each grid point, the inclination angle between the wall and the projection of the vorticity vector into the (x,z)-plane (see fig.4a), namely

$$\alpha = \tan^{-1}({}^V\overline{\omega_z}/{}^V\overline{\omega_x}) \tag{11}$$

has been calculated. The contribution from each grid point to the histogram was weighted with the magnitude of the projected vectors. The distributions and the fact, that the ratio of the peak values of the distributions to their minimum values is appreciably lower if one doesn't weight with the magnitude of the vorticity vector is in agreement with findings of Moin and Kim [12] in the case of the channel flow.

Influence of mean convection

Application of the standard solution method to the turbulent channel flow in a stationary reference frame led to some unexpected results, which showed up most spectacularly in a flattening of the mean velocity profile and in the shape of the longitudinal power spectra. Instead of slopes near -5/3 in the high wave number range we found shapes as depicted in figures 5a and a'. This behavior may also explain some of the deficiencies in our boundary layer results.

If we translate the coordinate system in the channel approximately with the mean profile average velocity, $u^G = u_m$, these anomalies mostly disappear. While the spectra adjust very fast (after about 0.2 problem times, see figures 5b,b') it takes over one problem time until a nearly stationary state is reached. It should be noted, that previous channel flow simulations [17,6] with similar SGS models and integration methods have been performed only in a moving reference frame. Because the moving coordinate system has essentially the effect of extracting most of the mean convection, we concluded, that the convection due to fluctuating velocities only and with it their nonlinear interaction is described with sufficient accuracy by the standard method, not however the transport due to mean convection, represented by the term $u^G \partial \phi / \partial x$. Note that in all cases the Courant numbers are well below one.

As one possibility to improve calculations in a stationary reference frame we split up the solution method into the following steps. First we apply (8.1), neglecting the mean convection terms and obtain v_i^*. Then we account for the mean convection by Fourier interpolation:

$$v_i^o(I\Delta x) = v_i^*(I\Delta x - \Delta t \, u^G(K)). \tag{12}$$

The spectra (see figures 5c,c') and the following more detailed discussion show that with this method we are able to make reasonable calculations in the non-moving coordinate system, too. Because the procedure allows u^G to be a function of z, improved spectra were received, compared to the simulations in the moving coordinate system, at every z-position ($u^G(z)$ was set equal to $\langle u \rangle (z)$). The drop off of the highest Fourier mode is a consequence of the fact, that this mode is not affected by the mean convection.

Channel flow results

The calculated mean velocity profile is given in fig.6a. Both, the maximum value of 26.3 and the difference between this value and the profile average value, which is 2.57, fit very well into the empirical formulae of Dean [2].
In fig.6b the computed resolvable rms velocity fluctuations and in fig.6c the contributions to the total turbulence kinetic energy are plotted together with the corresponding measurements of Comte-Bellot [4]. The agreement is satisfactory. The differences in the near wall region seem to be due to the applied boundary conditions (especially regarding the v-component) and to the truncation of Grötzbach's transport equation for the SGS energy where we loose terms which may possibly be important here. In agreement with Grötzbach [6] we notice that the w-rms-values are larger than the v-rms-values in the middle of the channel.
Profiles of the resolvable Reynolds shear stress, of the SGS part of the shear stress which is due to the inhomogeneous model and of the total Reynolds shear stress are displayed in fig.6d. These profiles indicate that the average Reynolds shear stress profile has attained the equilibrium shape that balances the longitudinal mean pressure gradient. For the given Reynolds number and grid resolution viscous stresses are not significant at the computed z-positions.
The rms value of the resolvable pressure fluctuations in the wall adjacent mesh cells is 2.9 . This value coincides with the experimental one.
In fig.6e the diagonal elements of the resolvable pressure-strain correlation tensor are drawn. These profiles show, as expected, that the longitudinal component of velocity fluctuations transfers energy to the cross-stream components [7].
The profiles of resolvable rms vorticity fluctuations in the channel (fig.6f) resemble those for the boundary layer (fig.1h). Here too, away from the wall the 'isotropy' of the vorticity fluctuations is evident.
Except for the wall adjacent regions our results for the resolvable pressure-strain correlation tensor and rms vorticity fluctuations compare qualitatively and also quantitatively well with the results of Moin and Kim [11].

The resolvable two-point correlation functions at $z = 1.0$ are plotted in figures 6g and h showing that, in general, for small separation distances, the correlation for the velocity in the direction of the displacement is larger than the corresponding transverse correlation. The pressure fluctuations are correlated over considerably greater distances in spanwise direction than they are in the longitudinal direction. Moreover, the pressure correlation is negative for large streamwise separations but is always positive for the spanwise separations. These results are confirmed by experiments and previous simulations (e.g., [4,5,6,11]).

Regarding the heat transport we consider two different cases. Temperature was treated as a passive scalar and therefore momentum transport remained unaffected. To show typical characteristics of the two cases figures 7a – f present profiles of some statistical quantities and of instantaneous fluctuations.

In case A a constant temperature gradient was imposed by keeping the temperatures of the lower and upper walls at fixed but different values. This leads to a constant heat flux profile between the two walls. The missing of the experimentally [13] found weak s-shape in the $\langle T \rangle$-profile (fig.7a) near the channel center line (which we get throughout in coarser grid simulations) may be due to the fact that the temperature has not yet reached a fully developed state.

In case B a constant volumetric heat source was applied and both walls were kept at the same temperature. Here the heat flux profile looks like that of the mean shear stress (fig.6d). Also the profiles for the mean temperature (fig.7a), the resolvable rms temperature fluctuations (fig.7b) and the resolvable temperature-velocity correlation coefficients (fig.7d) express the strong analogy which exists for this case between T and u. The good agreement between our temperature-normal velocity correlation coefficient and that of Grötzbach [6] for the same resolution (32x16x16) in fig.7d may be pointed out.

The contour lines of temperature fluctuations in figures 7e and 7f reflect the behavior of the rms values (fig.7b), i.e. maximum intensity in the core region and minimum values near the walls in case A and the contrary for case B.

CONCLUSIONS

Large-eddy simulation results have been presented for turbulent boundary layer and channel flow. The results for both cases show mostly reasonably good agreement with experiment.

In the case of channel flow we have demonstrated, especially by comparing spectra of velocity components, how a different treatment of the convection terms may influence the simulated fields. Treatment of mean convection by Fourier interpolation increased the accuracy of the simulations in a stationary reference frame considerably. Because this method necessitates periodic boundary conditions it cannot be used without modifications, if at all, for more general problems. It, however, shows the direction from where improvements of the numerical procedure can be expected. The other possibility to account for deficiencies of the numerical method by adjusting the SGS model is under investigation. In any case we notice a remarkable interdependence between numerical method, SGS model and resolved GS values especially in flow cases where strong mean convection is present.

Our treatment of the inflow boundary condition in the case of the turbulent boundary layer is one possible way which certainly can be improved. Other methods and the sensitivity of the simulated fields to an enlargement of the flow domain in the normal direction and to the outflow boundary conditions have still to be studied.

The Karhunen-Loeve expansion and the calculation of the distribution of the inclination angle of vorticity vectors should not only be seen as tools for the investigation of the existence of coherent structures in one specific flow case but also as a means to compare different flow cases or results computed with different numerical methods.

LITERATURE

[1] Andreopoulos,J., Durst,F., Zaric,Z., Jovanovic,J. Influence of Reynolds number on characteristics of turbulent wall boundary layers. Experiments in Fluids 2, 7 – 16, 1984.
[2] Bradshaw,P. Turbulence. Berlin, Heidelberg, New York: Springer, p.135, 1976.
[3] Cantwell,B.J. Organized motion in turbulent flow. Ann.Rev.Fluid Mech.13, 457 – 515, 1981.
[4] Comte-Bellot,G. Ecoulement Turbulent Entre Deux Parois Paralleles. Publications Scientifiques et Techniques du Ministere de l'Air No.419, 1965.
[5] Deardorff,J.W. A numerical study of three-dimensional turbulent flow at large Reynolds numbers. J.Fluid Mech.41, 453 – 480, 1970.
[6] Grötzbach,G., Schumann,U. Direct numerical simulation of turbulent velocity-, pressure-, and temperature-fields in channel flows. Proc. of the Symp. on Turbulent Shear Flows. Penn. State Univ., Apr.18 – 20, 1977.
[7] Hinze,J.O. Turbulence. New York: McGraw Hill, 1975.
[8] Klebanoff,P.S. Characteristics of turbulence in a boundary layer whith zero pressure gradient. NACA-TR-1247, 1955.
[9] Lumley,J.C. Stochastic Tools in Turbulence. New York: Academic Press, 1970.
[10] Moin,P. Probing turbulence via large-eddy simulation. AIAA-paper 84-0174, 1984.
[11] Moin,P., Kim,J. Numerical investigation of turbulent channel flow. J.Fluid Mech.118, 341 – 377, 1982.
[12] Moin,P., Kim,J. The structure of the vorticity field in turbulent channel flow. Part 1: Analysis of instantaneous fields and statistical correlations. J.Fluid Mech.155, 441 – 464, 1985.
[13] Page,F. Jr., Schlinger,W.G., Breaux,D.U., Sage,B.H. Point values of eddy conductivity and viscosity in uniform flow between parallel plates. Ind.Engng. Chem.44, 424 – 430, 1952.
[14] Schlichting,H. Boundary-Layer Theory. New York: McGraw Hill, 1968.
[15] Schmitt,L., Friedrich,R. Large-eddy simulation of turbulent boundary layer flow. In: Pandolfi,M., Piva,R. (eds.): Proc. of the Fifth GAMM-Conf. on Numerical Methods in Fluid Mechanics. Braunschweig, Wiesbaden: Vieweg. Notes on Numerical Fluid Mechanics 7, 299 – 306, 1984.
[16] Schmitt,L., Richter,K., Friedrich,R. A study of turbulent momentum and heat transport in a boundary layer using large-eddy simulation technique. In: Hirschel,E.H. (ed.): Finite Approximations in Fluid Mechanics. DFG – Priority Research Program, Results 1984 – 1985. Braunschweig, Wiesbaden: Vieweg. Notes on Numerical Fluid Mechanics, to appear.
[17] Schumann,U. Subgrid scale model for finite difference simulations of turbulent flows in plane channels and annuli. J.Comp. Phys.18, 376 – 404, 1975.
[18] Willmarth,W.W. Pressure fluctuations bencath turbulent boundary layers. Ann. Rev. Fluid Mech.7, 13 – 38, 1975.

ACKNOWLEDGEMENT. This work is supported by the German Research Society (DFG).

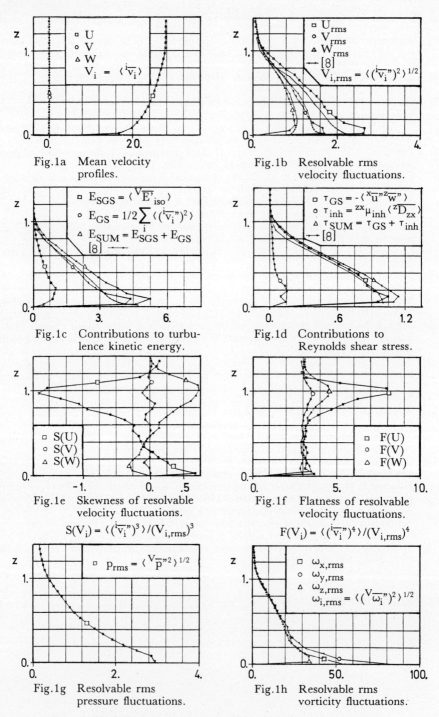

Fig.1 Turbulent boundary layer flow. Statistical mean values.

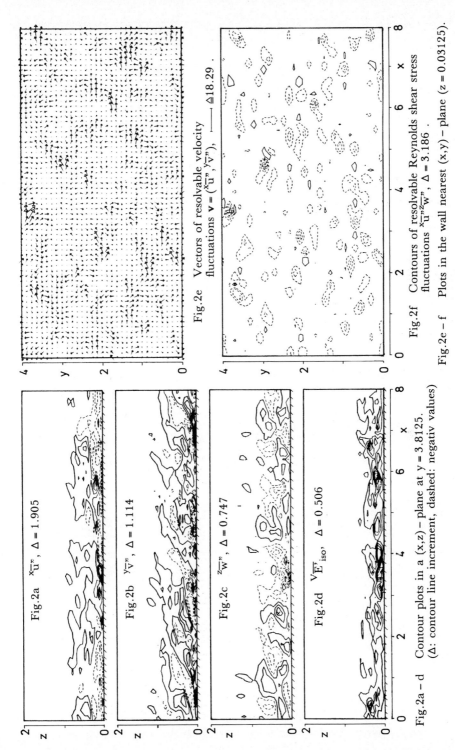

Fig.2e Vectors of resolvable velocity fluctuations $\mathbf{v} = (^x\overline{u}'', ^y\overline{v}'')$, —— ≙18.29.

Fig.2f Contours of resolvable Reynolds shear stress fluctuations $^x\overline{u}''^z\overline{w}''$, $\Delta = 3.186$.

Fig.2e – f Plots in the wall nearest (x,y) – plane (z = 0.03125). Instantaneous fluctuations.

Fig.2a $^x\overline{u}''$, $\Delta = 1.905$

Fig.2b $^y\overline{v}''$, $\Delta = 1.114$

Fig.2c $^z\overline{w}''$, $\Delta = 0.747$

Fig.2d $^v\overline{E}_{iso}$, $\Delta = 0.506$

Fig.2a – d Contour plots in a (x,z) – plane at y = 3.8125. (Δ: contour line increment, dashed: negativ values)

Fig.2 Turbulent boundary layer flow. Instantaneous fluctuations.

Table 1
Contribution of the dominant eigenfunction to turbulent kinetic energy and Reynolds shear stress.

$\lambda^{(1)}/E_t$	$P^{(1)}/P_t$	$\lambda^{(1)}/\lambda^{(2)}$
0.236	0.585	1.546

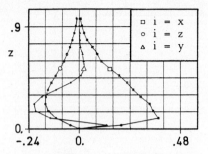

Fig.3a $(\lambda^{(1)})^{1/2}\phi_i^{(1)}$

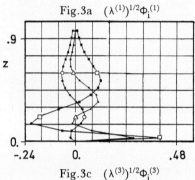

Fig.3b $(\lambda^{(2)})^{1/2}\phi_i^{(2)}$

Fig.3c $(\lambda^{(3)})^{1/2}\phi_i^{(3)}$

Fig.3a – c Components of the first three dominating eigenfunctions.

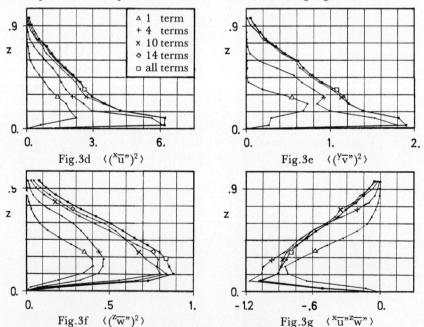

Fig.3d $\langle(^x\overline{u}")^2\rangle$

Fig.3e $\langle(^y\overline{v}")^2\rangle$

Fig.3f $\langle(^z\overline{w}")^2\rangle$

Fig.3g $\langle ^x\overline{u}"^z\overline{w}"\rangle$

Fig.3d – g Convergence of the Karhunen-Loeve expansion with increasing number of terms for the resolvable turbulence stresses.

Fig.3 Turbulent boundary layer flow. Karhunen-Loeve expansion.

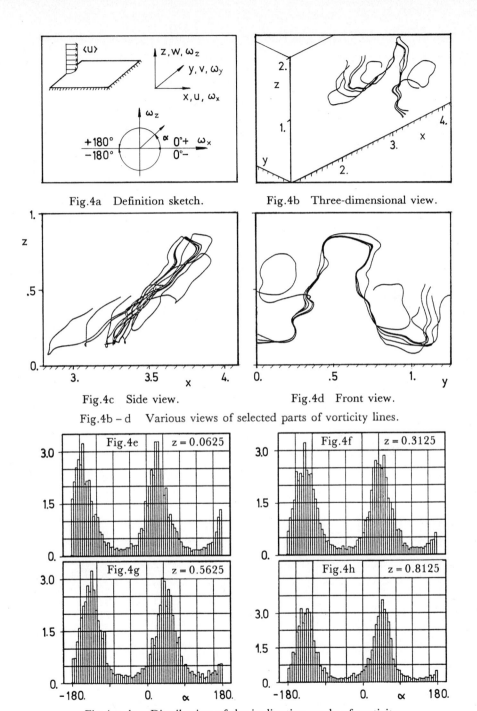

Fig.4a Definition sketch.
Fig.4b Three-dimensional view.
Fig.4c Side view.
Fig.4d Front view.
Fig.4b – d Various views of selected parts of vorticity lines.
Fig.4e – h Distribution of the inclination angle of vorticity vectors in (x,z)-planes at various z-positions.
Fig.4 Turbulent boundary layer. Evaluation of total vorticity.

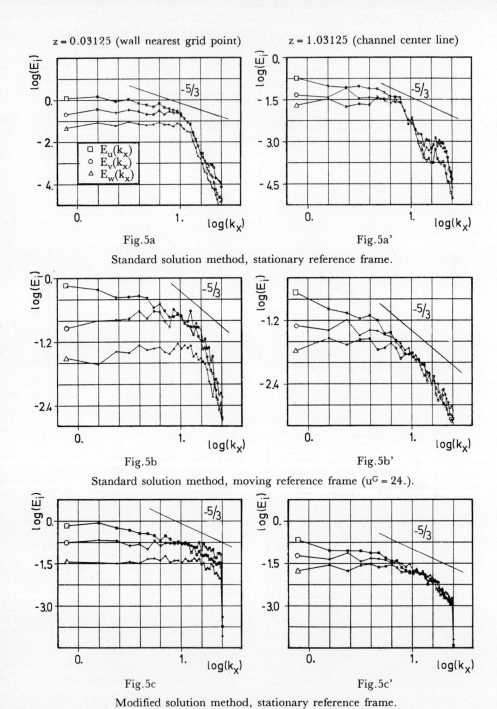

Fig.5 Turbulent channel flow. Influence of the treatment of the mean convection terms on the longitudinal power spectra.

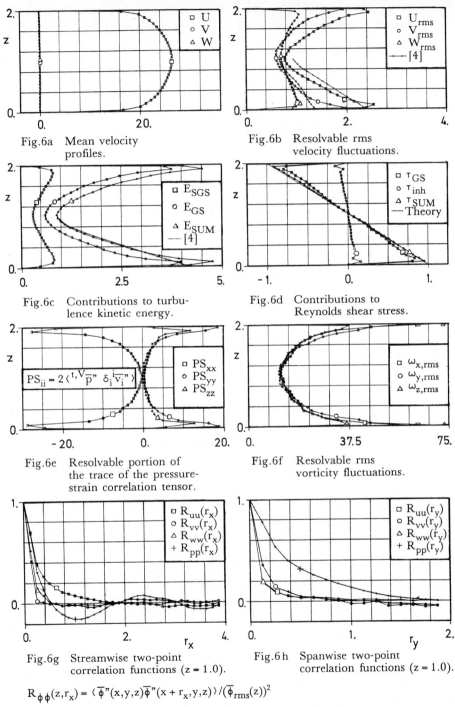

Fig.6 Turbulent channel flow. Statistical mean values.

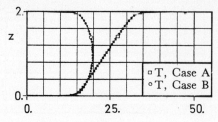
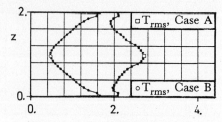

Fig.7a Mean temperature profiles.
$T \triangleq \langle {}^V\overline{T} \rangle$

Fig.7b Resolvable rms temperature fluctuations.
$T_{rms} = \langle ({}^V\overline{T}")^2 \rangle^{1/2}$

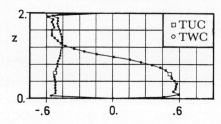
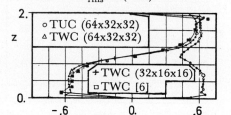

Fig.7c Case A

Fig.7d Case B

Fig.7c – d Correlation coefficients.
$TV_iC = \langle {}^V\overline{T}"{}^i\overline{v}_i" \rangle / (T_{rms} V_{i,rms})$

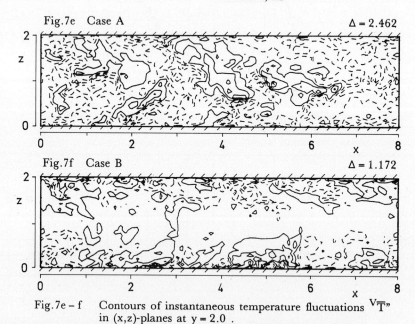

Fig.7e – f Contours of instantaneous temperature fluctuations ${}^V\overline{T}"$ in (x,z)-planes at y = 2.0 .

Fig.7 Turbulent channel flow with heat transport.
Case A: 'constant heat flux'.
Case B: 'constant volumetric heat source'.

NUMERICAL INVESTIGATION OF A VORTICAL STRUCTURE

IN A WALL-BOUNDED SHEAR FLOW

John Kim

NASA Ames Research Center, Moffett Field, California 94035

This is an abstract from Ref. [1]; for further details, readers are refered to the original article.

In order to gain a better understanding of the role of horseshoe vortices in turbulent transport processes, the temporal evolution of a channel flow containing a horseshoe vortex is investigated by integrating the time-dependent, three-dimensional, Navier-Stokes equations. A spectral method — Fourier series in the streamwise and spanwise directions and Chebychev polynomial expansion in the normal direction — is used for the spatial representation of the velocity field. The time advancement is made by a semi-implicit method: Crank-Nicolson scheme for viscous terms and Adams-Bashforth method for the non-linear terms. The resulting system of equations is solved by a spectral-tau method.

The computation is carried out with approximately 2×10^6 grid points ($128 \times 129 \times 128$, in $x, y,$ and z) for a Reynolds number of 3200, based on the centerline velocity and channel half-width (180 based on the wall shear velocity). The grid spacings in the streamwise and spanwise directions are $\Delta x^+ \approx 18$, and $\Delta z^+ \approx 6$ in wall units. Nonuniform meshes are used in the normal direction. The first mesh point away from the wall is at $y^+ \approx 0.05$, and the maximum spacing at the centerline of the channel is about 4.4 in wall units. No subgrid-scale model is used in the computation since the grid resolution is sufficiently fine to resolve all the essential turbulent scales. For a comparison, a similar calculation is carried out with approximately 4×10^6 grid points ($192 \times 129 \times 160$, in $x, y,$ and z) for a short period, and essentially no difference is found between the two results.

The initial vortical structure is constructed by applying a conditional sampling technique to the data base generated from a direct simulation of a turbulent channel flow

[2]; this results in a horseshoe-shaped vortical structure associated with the bursting process in the wall-bounded shear flow. The evolution of this vortical structure under the influence of the self-induced motion and the mean shear is investigated. It is shown that the initial sheet-like vortical structure rolls-up into a vortex tube when convected downstream (figure 1). Turbulence characteristics associated with the vortex are also investigated. The production of vorticity due to vortex stretching is high inside the vortex legs, and also substantial in the tip region and above the legs. The stretching due to the induced motion by the legs is found to create a secondary structure above the legs. High Reynolds shear stress is produced near the tip of the vortex due to the self-induced motion.

REFERENCES

[1]   Kim, J. : "Evolution of a vortical structure associated with the bursting event in a channel flow", Turbulent Shear Flows V, ed. by F. Durst, B. E. Launder, F. W. Schmidt and J. H. Whitelaw, Springer-Verlag Berlin Heidelberg (1986); also in Proceedings of the Fifth Symposium on Turbulent Shear Flows, Cornell University, Ithaca, New York, August 7-9, 1985.

[2]   Kim, J., Moin, P. : "The structure of the vorticity field in turbulent channel flow. Part 2. Study of ensemble-averaged fields", J. Fluid Mech., **162** (1986) pp. 339-363.

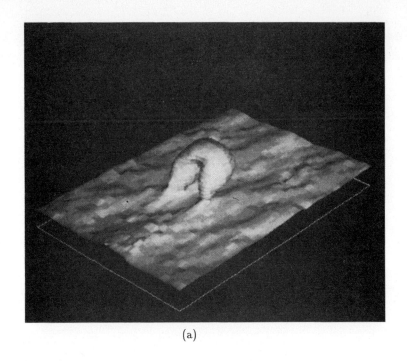

(a)

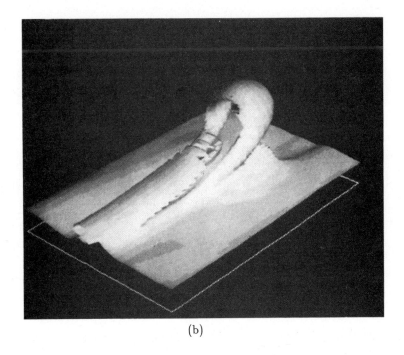

(b)

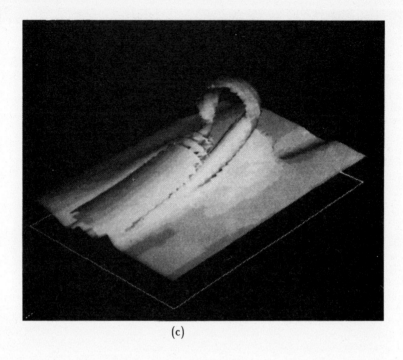

(c)

Fig. 1 Evolution of a vortical structure in turbulent channel: (a) $t^* = 0$; (b) $t^* = 17.2$; (c) $t^* = 28.8$. Here t^* is a dimensionless time nondimensionalized by the channel half-width and the centerline velocity. The surface shown in the figure represents a surfaces of constant enstrophy. This is a perspective view of a three-dimensional field. The mean flow direction is from lower left to the upper right, and the box shown in the figure represents a portion of the lower wall of the channel. The streamwise and spanwise extents of the wall shown are 3.2 δ and 2.1 δ, respectively, or 575 and 375 in wall units.

RECENT RESULTS ON THE STRUCTURE OF TURBULENT SHEAR FLOWS USING SIMULATION DATABASES

Parviz Moin

NASA Ames Research Center, Moffett Field, California 94035

During the past year the use of large eddy and direct simulation databases has made possible significant advances in our understanding of the organized structures in turbulent shear flows. Here, I summarize some of these contributions, and cite the references where they are published in their entirety.

In addition to the investigation of the effects of curvature on turbulence, the results of the direct simulation of curved channel flow of Moser & Moin [1] were used to study the *near wall* coherent structures. In this simulation the computational grid resolution was sufficient to reproduce the alternating high and low speed streaks with mean spacing of 100 wall units. However, unlike the streamwise velocity, the pattern of contours of the normal velocity component in the wall region does not show significant elongation; rather, it is dominated by small regions of intense fluctuations (see Fig. 1). Here, the flow is in the θ-direction, r is the direction normal to the walls, and z is the spanwise direction. In Fig. 2 one of the regions of intense velocity fluctuation (the framed area in Fig. 1) is enlarged. The velocity vectors in (r, z) planes passing through points marked **B**, through **E** in Fig. 2 are shown in Fig. 3. A vortex inclined to the wall is clearly discernable. It is emphasized that only a single vortex is observed near the wall. Further downstream and away from the wall, this vortex is joined by another vortex of opposite sign. In fact, in a random sampling of (r, z) planes, significantly more solitary vortices than vortex pairs were observed. These patterns and the spanwise two-point correlation profiles indicate that the wall layer is *not* composed of elongated pairs of counter-rotating vortices, as is commonly hypothesized.

One of the more significant contributions of this work was the first demonstration of *direct* simulation of a fully developed, statistically stationary, turbulent flow. Although the computational resolution was not sufficient to resolve the Kolmogorov scale, scales with appreciable contribution to turbulent dissipation spectra were resolved.

An investigation into the existence and frequency of occurrence of hairpin vortices in turbulent channel flow was carried out by Moin & Kim [2]. Using statistical analysis and visualization of the instantaneous vorticity fields in three dimensional space it was shown that turbulent channel flow consists of a large number of horseshoe (or hairpin) vortices inclined to the wall. The hairpins are formed from the roll-up of sheets of spanwise vorticity by random velocity fluctuations and stretching by mean rate of strain. The hairpins do not have legs elongated in the streamwise or spanwise direction as has been proposed by some investigators. In the second part of this investigation, [3], it was shown by using conditional sampling techniques that the bursting process is associated with hairpin vortices, thus, establishing their importance in the turbulence production mechanism.

In the above study it became apparent that the tip of the hairpins progressively assume circular shapes and that the computational resolution may not be adequate to resolve possible pinching of the tip portion into a ring vortex. To study the time evolution of hairpin vortices, Moin, Leonard & Kim [4] used a three-dimensional vortex tracking method. The calculations were initialized with an isolated parabolic shaped filament of concentrated vorticity resembling a stretched hairpin vortex (Fig. 4a). Due to its high induced velocity, the tip region lifts from the plane where the vortex was placed initially, resulting in high curvature of the filament in the (x, y) plane. The subsequent induced motions lead to a nearly circular shaped tip region and ultimately generate a ring vortex (Fig. 4b). This result provides a mechanism for the formation of ring vortices which have been observed by Falco [5] in turbulent boundary layers.

Our observations of the vortical structures in the channel flow led us to conjecture that hairpin vortices are the characteristic structures not only in wall-bounded flows but in *all* turbulent shear flows. Moin, Rogers and Moser [6] examined the structure of vorticity field in homogeneous turbulent shear flow using the same techniques as in the channel. Rogallo's [7] computer program with up to 128x128x128 points was used. The results conclusively showed that homogeneous turbulent shear flow consists of coherent hairpin vortices, verifying the above assertion. The presence of mean shear causes a remarkable organization of the vorticity field in turbulent flows.

Finally, the characteristic eddy decomposition theorem was applied to turbulent channel flow database. This decomposition also known as Lumley's orthogonal decomposition [8] provides a quantitative definition of coherent structures in turbulent flows as well as unambiguous determination of their contribution to turbulence stresses. It is a mathematically elegant procedure for identification of coherent structures and representation of the entire flow field in terms of these eddies. This is an ideal application of the simulation databases because of the large magnitude of the required input data. In this decomposition the instantaneous velocity field is decomposed into a series of deterministic functions (eddies) with random coefficients

$$u_i = \sum_n a_n \phi_i^{(n)}(\mathbf{x}) \tag{1}$$

Given an ensemble of realizations of the velocity field $u_i(\mathbf{x},t)$ the deterministic vector functions or eddies, $\phi_i^{(n)}(\mathbf{x},\mathbf{t})$, are chosen such that they have the highest possible root-mean square correlations with the members of the ensemble. It can be shown that ϕ_i's are the eigenfunctions of a Fredholm eigenvalue problem with the correlation tensor R_{ij} as the kernel [8].

Moin [9] applied a two-dimensional variant of this technique to the channel flow database. It was shown that the dominant eigenfunction contributed about 30% to total turbulent kinetic energy and its contribution to Reynolds shear stress was about 70%. When the problem was formulated for the wall region alone, it was possible to recover virtually all the turbulent kinetic energy and Reynolds shear stress with only 5 terms in the expansion of equation (1). In the directions of flow homogeneity this expansion is combined with the shot-noise expansion [10] to yield the characteristic eddy that is sprinkled in the flow. From second order statistics the characteristic eddy can be obtained to within a phase factor. This phase factor which is essential for determination of the shape of the eddy is recovered from third order statistics. The bi-spectra of the coefficients a_n were calculated from the database, from which the unknown phase angle was determined and the characteristic eddy was constructed. In the wall region the dominant eddy depicts the sweep event (i.e., high speed fluid moving towards the wall); and in the core region it displays the ejection event. These findings are consistent with the measurements of Wallace, Eckelmann and Brodkey [11], using quadrant analysis.

References

[1] MOSER, R. D. & MOIN, P. Direct numerical simulation of curved turbulent channel flow. *NASA TM 85974* (1984). Also, *Report TF-20, Department of Mech. Engng., Stanford Univ., Stanford Calif.*

[2] MOIN, P. & KIM, J. The structure of vorticity field in turbulent channel flow. Part 1. Analysis of the instantaneous fields and statistical correlations. *J. Fluid Mech.* **155**, 441 (1985).

[3] KIM, J. & MOIN, P. The structure of vorticity field in turbulent channel flow. Part 2. Study of ensemble-averaged fields. *J. Fluid Mech.* **162**, 339 (1986).

[4] MOIN, P. LEONARD, A. & KIM, J. Evolution of a curved vortex filament into a vortex ring. *NASA TM 86831* (1985).

[5] FALCO, R. E. Coherent motions in the outer region of turbulent boundary layers. *Phys. Fluids* **20**, S124 (1977).

[6] MOIN, P., ROGERS, M. & MOSER, R. D. Structure of turbulence in the presence of uniform shear. Proc. of the Fifth Symposium on Turbulent Shear Flows, Cornell University, Ithaca, New York, August 7-9, (1985).

[7] ROGALLO, R. S. Numerical experiments in homogeneous turbulence *NASA TM 81315* (1981).

[8] LUMLEY, J. L. The structure of inhomogeneous turbulent flows. In *Atmospheric Turbulence and Radio Wave Propagation*, ed. A. M. Yaglom & V. I. Tatarsky, pp. 166. NAUKA, Moscow (1967).

[9] MOIN, P. Probing turbulence via large eddy simulation. *AIAA paper 84-0174* (1984).

[10] RICE, S. O. Mathematical analysis of random noise. Bell System Tech. J. **23**, 282 (1944).

[11] Wallace, J. M., Eckelmann, H. & Brodkey, R. S. The wall region in turbulent shear flow. *J. Fluid Mech.* **54**, 39 (1977).

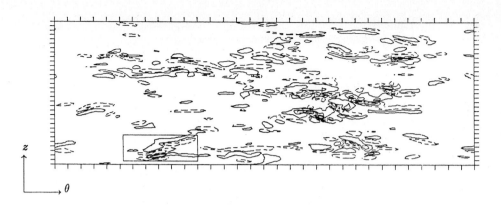

Figure 1 Contours of the normal (radial) velocity v in the (θ, z)-plane near the concave wall, $y^+ = 6.14$. Negative velocities (in the direction away from the concave wall) are contoured by dashed lines.

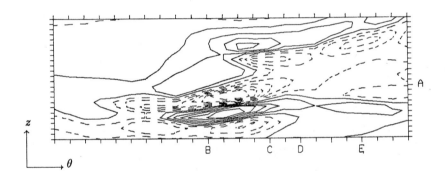

Figure 2 Contours of the radial component of velocity, v, in the (θ, z)-plane near the concave wall, $y^+ = 6.13$. Enlargement of the region marked with a box in Figure 1. The domain is 135 wall units in the z direction and 404 wall units in the θ direction.

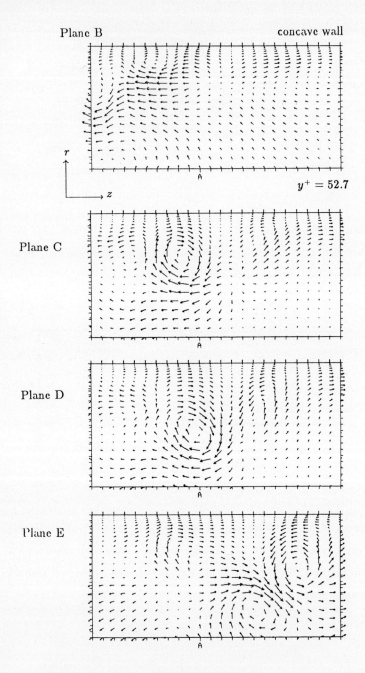

Figure 3 Velocity vectors projected into (r,z)-planes at the θ locations marked "**B**" through "**E**" in Figure 2. The domain is 135 wall units in the z direction. Point **A** is the reference point between Figures 2 and 3.

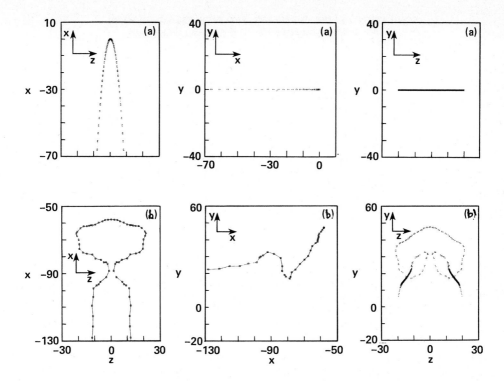

Figure 4 Evolution of the parabolic vortex filament $x = -az^2$ ($a = 1$). Plan views are in the left column, the middle column shows the side views and the end views are shown in the right column. Non-dimensional time, $ta^2\Gamma =$ (a)0; (b)6.5×10^4. The symbols indicate the location of the node points and Γ is the circulation.

SIMULATION OF THE TURBULENT RAYLEIGH-BENARD PROBLEM USING A SPECTRAL/FINITE DIFFERENCE TECHNIQUE

T. M. Eidson
George W. Woodruff School of Mechanical Engineering
Georgia Institute of Technology
Atlanta, Georgia 30332 U.S.A.

M. Y. Hussaini
Institute for Computer Applications in Science and Engineering
NASA-Langley Research Center
Hampton, Virginia 23665 U.S.A.

T. A. Zang
NASA Langley Research Center
Hampton, Virginia 23665 U.S.A.

SUMMARY

The three-dimensional, incompressible Navier-Stokes and energy equations with the Boussinesq assumption have been directly simulated at a Rayleigh number of 3.8×10^5 and a Prandtl number of 0.76. In the vertical direction, wall boundaries were used and in the horizontal, periodic boundary conditions were applied. A spectral/finite difference numerical method was used to simulate the flow. At these conditions the flow is turbulent, and a sufficiently fine mesh was used to capture all relevant flow scales. The results of the simulation are compared to experimental data to justify the conclusion that the small scale motion was adequately resolved.

INTRODUCTION

Direct simulation of turbulent fluid flows is now possible with the large vector computers that have become available [1,2]. Prediction of low-order flow statistics is definitely within current capabilities, and some results have already been published which show predictions of small scale turbulent features which are consistent with experimental observations [3-5]. The current study was undertaken to explore the quality of information that can be extracted from a direct flow simulation (DFS) of turbulence on a sufficiently fine mesh.

The turbulent Rayleigh-Benard problem (natural convection) was chosen for study since it is a simple turbulent flow for which a good body of experimental measurements exists. Moreover, some DFS and large-eddy simulations (LES) of this problem have been published albeit on coarser meshes. While experimental data do exist, measurements of velocity, where no mean flow exists, are difficult. Hence, there is much to be learned about turbulent natural convection from an accurate simulation.

The two requirements for conducting such a study are a high-speed computer and an efficient, accurate flow simulation code. The CYBER-205 computer with a 16 mega-word memory provides sufficient computation power. This current code has been extensively tested, and various versions of it have been used to study transition in channel flow [6]. The version used in this study includes the addition of the energy equation and a modified vertical momentum equation that includes buoyancy consistent with the Boussinesq assumption.

A simulation of a turbulent flow was then conducted, and these data as well as a discussion of the code will be presented in this paper. Though the overall goal of this work is an in-depth examination of the quality and type of information that can be extracted from such a simulation, the purpose of this paper is to document the basic simulation. The simulation results will be compared with experimental mean measurements as well as previous DFS results. The increased resolution of this work over previous DFS resulted in an improvement in the prediction of the Nusselt number; it was sufficiently close to experimental results to suggest that in addition to a good prediction of the large-scale flow, the small-scale features are accurately represented. Grotzbach discusses this connection extensively [7,8]. Comparisons with experimental data which are more dependent on the small scale components of the flow will also be presented to justify further this conclusion.

RAYLEIGH-BENARD PROBLEM

The Rayleigh-Benard problem is a simple geometry, laboratory-type problem used to study natural convection (figure 1). Chandrasekhar [9] and Busse [10] have described the basic problem and discuss both the stability analysis and some experimental results. Krishnamurti [11,12] has summarized much experimental data and developed a map showing the qualitative flow at different values of Ra and Pr, the principal independent problem parameters (defined below). For Pr = 0.76 (air) and Ra = 3.8×10^5 the motion is turbulent, although it should possibly be qualified as low Reynolds number turbulence. Several experimental studies [13-18] and numerical simulations [7,19,20] (both LES and DFS) have been completed in the qualitatively similar Pr-Ra region. The flow at these values of Pr and Ra consists of a core flow (a horizontal layer in the middle 80% of the fluid layer) and a boundary region near each plate. The turbulence is statistically homogeneous in the horizontal directions for both layers. In the core the vertical variation of most statistical quantities is small. In the boundary layer there is a transition from molecular dominated physical processes near the wall to the fully turbulent core flow.

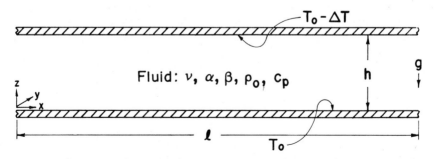

Figure 1. Geometry and Parameters of the Rayleigh-Benard Problem

This flow is described by the incompressible Navier-Stokes equations modified to include the effect of temperature-induced density variations on the buoyancy force (Boussinesq assumption) plus the temperature equation. These equations, when non-dimensionalized by α, h, and ΔT, are

$$\frac{\partial u_i}{\partial t} + \frac{\partial (u_i u_j)}{\partial x_j} = -\frac{\partial P}{\partial x_i} + Pr \frac{\partial^2 u_i}{\partial x_j^2} + Pr\, Ra\, T \delta_{i3}\,, \qquad (1a)$$

$$\frac{\partial T}{\partial t} + \frac{\partial (T u_j)}{\partial x_j} = \frac{\partial^2 T}{\partial x_j^2} + u_3\,, \qquad (1b)$$

and

$$\frac{\partial u_j}{\partial x_j} = 0. \qquad (1c)$$

The temperature and pressure in equations (1a,b,c) are the difference between the actual temperature, T_a, and pressure, P_a, and the values due to the static temperature gradient only. These are defined as follows:

$$T_a(\underline{x},t) = T_o - x_3 + T(\underline{x},t)$$
$$= T_o + T_r(\underline{x},t)$$

and

$$\frac{\partial P_a(\underline{x},t)}{\partial x_3} = -\frac{gh^3}{\alpha^2} - Pr\, Ra\, x_3 + \frac{\partial P(\underline{x},t)}{\partial x_3}.$$

The dependent variables in the problem are the velocity components, u_i or (u,v,w), the temperature, T, and the fluid pressure, P. The independent variables are the spacial coordinates, x_i or $\underline{x} = (x,y,z)$, and time, t. The indices i = 1 and i = 2 signify the horizontal directions, and i = 3 denotes the vertical direction. The problem parameters are the thermal diffusivity, α; the kinematic viscosity, ν; the acceleration of gravity, g; the reference fluid density, ρ_o; the coefficient of thermal expansion, β; a reference temperature (the temperature of the lower plate), T_o; and the temperature difference between the two plates, ΔT. The dependent and independent variables have been non-dimensionalized by α/h (velocity), ΔT(temperature), $\rho_o \alpha^2/h^2$(pressure), h (coordinates), and h^2/α (time). The Rayleigh number, $Ra = g\beta \Delta T h^3/\nu\alpha$, and Prandtl number, $Pr = \nu/\alpha$, are the principal non-dimensional problem parameters.

NUMERICAL METHOD

The Fourier finite difference algorithm developed by Moin and Kim [21] for their large-eddy simulations of turbulent channel flow has been applied, with modifications, to the present direct simulation of turbulent Rayleigh-Benard flow. This Rayleigh-Benard Fourier-finite difference Method (RBFFDM) is an unsplit method on a grid staggered (for the pressure variable and the continuity equation) in the vertical (z) direction. Fourier collocation is used for the spacial discretization in the x and y directions whereas in the vertical direction second-order finite-differences are employed on the non-uniform grid,

$$z_k = (1 - \cos(k\pi/N_z))/2, \qquad k = 0, 1, \ldots, N_z.$$

The time discretization is Crank-Nicolson for the viscous and conductive terms and backward Euler for the pressure gradient term. The advection and buoyancy terms are handled by a third-order Adams-Bashforth method.

The implicit part of the algorithm requires, for each pair of horizontal Fourier wavenumbers, the solution of 2 real, block tridiagonal systems (involving the velocities and pressure) and, independently, 2 real, scalar tridiagonal systems (for the temperature). The block-tridiagonal equations were scaled as described by Zang & Hussaini [6] for their Fourier-Chebyshev version of the corresponding channel flow algorithm. Pivoting has proven to be unnecessary for this system. The block-tridiagonal solution algorithm takes advantage of the many zero elements which occur. Vectorization of this phase of the algorithm is achieved by solving for many pairs of Fourier wavenumbers at the same time.

Equations (1a,b,c) were solved on the region, $0 \leq x \leq A_x$, $0 \leq y \leq A_y$, $0 \leq z \leq 1$. Under the present scaling of the vertical direction, the lengths A_x and A_y correspond to the aspect ratios of the two horizontal directions to the vertical one. The boundary conditions at the lower and upper walls, $z = 0$ and $z = 1$, are the conventional no-slip and no temperature jump conditions. In the horizontal directions, periodic boundary conditions are assumed. These aspect ratios must be large for reasonable correspondence with experiments. The aspect ratios are related to the resolution by $A_x = N_x \Delta x$ and $A_y = N_y \Delta y$. The computer memory limitation places an upper bound on $N = N_x N_y N_z$; therefore, to have a large A_x, along with sufficiently small Δx_i, some compromise is necessary. The values chosen, $A_x = 4$ and $A_y = 2$, allow the available computer memory to be used for better small scale resolution. This will be further discussed in Section IV.

The RBFFDM code has been implemented on a CDC Cyber 205 with 2 pipes and 16 million 64-bit words of main memory. For each grid point 13 variables were stored. Additional storage equivalent to 7 variables per grid point was used to facilitate vectorization. A total of 11 million words was used for the simulation on the 128 × 64 × 64 grid. Vector lengths for the explicit portion of the algorithm were between 4 and 40 thousand. Typically, one-fourth of the implicit equations were solved together. The vector lengths here were roughly 1,000. The linked triad feature was heavily exploited. A single time-step required 6.8 seconds of CPU time and no I/O time since the job was run entirely within the central memory. The sustained speed of the calculation was 100 MFLOPS.

The start-up phase of the calculation took 2100 time-steps and the data collection an additional 5600 steps. A total of 12 hours of CPU time was required for the data collection. This includes the time for some preliminary diagnostics. The Courant number, defined as the maximum over the grid of the quantity

$$\left[\frac{|u|}{\Delta x} + \frac{|v|}{\Delta y} + \frac{|w|}{\Delta z}\right]\Delta t \,,$$

ranged between 0.19 and 0.26 and averaged 0.23 for the data collection phase.

COMPARISON WITH PREVIOUS WORK

The turbulent flow of the Rayleigh-Benard problem is assumed to be homogeneous in both horizontal directions as well as statistically steady in time (after a start-up period). Experimental data are usually presented using some combination of a long-time average as well as a spacial average in one or both horizontal directions. The simulation results presented as horizontal averages, $\langle \rangle$, have been averaged over both the x and y directions (except for the 1-D x-spectra which were averaged in y only). In addition, they have been time averaged over a time period equal to $10/W_c$, where $W_c = (Nu\ Pr\ Ra)^{1/3}$. (See Deardorff [22] for a discussion of W_c, the scaling velocity for the large eddies.) This period, which should consist of several large eddy turn-over times, was found adequate by Eidson [19]. Volume averages were time averaged as well. All the simulation results presented below are both horizontally and temporally averaged unless otherwise specified.

Prediction of the average vertical heat flux, the Nusselt number in non-dimensional form, is an important result of any natural convection study. Previous simulations have predicted values of the Nusselt number, Nu, which are slightly higher than those measured experimentally. Grotzbach [8] has discussed extensively this discrepancy and has shown that inadequate resolution is partially responsible. In table 1, the results from both simulations and experiments are shown. The prediction of the present study lies at the the upper range of experimental measurements and below the Nusselt numbers predicted by previous, coarser-grid simulations. In the present simulation, Nu was calculated at each z level. The average value is reported in table 1. The variation with z was small (approximately ± 0.1) except very near the lower wall where Nu increased to 7.0.

The aspect ratio of the horizontal to vertical boundary lengths also is known to affect Nu [8,15]. The values of A_x = 4 and A_y = 2 are smaller than the values of 4 to 7 suggested by experimentalists as the minimum for removing significant side boundary effects. Although the aspect ratio effect is not negligible, especially on Nu [8], a large aspect ratio was foregone in favor of better small scale resolution in view of the goal of this study: to resolve eddies down to nearly the dissipation scales.

The relative temperature, T_r, vertical profile is compared to data of studies by Chu and Goldstein [16] in figure 2. A line with slope, $\partial T_r/\partial z$ = Nu, is drawn in this figure. From this one can estimate the conductive layer thickness, δ_c. This will be used later in a more extensive examination of the temperature data near the wall.

In figures 3 and 4, the vertical dependence of the velocity and temperature RMS values are compared to the simulation of Grotzbach [7] and the measurements of Deardorff and Willis [13] (slightly larger Ra). In table 1 a comparison for the centerline values is presented for

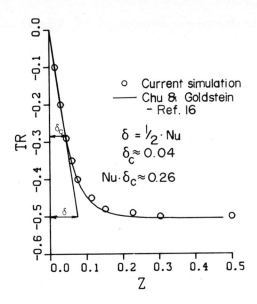

Figure 2. Estimation of the region for which the relative temperature varies linearly with z.

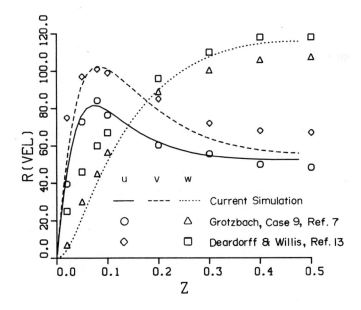

Figure 3. Comparison with experiments and previous simulations of the RMS of the velocity components.

Table 1. Horizontal Averages at the Vertical Center

Direct Simulation	Grid	Ra	Nu	u_{RMS}	w_{RMS}	$T_{r_{RMS}}$
Current Study	128×64×64	3.8×10^5	6.6	52	116	0.08
Grotzbach		3.8×10^5				
(case 7)	16×16×16		7.8	--	109	--
(case 9)	32×16×16		7.4	50	107	0.08
(case 14)	64×32×32		6.9	--	109	0.08
Eidson	64×64×16	3.8×10^5	8.1	70	112	0.11
Experiment						
Deardorff & Willis		6.8×10^5	5.8	65	120	0.08
Carroll		3.8×10^5	5.9*	--	--	0.05*
Fitzjarrold		3.8×10^5	6.2*	80*	120*	0.05*
Chu & Goldstein		3.8×10^5	6.5*	--	--	--
Goldstein & Chu		3.8×10^5	5.4*	--	--	--

*Calculated from curve fit of data over a range of Ra.

slightly the predicted levels of the RMS temperature over previous simulations. However, no systematic change in the RMS velocity levels with the improved spacial resolution was observed. Both trends are consistent with Grotzbach's results. Considering the variation in experimental values and the uncertainty in aspect ratio effects, the results appear quite satisfactory.

The wT correlation coefficient, C(wT), is constant for $0.2 < z < 0.8$, giving a value of 0.71 (figure 5). Both Grotzbach [7] and Eidson [20] previously obtained a value of approximately 0.67. Deardorff and Willis measured approximately 0.60 for a slightly higher Ra, but C(wT) should decrease with increasing Ra. Near the wall all four studies differ. Deardorff and Willis measured a significant drop as the wall was approached, but warned that since the numerator and denominator of C(wT) become small, their results are uncertain. The increase in C(wT) near the wall found in the current study was not observed in either previous simulation, but neither of these had sufficient resolution in this region. Note that only the data of Grotzbach's case 9 with 16 vertical grid points was available for comparison. Also, horizontal averages of the uv, uw and vw correlations were calculated. They were all approximately zero as would be expected for turbulence homogeneous in the horizontal directions.

In figure 6, the several terms in the horizontally averaged kinetic energy equation are plotted versus z and compared with experimental data at a slightly larger value of Ra. These terms are

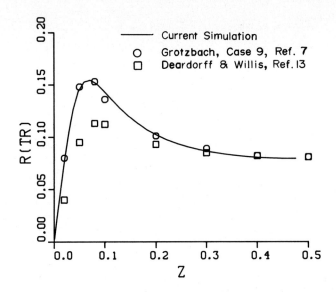

Figure 4. Comparison with experiments and previous simulations of the RMS of relative temperature.

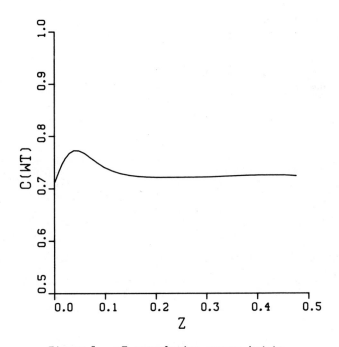

Figure 5. wT correlation versus height.

$$\frac{\partial \langle E \rangle}{\partial t} = \Pr \operatorname{Ra} \langle wT_a \rangle - \frac{\partial}{\partial z} \langle w(E + P_a) \rangle$$

$$\text{Production} \qquad \text{Diffusion}$$

$$- \Pr \left\langle \frac{\partial u_i}{\partial x_j} \frac{\partial u_i}{\partial x_j} \right\rangle + \Pr \frac{\partial^2 \langle E \rangle}{\partial z^2},$$

$$\text{Dissipation} \qquad \text{Molecular Transfer}$$

where

$$E = \frac{1}{2} u_i u_i .$$

Since the production term in the core is equal to (Pr Ra Nu), the experimental data of Deardorff & Willis [13] at Ra = 6.3×10^5 and Nu = 5.7 would be expected to be about 40% higher than the current simulation. For clarity, only a few key values from Deardorff and Willis are shown to demonstrate the general agreement between experiment and simulation. The four terms are all essentially constant and approximately 40% below the experimental data for the core region. In the boundary region the variation with z is similar for both experiment and simulation. The diffusion term (for the experiment and simulation) peaks at a z value of 0.035, and the molecular diffusion term of the simulation peaks at a slightly higher z. The experimental point plotted at a low z value is at these peaks for these two terms. Actually, one would expect the peaks for the lower Ra to be at a slightly larger z since the boundary region thins with increasing Ra. From a closer inspection of the data, the molecular dissipation also can be seen to have a small positive value between $z = 0.1$ and $z = 0.2$ as was found by Deardorff and Willis. The production and dissipation terms have the same general shape in both studies. The volume averages of the production and dissipation terms were 1.641×10^6 and 1.637×10^6, respectively. Also, the volume average of the molecular transfer term was 9.0×10^3.

One advantage of the simulation technique is that quantities which are experimentally difficult to measure can be easily calculated. Figures 7 to 9 give three examples. In figure 7 the absolute value of each of the three components of vorticity is shown. As expected, the x and y components are large near the wall due to the creation there of a boundary layer by the large eddies. Near the center the flow is more isotropic. The ratio of the volume average of the horizontal to vertical vorticity was 5.0 and 4.3 for the x and y directions, respectively. This is higher by a factor of 2 than in the previous DFS by Eidson [19]. Figure 8 shows that the only significant velocity skewness is for the w component near the wall. A negative value for the w component is reasonable since the fluid particles with negative velocity (near the lower wall) come from the core where there is more velocity variation due to the turbulent cascading process. The particles with a positive velocity originate near the wall where the motion is damped and they have a more uniform value. Recall that the horizontal average of the w velocity equals zero. The flatness profile of the velocities is shown in figure 9. For the horizontal components, these approach 3 in the core, a value which is similar to that in channel flow turbulence away from the wall [21]. The larger flatness factor near the wall suggests that the flow is more intermittent in this region.

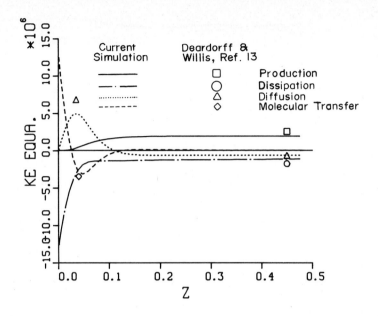

Figure 6. Terms in the kinetic energy equation versus height.

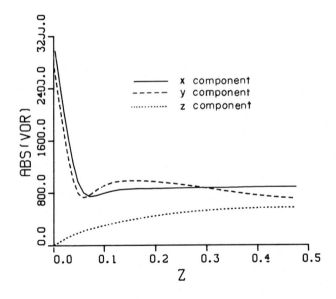

Figure 7. RMS of the three vorticity components versus height.

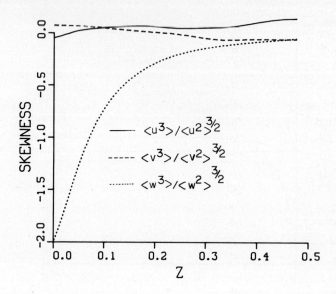

Figure 8. Skewness factor of the three velocity components versus height.

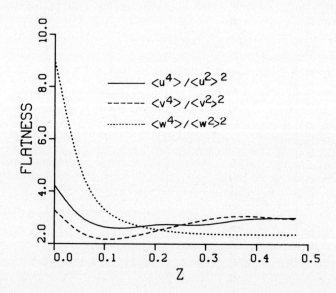

Figure 9. Flatness factor of the three velocity components versus height.

The wT cross-spectra for several z values are plotted in figures 10 to 13. These are 1-D spectra in the x-direction. They have been normalized so that they integrate to $\langle wT \rangle$. The wavenumber, κ, is defined such that $2\pi\kappa x$ is the argument of the periodic expansion functions. Notice that the spectra decrease with κ except near the wall where they are roughly constant with κ. This trend was found also in the u, w and T spectra but not in v. Near the center the most energy would be expected in the largest available scales for the small values of A in this study. Near the wall, smaller horizontal structures would be expected, and thus significant energy at a wavenumber greater than the minimum is a reasonable result. Since this is low Ra turbulence, an equilibrium region would not be expected, and indeed none was found here. The drop-off with wavenumber is rapid due to the importance of viscous effects at all the scales.

To evaluate the simulation further, a more sensitive measure of the small scale features of the flow field, especially in the wall region, was sought. Carroll [14] has made direct instantaneous measurements of the temperature and the vertical temperature gradient and calculated the mean, standard deviation, and skewness. Carroll's data are compared with the results of this simulation in figures 14 to 20. The data of Carroll in these figures are a mean curve (drawn by Carroll) through the data points. The bars show the range of data scatter. The mean of the relative temperature (simulation data shown in figure 2) and its vertical derivative (figure 18) show good agreement with Carroll's data. The higher order statistics give a variation with z which is similar to the experimental results although the magnitudes differ in some cases. In table 1, a significant variation in experimental measurements of the RMS of T_r, $R(T_r)$, between different experimental studies is apparent. Carroll used a stationary probe, and this may have reduced the fluctuations coming from the direction of the probe and lowered the statistical levels. Also Carroll measured the same RMS temperature level for Nu ranging from 5.5 to 14.0. Both experiments [13,15] and simulations [20] have shown that $R(T_r)$ increases with decreasing Nu (or Ra). Since the magnitude of the data is in question in some cases, the conclusions drawn from the simulation/experiment comparison are based mainly on changes of slope with vertical distance from the wall. These variations correspond to differences in the dominant physics in the different layers, and the simulation was able to predict these layers in agreement with the experimental results.

A four-layer (or three-layer with the middle divided into 2 sub-layers) model is consistent with both the experimental and simulation results. These are:

(i)   a conduction dominated layer $0 < Nu*z < \delta_1^*$

$0.1 < \delta_1^* < 0.3$

(ii)   an inner boundary layer $\delta_1^* < Nu*z < \delta_2^*$

$\delta_2^* \approx 0.5$

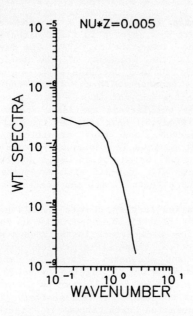

Figure 10. One-dimensional wT cross spectra (averaged in y only) versus x wavenumber.

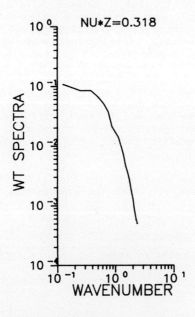

Figure 11. One-dimensional wT cross spectra (averaged in y only) versus x wavenumber.

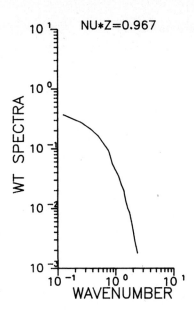

Figure 12. One-dimensional wT cross spectra (averaged in y only) versus x wavenumber.

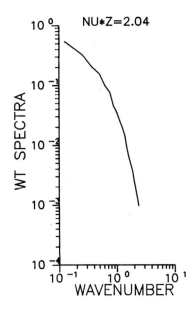

Figure 13. One-dimensional wT cross spectra (averaged in y only) versus x wavenumber.

(iii) an outer boundary layer $\quad \delta_2^* < $ Nu*z $ < \delta_3^*$

$\delta_3^* \approx 1.5$

(iv) the core $\qquad\qquad\qquad\qquad \delta_3^* < $ Nu*z $ < $ Nu/2 .

These layers are all well resolved by the simulation as they contain 8, 4, 9, and 11 grid points respectively (assuming $\delta_1^* = 0.2$). Determination of δ_1^* is uncertain. As was mentioned previously (figure 2), divergence of the temperature profile from a straight line with a normalized slope equal to Nu (the temperature gradient at the wall) gives $\delta_1^* = \delta_c^* = 0.26$. Another measure is that the wT correlation rises from 1% to 10% of its core value near Nu*z ≈ 0.1.

The inner boundary layer is characterized by the transition from molecular to turbulent dominated processes. Both experiment and simulation show that $\partial T_r/\partial z$ has been reduced by 50% between Nu*z of 0.4 and 0.5 (figure 18). In the inner boundary layer, $R(T_r)$ reaches a maximum (figure 14), the skewness factor for T_r, $S(T_r)$, changes sign (figure 16), and the flatness factor for T_r, $F(T_r)$, reaches a minimum (figure 17). The skewness, $S(T_r)$, continues to increase reaching a positive maximum in the outer boundary layer. The above trends, shown in both studies, suggest significant changes in the flow character in these layers.

In the outer layer, the flow transitions to a core region where the large eddies (the most efficient heat transfer agents) carry the heated fluid from the bottom to the top. At Nu*z ≈ 1.5, the temperature gradient, $\partial T_r/\partial z$, has been reduced to 1% of its maximum value (figure 18). The skewness, $S(T_r)$, decreases in the outer layer and approaches zero near the center of the core (figure 16). Also note, in the outer layer $R(T_r)$ varies as $z^{-1/3}$ compared with z^{+1} in the inner layer, a trend which is found in both studies (Figure 15). The simulation predicts a transition to a constant level for $R(T_r)$ in the center in agreement with Deardorff and Willis [13] (Figure 4). However, Carroll's data suggest that $R(T_r)$ varies as $z^{-1/3}$ in the core as well as the outer layer.

A possible discrepancy between the simulation results and Carroll's data is seen in the $R(\partial T_r/\partial z)$ plot (figure 19). At values of Nu*z between 0.4 and 1.0 the simulation shows a hump. However, Carroll's choice of an "average curve" has significant scatter only on top of the curve. Re-drawing the data fit through this scatter would result in a similar hump in the outer boundary layer region. Another example, where the simulation results suggest that the "averaged curve" through sparse, scattered data is not correct, is Carroll's choice for the temperature skewness profile for Nu*z < 0.1 (figure 16). He suggests that the curve should continue decreasing to $S(T_r) = -1.5$ and then rise to zero at Nu*z ≈ 0.01. However, only a few data points which gave significant scatter were measured in this region. From an examination of the data in Carroll's paper, it is not inconsistent to draw the experimental data

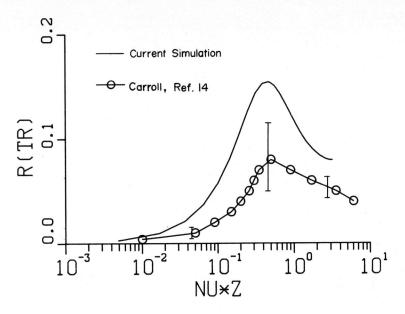

Figure 14. Comparison with experiment of the RMS relative temperature versus normalized height.

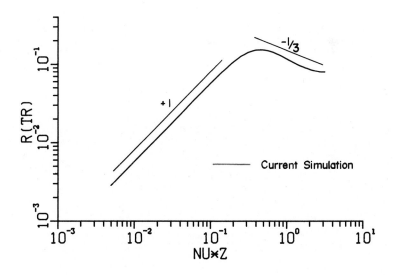

Figure 15. RMS relative temperature versus normalized height. The experimentally determined slopes of Carroll are shown for comparison.

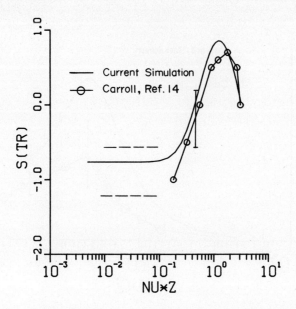

Figure 16. Comparison with experiment of the skewness of the relative temperature versus normalized height.

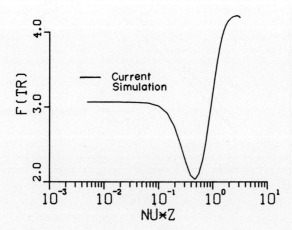

Figure 17. Flatness factor of the relative temperature versus normalized height.

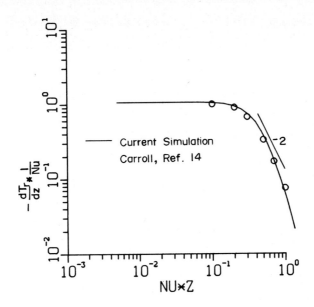

Figure 18. Comparison with experiment of the normalized relative temperature derivative versus normalized height. The slope shown is from the theory of Malkus.

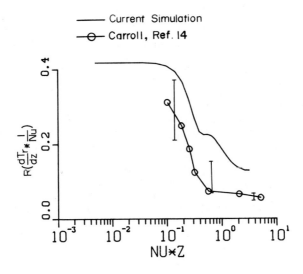

Figure 19. Comparison with experiment of the RMS of the normalized relative temperature derivative.

fit to turn between Nu*z = 0.1 and 0.2 and be parallel to the simulation results somewhere between the dashed lines shown in figure 16. One discrepancy that cannot be explained by data scatter is the comparison between studies of $S(\partial T_r/\partial z)$ away from the wall. The simulation predicts that $S(\partial T_r/\partial z)$ returns to zero in the outer layer and remains zero in the core (figure 20). However, the experiments show a return to negative skewness in the core.

The final comparison is for the temperature gradient versus Nu*z near the wall (figure 18). Various power laws have been hypothesized which predict a linear slope for the data plotted in log-log form. Carroll [14], Businger [23], and Monin and Yaglom [24] discuss the various theories. Carroll's data show that only for the range 0.4 < Nu*z < 1.0 does a simple power law seem reasonable and even then the data have a slow change in slope in this region. The current simulation fits Carroll's data nicely and thus also predicts a slope of −2 in this range of z. The value of −2 is predicted by the theory of Malkus [25].

From these comparisons, it can be seen that the simulation data represent a turbulent realization which has a good agreement with experimental, averaged measurements. The agreement, at least in prediction of sub-layers using the variance and skewness data, shows that the small scale turbulent features are accurately represented near the wall. This is important because a complete understanding of the Rayleigh-Benard convection requires studying the thin boundary layer which exists as a result of the large scale motion in the core. From isothermal, flat plate boundary layer studies, it is known that significant small scale events occur in this type of region which are important to the global flow.

CONCLUDING REMARKS

From the comparison with the gross properties of the experimental data, it can be concluded that the completed computer simulation is a reasonably accurate realization of a turbulent flow even down to the dissipation scales. Moreover, the detailed database has enabled us to identify a feature in the skewness of the vertical temperature derivative which has been overlooked in the experiment due to data scatter.

Obviously, a far more detailed analysis of the simulation results is in order. This simulation is of sufficient accuracy to study the small scale structure of a natural convection flow in a similar manner to the work previously done for channel flows [4,21]. Moreover, it can be used to help interpret experimental measurements which are typically limited to a few spacial points and fewer than all three velocity components. Finally, we feel this database can be used to evaluate turbulent models and other theoretical ideas.

ACKNOWLEDGMENT

The authors would like to express their appreciation to Control Data Corporation for the use of a CYBER-205 at the Arden Hills Plant in Minnesota during the initial phase of this study, and to J. Lambiotte and G. Tennille for their assistance in running the code.

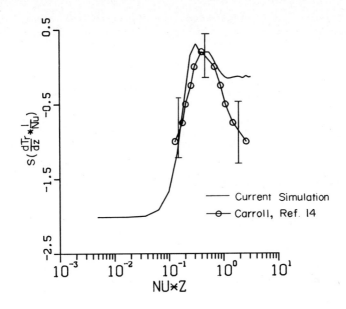

Figure 20. Comparison with experiment of the skewness factor for the normalized relative temperature derivative.

REFERENCES

[1] Rogallo, R. S. and Moin, P. [1984] Numerical simulation of turbulent flows. **Ann. Rev. Fl. Mech.**, **16**, 99-137.

[2] Schumann, U., Grotzbach, G. and Kleiser, L. [1980] Direct numerical simulation of turbulence. **Prediction Methods for Turbulent Flows**, Hemisphere Publ. Corp., (W. Kollman, ed.), 124.

[3] Brachet, M. E., et al., [1983] Small-scale structure of the Taylor-Green vortex. **J. Fluid Mech.**, **130**, 411-453.

[4] Moin, P. and Kim, J. [1985] The structure of vorticity field in turbulent channel flow. Part 2. Study of ensemble-averaged fields. **J. Fluid Mech.** To appear.

[5] Moser, R. D. and Moin, P. [1984] Direct numerical simulation of curved turbulent channel flow. NASA TM 85974. Also Report TF-20, Department of Mechanical Engineering, Stanford University.

[6] Zang, T. A., and Hussaini, M. Y. [1985] Numerical experiments on subcritical transition mechanisms. AIAA Paper 85-0296.

[7] Grotzbach, G. [1982] Direct numerical simulation of laminar and turbulent Benard convection. **J. Fluid Mech. 119**, 27.

[8] Grotzbach, G. [1983] Spacial resolution for direct numerical simulation of the Rayleigh-Benard convection. **J. Comp. Phys. 49**, 241.

[9] Chandrasekhar, S. [1961] **Hydrodynamic and Hydromagnetic Stability.** Clarendon Press, Oxford.

[10] Busse, F. H. 1981 Transition to turbulence in Rayleigh-Benard convection. **Topics in Applied Physics. Vol. 45: Hydrodynamic Instabilities and the Transition to Turbulence.** (H. L. Swinney & J. P. Gollub, eds.), Springer-Verlag.

[11] Krishnamurti, R. [1970a] On the transition to turbulent convection. Part 1. The transition from two- to three-dimensional flow. **J. Fluid Mech. 42**, 295.

[12] Krishnamurti, R. [1970b] On the transition to turbulent convection. Part 2. The transition to time-dependent flow. **J. Fluid Mech. 42**, 309.

[13] Deardorff, J. W., and G. E. Willis [1967] Investigation of turbulent thermal convection between horizontal plates. **J. Fluid Mech. 28**, 675.

[14] Carroll, J. J. [1976] The thermal structure of turbulent convection. **J. Atmos. Sc. 33**, 642.

[15] Fitzjarrald, D. E. [1976] An experimental study of turbulent convection in air. **J. Fluid Mech. 73**, 693.

[16] Chu, T. Y. and R. J. Goldstein [1973] Turbulent convection in a horizontal layer of water. **J. Fluid Mech. 60**, 1, 141.

[17] Goldstein, R. J. and T. Y. Chu [1969] Thermal convection in a horizontal layer of air. **Prog. Heat and Mass Transfer. 2**, 55.

[18] Denton, R. A. and I. R. Wood [1979] Turbulent convection between two horizontal plates. **Int. J. Heat Mass Transfer. 22**, 1339.

[19] Eidson, T. M. [1982] Numerical simulation of the turbulent Rayleigh-Benard problem using subgrid modeling. Ph.D. Thesis, University of Michigan.

[20] Eidson, T. M. [1985] Numerical simulation of the turbulent Rayleigh-Benard problem using subgrid modeling. **J. Fluid Mech. 158**, 245.

[21] Moin, P. and Kim, J. [1982] Numerical investigation of turbulent channel flow. **J. Fluid Mech. 118**, 341.

[22] Deardorff, J. W. [1970] Convection velocity and temperature scales for the unstable planetary boundary layer and for Rayleigh convection. **J. Atmos. Sci. 27**, 1211.

[23] Businger, J. A. [1973] Turbulent transfer in the atmospheric surface layer. **Workshop in Micrometeorology,** American Meteorology Society, 67.

[24] Monin, A. S. and Yaglom, A. M. [1971] **Statistical Fluid Mechanics,** 1 MIT Press.

[25] Malkus, W. V. R. [1954] The heat transport of thermal turbulence. **Proc. Roy. Soc. (London), A.** 225 196.

[26] Grotzbach, G. [1980] Uber das raumliche Auflosungsvermogen numerischer Simulationen von turbulenter Benard-Konvektion. Kernforschungszentrum Karlsruhe. KfK 2981 B.

APPLICATION OF THE TURBIT-3 SUBGRID SCALE MODEL TO SCALES
BETWEEN LARGE EDDY AND DIRECT SIMULATIONS

Günther Grötzbach
Kernforschungszentrum Karlsruhe GmbH
Institut für Reaktorentwicklung
Postfach 3640, 7500 Karlsruhe
Federal Republic of Germany

SUMMARY

A method is presented to calculate the coefficients of subgrid scale models. The theory accounts for all details of the finite difference scheme as it is actually used in a simulation code. The dominating model coefficients are found to depend on local flow parameters. This method as it is implemented in the TURBIT-3-code makes the subgrid scale model selfadaptive to all scales from direct to large eddy simulations. This feature is verified by simulations for turbulent liquid metal flows in annuli. According to the domination of large scales due to the large conductivity of the fluid the theory automatically switches the subgrid scale heat flux model gradually or totally off. It is necessary for successful verification of the results to really use the predicted radially space-dependent coefficients at all intermediate scales between large eddy and direct simulations. For an internally heated horizontal convection layer the predicted model coefficients are compared to direct simulations on grids with different resolution. The agreement of the maximum grid widths allowable for a direct simulation found on both ways shows that the theory to calculate the model coefficients can also be used to check the spatial resolution capabilities of grids in advance to direct numerical simulations.

1. INTRODUCTION

Subgrid scale models have been developed by various authors and have been used for large eddy simulations. Overviews for models and applications are given e.g. in /1, 2/. In most simulations coefficients are applied which are independent of the spatial variables. The reliability of the models seems to be sufficient because no major problems arose from different applications. The only blemish, e.g. for the widely used Smagorinsky-type model, is that the coefficients require some adjustment to the problem considered. As a consequence successful subgrid scale models for the transition range between large eddy simulations with space independent coefficients and direct simulations with vanishing subgrid scale terms are missing /1/. This restricts the applicability of existing subgrid scale models to codes using wall functions, because a finer resolution of the viscous sublayer combined with linear wall approximations would locally require an adequate transition to vanishing subgrid scale fluxes near the walls.

Indeed, problems get obvious from applications of large eddy simulations resolving the viscous sublayer directly. E.g. in /3/ a subgrid scale model has been used which should be suited for wall bounded flows. The subgrid scale fluxes have been split into inhomogeneous and isotropic parts. The first part was represented by a modified Prandtl-van Driest mixing length model, the second one by a Smagorinsky model with a space-independent coefficient. Despite using a very fine grid this sophisticated model gave too strong a damping for turbulence near the walls.

In this paper we will briefly present a subgrid scale (SGS) model which allows for a more accurate application to the transition range between large eddy simulations (LES) and direct numerical simulations (DNS). This model as it is used in the TURBIT-3 code /4, 5/ differs considerably in several features from most of the published ones:

- The finite difference scheme is formally deduced by applying Schumann's method of volume integration using the Gaussian theorem /6/. This allows to use highly anisotropic grids in which the mesh widths may differ by a factor of 10 or more.
- The SGS-fluxes are given as values averaged over the different mesh cell surfaces. Therefore the model accounts for anisotropy of the grids and to some extent also for anisotropy of turbulence.
- Most of the coefficients depend on the local discretization and are therefore space-dependent. These are calculated for each new grid or new flow parameters assuming isotropy of the subgrid scales and the validity of Kolmogorov's energy spectra.

One objective of this paper is to present the method for the calculation of most of the coefficients. The method is an extension of that introduced in /6/. It enables us to fit only once the remaining coefficients which cannot be calculated /7/, and to use a fixed set of correction factors for all applications. Another objective is to show with applications of the TURBIT-3 code that the method to calculate the SGS-coefficients for each simulation is the key to successful simulations for channel flows with large Reynolds numbers, but very small Prandtl numbers. The length scales in the temperature fields of these flows are in the intermediate range between direct and large eddy simulations. Finally it will be shown that the method for calculation of the coefficients may also be used to test the spatial resolution capabilities of grids in advance of using them in direct numerical simulations of turbulent flows.

2. THEORETICAL MODEL

The principles of the method to calculate the coefficients of subgrid scale models will be shown for the example of the SGS-heat flux model of the TURBIT-3-code /5/.

2.1 SUBGRID SCALE HEAT FLUX MODEL

Starting from the time-dependent, three-dimensional thermal energy equation, the subgrid scale heat fluxes are deduced with Schumann's method /6/. That means, the convective terms are integrated over mesh cell volumes $V = \Delta x_1 \Delta x_2 \Delta x_3$. With the Gaussian theorem the partial derivative is directly transformed into a finite difference operator for the convective fluxes averaged over the mesh cell surfaces $^jF = V/\Delta x_j$:

$$1/V \int_V \partial/\partial x_j (u_j T) dv = \delta_j \overline{{}^ju_j T}$$
$$= \delta_j (\overline{{}^ju_j}\,{}^j\overline{T} + \overline{{}^ju'_j T'}). \qquad (1)$$

Here u_j is a velocity component, T is the temperature, and δ_j is the central finite difference operator in the j-direction. The Einstein summation rule is applied to all terms bearing the same subscript twice.

Thus one gets for each of the three different surfaces of a mesh cell a different convective heat flux term. After splitting off the resolvable parts one has to introduce models for the subgrid scale parts $\overline{{}^ju'_j T'}$.

Gradient diffusion is assumed with effective thermal conductivities ^ja_T. In addition each SGS flux is formally split into an isotropic part which is proportional to the gradient of the local and instantaneous temperature fluctuations and into an inhomogeneous part accounting for wall effects. The latter is proportional to the gradient of the time mean value $\langle T \rangle$ of the temperature field.

$$\overline{{}^ju'_j T'} = -\,{}^ja_T\,\delta_j\,({}^j\overline{T} - \langle {}^j\overline{T} \rangle) - {}^ja^*_T\,\delta_j\,\langle {}^j\overline{T} \rangle. \qquad (2)$$

This formal splitting allows for strictly applying the mathematical tools for isotropic turbulence to the calculation of the coefficients of the isotropic part of the model. The expression used for the eddy conductivity in the isotropic part is a type of a Prandtl-Kolmogorov or energy-length-scale model:

$${}^ja_T = C_{T2}\,{}^jC_T\,({}^jF\,\overline{{}^jE'})^{1/2}. \qquad (3)$$

The length scale is determined from the mesh cell surface over which the average of the convective flux (2) has been taken, $^jF^{1/2}$. The energy $\overline{{}^jE'}$ is the subgrid scale part of the kinetic energy of turbulence. This is determined from an additional three-dimensional time-dependent transport equation. Details of the equation, of the anisotropy factor ^jC_T, which is equal to one for $\Delta x_1 = \Delta x_2 = \Delta x_3$, and of the model for the inhomogeneous part of eq. 2 are given in /5, 7/. The important coefficient is C_{T2}. It is calculated in the following chapter.

2.2 CALCULATION OF COEFFICIENTS

The pioneer of the method to calculate the coefficient of a subgrid scale model is Lilly /8/. He assumed that turbulence at subgrid scales is isotropic for large Reynolds numbers. Then the coefficient of the Smagorinsky model can be calculated using the theoretical tools for isotropic turbulence and assuming the validity of the Kolmogorov energy spectrum $E(k)$:

$$E(k) = \alpha\,\langle \varepsilon \rangle^{2/3}\,k^{-5/3}. \qquad (4)$$

k is the wavenumber and ε is the dissipation of the kinetic energy of turbulence. The only empirical parameter in this theory is the Kolmogorov constant $\alpha = 1.5$. It has been found from several applications of the Smagorinsky model that the coefficients calculated with this theory have to be fitted to the special problem considered.

Schumann /6, 9/ extended Lilly's theory by including not only the local grid widths and the actually used volume and surface averaging operators, but also including all details of the actually used finite difference scheme. Especially all approximations required to code the volume averaged conservation equations on a staggered grid are taken into account. So, e.g., different discretizations for the deformation tensor squared result in very different model coefficients /9/. As these discretizations similarly give very different results in the numerical simulations, most of the numerical deficiencies due to discretizing the averaged equations are compensated automatically with this method.

To extend the applicability of Schumann's method to small Reynolds number flows, this author considered the influences of the deviations of real energy spectra from the Kolmogorov spectrum /10/. Applying a combined von-Karman-Kolmogorov-Pao-spectrum, corrections have been deduced for the coefficient in the subgrid scale shear stress model and in the dissipation model in the equation for $\sqrt{jE'}$. These factors are not applied in the code, but have been used to develop extensions of the model. The subgrid scale kinetic energy equation was modified to include the production due to the inhomogeneous part of the subgrid scale model. Accordingly, additional dissipation terms were necessary to make the model consistent and to make it suitable for inhomogeneous flows with small Reynolds numbers. Finally from applications of this extended model to low Peclet number flows it became obvious that at scales between direct and large eddy simulations the calculated model coefficients must also depend on local flow parameters. The extended method developed to calculate the coefficients for applications to all scales will be given below on the example of the isotropic subgrid scale heat flux model (2, 3).

A definition equation for the coefficient C_{T2} of the subgrid scale heat flux model can be deduced from the balance equation for thermal variances in the subgrid scales, $\overline{vT'^2}$. For stationary and isotropic turbulence production and dissipation ε'_T are balanced:

$$-\overline{ju'_jT'}^j \, \delta_j \overline{T}^j - \overline{\varepsilon'_T}^v = 0 . \tag{5}$$

The overbar ___j denotes averaging over values defined in neigbouring mesh cells. Introducing the isotropic part of (2) and (3) in the production term results in an equation for C_{T2}:

$$C_{T2} = \frac{\langle \varepsilon_T \rangle - \langle \overline{D_{Tj}}^2 \rangle / (Re_o Pr)}{\gamma_T \, {}^jC_T \, {}^jF^{1/2} \langle \overline{jE'} \rangle^{1/2} \langle \overline{D_{Tj}}^2 \rangle} . \tag{6}$$

The numerator is rewritten in terms of the difference of total dissipation of temperature variances, ε_T, minus the resolved part of the dissipation. Re_o is the shear Reynolds number, $Re_o = u_\tau D/\nu$, and Pr the Prandtl number, $Pr = \nu/a$, resulting from the normalization process. ν and a are the diffusivities for momentum and heat. The temperature deformation squared is used in its actual finite difference representation including all surface averages and interpolations:

$$\langle \overline{D_{Tj}}^2 \rangle = \langle \delta_j \overline{T}^j \, \delta_j \overline{T}^j \rangle . \tag{7}$$

The denominator of (6) represents the production term. To avoid the calculation of triple correlations the time average has been split into that of the SGS-energy and of the temperature deformation squared. This assumption requires to introduce an unknown correction factor, γ_T, which is greater than one.

For isotropic turbulence the statistical mean values do not depend on the location considered, but only on the distance between locations. Therefore, the rewrite of (7), e.g. for $j = 1$, is:

$$\langle \overline{D}^2_{T1} \rangle = -1/\Delta x_1^2 \ (\langle \overline{T}\overline{T}(3/2 \ \Delta x_1, \ 0, \ 0)\rangle - \langle \overline{T}\overline{T}(\Delta x_1/2, \ 0, \ 0)\rangle). \tag{8}$$

Two-point correlations for nonaveraged variables can be calculated if one assumes for all wave numbers the validity of the Batchelor spectrum for the Energy E_T of the temperature fluctuations:

$$E_T(k) = \beta \ \langle \varepsilon \rangle^{-1/3} \ \langle \varepsilon_T \rangle \ k^{-5/3}. \tag{9}$$

For the Batchelor constant we use $\beta = 1.3$. The large uncertainty of this value is of minor importance because it is finally absorbed in the correction factor γ_T. This results in the following expression for the two-point correlation

$$R_{TT}(\underline{r}) = \langle T(\underline{x}) \ T(\underline{x}+\underline{r}) \rangle \tag{10}$$
$$= -9/20 \ \beta \ \Gamma(1/3) \ \langle \varepsilon_T \rangle \ \langle \varepsilon \rangle^{-1/3} \ r^{2/3} + R_{TT}(\underline{0}),$$

where $\Gamma(x)$ = gamma function and $r = |\underline{r}|$.

Two-point correlations of averaged variables as in equation (8) can be obtained by integrating over the mesh cell surfaces considered:

$$\langle {}^i\overline{T}(\underline{x}) \ {}^j\overline{T}(\underline{x} + \underline{r})\rangle = ({}^iF {}^jF)^{-1} \iint_{{}^iF} \iint_{{}^jF} R_{TT}(\underline{y} - \underline{x} - \underline{r}) \ dF(\underline{y}) \ dF(\underline{x}). \tag{11}$$

For most distance vectors $\underline{r}$ and orientations of the averaging surfaces ${}^iF$ these integrals can only be evaluated by numerical integrations. In some cases the integration range contains singularities. This would cause a tremendous numerical effort to evaluate the interlocked integrals if one would not use the weighting function formalism provided by Uberoi & Kovasznay /6, 11/. We give the final result which is deduced directly in a lengthy procedure /10/ without any further approximations. Equation (6) is replaced by:

$$C_{T2} = \frac{1/2 - \beta/(Re_o Pr) \ \Gamma(1/3) \ 9/20 \ h^{-4/3} \ \langle \varepsilon \rangle^{-1/3} \ DT2(\underline{\Delta x})}{\beta \ \alpha^{1/2} \ (9/20 \ \Gamma(1/3))^{3/2} \ f_2(\underline{\Delta x}) \ \gamma_T} \tag{12}$$

where

$$h = (\Delta x_1 \ \Delta x_2 \ \Delta x_3)^{1/3}. \tag{13}$$

The functions DT2 and f_2 contain the integrals discussed above depending only on geometrical details of the finite difference grid. They are of order one.

For high Reynolds number flows and coarse grids the term containing the dissipation can be neglected. Then C_{T2} does not depend on the mean grid width h, but it depends only on the anisotropy of the grid and on the spectral constants α and β. Therefore, C_{T2} is nearly space-independent at such applications. For low Reynolds number or low Peclet number flows, $Pe_O = Re_O Pr$, or for very fine grids, the term containing the dissipation cannot be neglected and assumptions concerning the local dissipation of kinetic energy must be made. This dependence on local flow parameters enforces a strong space-dependence of C_{T2}. This term can reduce C_{T2} to vanishing values. It gives automatically the transition to a direct simulation scheme with zero coefficients. Thus, the numerator of (6, 12) can also be used to check whether grids are fine enough for direct simulations or not /12/.

The dissipation profile has to be approximated appropriately for each type of flow to calculate the initial value of C_{T2} (see next chapter). In principal it could be updated after some simulation time by evaluating the actual dissipation from the simulation results. Such an improvement was not necessary in our simulations because the dissipation profile could be estimated with sufficient accuracy in advance of simulations for all flows we considered.

The influence of the correction factor γ_T on the temperature statistics has been investigated in a sensitivity study in 1976 /7, 10/. As no considerable effect could be detected for all Prandtl numbers less than one, γ_T has arbitrarily been chosen to be 1.4. This is the value also used in the subgrid scale shear stress model where an analogous coefficient C_2 exists. These fixed values for both correction factors are used throughout all of our simulations for all types of flows.

All other coefficients in the isotropic part of the model like ^jC_T including those in the subgrid scale energy equation are calculated in an analogous manner. All are newly calculated for each new application of the code to a new flow with other parameters or other grid widths.

3. NUMERICAL RESULTS

In this chapter we apply the theory to calculate the coefficients in two different ways to two different flow problems. In the first example calculated coefficients are used for large eddy simulations of the pressure gradient driven flow through annuli. In the second example calculated coefficients are used to check the spatial resolution capabilities of grids in advance of direct simulations of a natural convection problem.

3.1 SIMULATIONS FOR SMALL PECLET NUMBER FLOWS

Several nonbuoyant annular flows are considered with periodic boundary conditions in the circumferential ($\varphi = x_2$) and in the axial direction ($z = x_1$) which is the direction of the mean pressure gradient. The periodicity lengths used here are π and 3.2 times the channel width D. The grids use $N_1 \times N_2 \times N_3 = 32 \times 32 \times 16$ mesh cells. The fluid is heated at the inner wall by prescribing a constant wall heat flux q_w. The outer wall is adiabatic. The fluids used in the simulations have a Prandtl number of 0.007, which represents e.g. liquid sodium, and 0.0214 which represents e.g. liquid mercury. As the Reynolds numbers $Re = u_b d_h/\nu$ are about 10^5 the large eddy simulation scheme of the TURBIT-3 code is activated by the theory to calculate the coefficients. The wall layers of the velocity field are modelled as in /7/. For the temperature field linear wall approximations are applied.

For a qualitative investigation of simulation results contour-line plots of the resolved instantaneous temperature fluctuations are shown in Fig. 1 for the two different Prandtl numbers. The isolines look random as expected for turbulent flows. The temperature fluctuations are larger near the heated wall than near the adiabatic wall. The amplitude of the fluctuations indicated by the contourline increment Δ is larger in the mercury flow, and the location of its maximum is closer to the wall. Both sections show larger scales in the centre of the channels, whereas the scales are smaller near the walls. In general, large scales are dominant because the large thermal conductivity of liquid metals filters the small scale variations out.

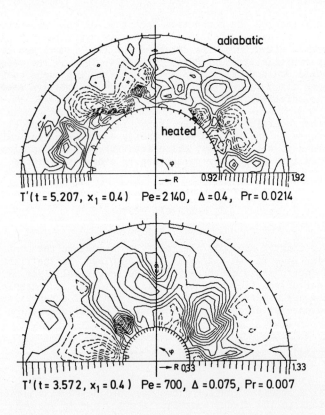

Fig. 1: Instantaneous resolved temperature fluctuations $\overset{V}{T}{}' = \overset{V}{T} - \langle \overset{V}{T} \rangle$ in two annuli with different ratios of radii. Solid lines represent positive values, dashed lines negative values. The temperature fields are normalized by $T^* = q_w/(\rho\, C_p\, u_\tau)$.

Some conclusions on the resolution capabilities of the grids used can be deduced from this figure. For liquid sodium the scales in the temperature field are even near to the heated wall much larger than the grid widths. Whereas for liquid mercury it is obvious that at many places the temperature variations have scales of about the mesh cell sizes indicated at the boundaries. Thus one expects, that a SGS-heat flux model is required at the Peclet number of Pe = Re Pr = 2 140 of the mercury flow, but no SGS-heat flux model has to be used at the much smaller Peclet number of the sodium flow, Pe = 700.

The coefficient C_{T2} of the subgrid scale heat flux model used in these simulations is calculated on the basis of the universal dissipation profile. This is deduced assuming equality of production and dissipation of kinetic energy. Application of the Prandtl mixing length model and the universal logarithmic velocity profile furnishes:

$$\varepsilon = (\kappa y)^2 \; |\partial \langle u_1 \rangle / \partial x_3|^3 = 1/|\kappa y|. \tag{14}$$

Where $\kappa = 0.4$ is von Karman's constant and y is the distance to the next wall.

The radial profiles of C_{T2} calculated from equations (12, 14) are mainly a result of the dissipation profile (Fig. 2). Only the grid used for Pe = 3250 is non-equidistant near the walls. The larger values of C_{T2} for all grids near the outer wall are due to the larger mesh size there. The influence of the Peclet number is as expected: The predicted C_{T2} values decrease for decreasing Peclet numbers. For the mercury flow given in Fig. 1 (Pe = 2140) a strong radial variation of C_{T2} follows. This

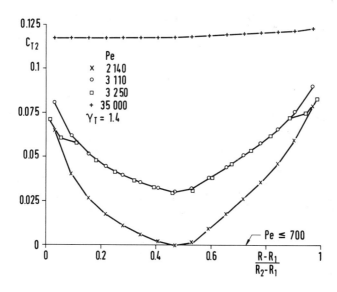

Fig. 2: Radial distribution of the calculated coefficient C_{T2} in annuli. R_1 is the radius of the inner wall, R_2 that of the outer wall.

computation is done for the transition range between a pure direct simulation of the temperature field and a pure large eddy simulation with a nearly constant coefficient as it follows for large Peclet numbers (e.g. for Pe = 35 000 in Fig. 2). For the sodium flow with Pe = 700, the coefficient C_{T2} is zero all over the channel. This means that for Pe = 700 a direct simulation of the temperature field is performed without a subgrid scale heat flux model. Thus in the thermal energy equation a single open parameter does not exist.

In a sensitivity study for liquid metal flows in the transition range between DNS und LES a considerable insensitivity of the results to changes of the correction factor γ_T was observed /13/. In varying γ_T from 0.2 to 100., no remarkable influence on the temperature fields could be detected for all meaningful values ($\gamma_T \geq 1$). Therefore, it is not of importance that for these Peclet numbers the real temperature spectra do not really follow eq. (9), but are much steeper, and that the Batchelor constant β is not well known.

To verify the simulations the calculated maxima of the radial profiles of the temperature-root-mean-square values are compared to experimental data in Fig. 3. For reference of the experimental data see /13/. Unfortunately, all experimental data scatter considerably. It has been shown by Lawn /14/ that most of these experimental data sets are not consistent. Obviously it is very hard to perform accurate measurements in

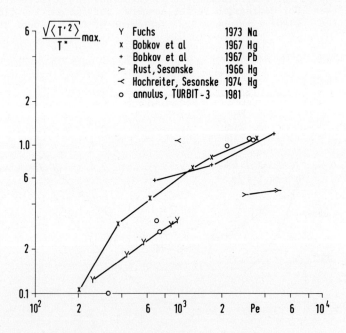

Fig. 3: Peaks of radial temperature RMS-value profiles. The experimental data are for pipes. The calculated data for Pe < 1000 are for liquid sodium in annuli with $R_1/R_2 = 0.25$; the others are for liquid mercury in annuli with $R_1/R_2 = 0.479$.

liquid metal flows. Nevertheless this comparison gives important results: None of the calculated data is out of range, despite the fact that according to Fig. 2 some simulations use no SGS-heat flux models, some use models with strongly varying C_{T2}-profiles in the radial direction, and that the position of T'_{max} moves with increasing Peclet number closer to the wall, physically and in the calculation, see e.g. Fig. 1. This latter effect acts against the C_{T2}-profile which gives larger damping values near the heated wall.

To confirm this more qualitative verification of our simulations using calculated non-constant coefficients, the radial heat flux cross-correlation coefficient is given in Fig. 4 for all simulated liquid metal flows in annuli. The calculated correlation coefficients are about 0.45. For each Prandtl number they decrease with increasing Peclet number. This is mainly due to flattening of the local T'_{max}-values by volume averaging of the basic equations, because the T'-distribution becomes steeper at larger Peclet numbers. For such liquid metal flows we have no experimental data available for comparison. The values calculated for these Peclet numbers are about the same as those calculated and measured for Prandtl numbers around one. Thus the numerical data seem to be realistic and consistent.

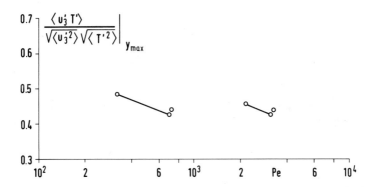

Fig. 4: Resolved radial heat flux correlation coefficient at position y_{max} of the maximum of the RMS temperature values.

Both types of results could not be obtained as reasonable as shown here without assuming that C_{T2} depends on the radial dissipation profile. Especially the calculated radial profiles of mean temperature and eddy conductivity could only be verified in /13/ by using the space-dependent C_{T2}-profiles. That means also that non-constant turbulent Prandtl-numbers $Pr_{SGS}^{T2} = C_2/C_{T2}$ have to be used for all simulations of liquid metal flows with scales in the transition range between LES and DNS because at large Reynolds numbers C_2 is nearly space-independent, but not C_{T2}.

This experience seems to be in contradiction to the considerable insensitivity of the statistics of the simulated temperature fields against changes of the correction factor γ_T found with this subgrid scale model at molecular Prandtl-numbers less than one. The solution to this problem is found in the fact that the physical basis of the subgrid scale flux assumption, eqs. (2,3), has been extended by the theory to calculate the coefficient C_{T2} in such a manner that flow dependent information is introduced in the coefficient by choosing an adequate dissipation profile. Following our experience, this results in radial distributions of the coefficient C_{T2} which need no readaptation to the special problem considered. Inadequate models usually do not show such a good natured behavior. Thus, the large insensitivity against changes of γ_T found is a qualitative confirmation for the total model including the calculation of C_{T2}. Moreover it indicates that for such applications it is more appropriate to use adequate radial profiles for the coefficients than to use optimized or readapted but radially constant values.

3.2 PREDICTION OF RESOLUTION LIMITS FOR DIRECT SIMULATIONS

A series of direct numerical simulations on different grids is used to check the resolution limits predicted by the theory for calculation of the subgrid scale coefficient C_{T2}. For this test we do not use the predicted subgrid scale coefficients in the simulations, but we enforce them to be zero. At the end of the simulations we compare the resolution limit found with that one resulting from the predicted coefficients.

A plane horizontal channel is considered with periodic boundary conditions in both horizontal directions. Both periodicity lengths are 2.8 times the distance D between the walls. The fluid with $Pr = 6$, which represents e.g. molten fuel of nuclear reactors, is internally heated by a homogeneously distributed volumetric heat source Q. The heat is removed from the channel by prescribing equal wall temperatures, T_w, at both walls. The Rayleigh number chosen is $Ra = g\beta QD^5/(\nu a \lambda) = 4 \times 10^6 = 107 \times Ra_{cr}$, where β = volume expansion coefficient and λ = thermal conductivity. For these simulations linear wall approximations are applied to all variables. Thus, the spatial resolution capability of the grid remains the only model parameter to be adapted.

The convection phenomena are investigated on the basis of vertical and horizontal sections through the temperature field at different times, Fig. 5. In the vertical sections one finds the internally heated hotter fluid bounded by the colder walls. The lower boundary layer near $x_3 = 0$ is stably stratified. The core is weakly stable or neutral. The upper boundary layer near $x_3 = 1$ is the only unstable one which drives the flow. Here the cold plumes develop, fall downwards and sometimes deeply penetrate the lower boundary layer. The interferogram from the experiment by Jahn /15/ shows about the same typical length scales and about the same position of the temperature maximum.

The horizontal sections are for a plane near the temperature maximum T_{max}. Most of the cold plumes form open cells with a hot, isothermal core. The plumes start locally. With increasing time the initially round plumes grow to plane plumes and merge with other ones. Sometimes closed cells are formed. These then contract to knots. The evaluation of sections at other heights shows that jets plunging downwards from knots have the largest kinetic energy so that some of these reach the lower wall. In contrast to the jet velocities the velocity in the connecting cell or spoke structure is much smaller. Therefore the cell structure can only be found in the

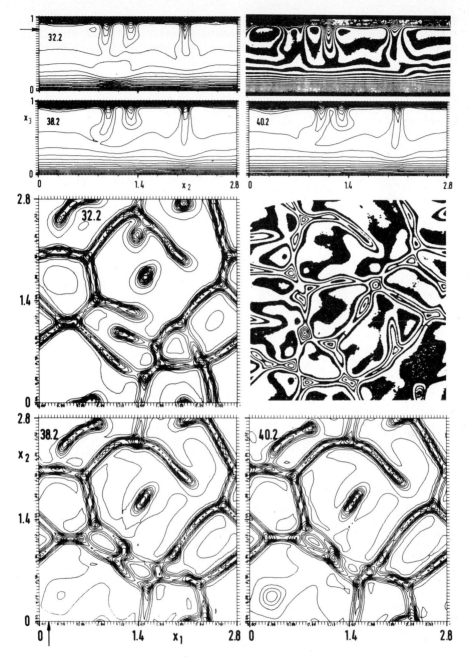

Fig. 5: Vertical and horizontal sections, x_2-x_3 and x_1-x_2 planes, of calculated temperature fields at several times and interferograms from experiments by Jahn. The arrows denote the origins of the other sections. Δ(vert.) = 0.08, Δ(horiz.) = 0.0625. The temperature fields are normalized by the initial value of $\Delta T_o = \langle T_{maxo} - T_w \rangle$.

upper third of the channel including the upper boundary layer /5/. All the features of this flow show that fully turbulent flow is simulated as it was to be expected at this Rayleigh number. For lower Rayleigh numbers in the transition or laminar regime one would find irregular and regular hexagonal cells /5/.

The important result of this simulation is that most scales of the temperature fields are larger than the grid widths which result from the discretization by 64 x 64 x 32 mesh cells. But there exist also a few structures with scales which are comparable to the grid widths.

A first check on whether the resolution of the grid used is sufficient or not is performed investigating the predictions for the subgrid scale heat flux coefficient C_{T2} in Fig. 6. The dissipation profile re-

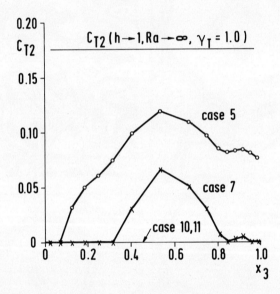

<u>Fig. 6:</u> Vertical distribution of C_{T2} for an internally heated fluid layer with Ra = 4 x 10^6. The grids are specified in the text. γ_T = 1.

quired in equation (12) is approximated here by a linear distribution between zero near the lower wall and the value ε_{max} at the upper wall following from dimensional analysis:

$$\varepsilon_{max} \approx (Nu_2 - Nu_1)((Nu_1 + Nu_2)/(RaPr))^{1/2} \qquad (15)$$

The Nusselt numbers at bottom, Nu_1, and at top, Nu_2, are estimated from the empirical correlations given by Jahn /15/. The velocity scale is $u_o = (g\beta(T_{maxo} - T_w)D)^{1/2}$. This assumption for the dissipation profile had to be verified by using the results of the simulations because no experimental results for turbulence data exist for this type of flow.

The different curves in Fig. 6 are for the same channel and for the same Rayleigh number, but for different node numbers. We use 16^3 nodes in case 5, $32^2 \times 16$ in case 7, 32^3 in case 10, and $64^2 \times 32$ in case 11. In all cases the vertical grid width is finest at the top of the channel ($x_3 = 1$), and coarsest in the middle. Accordingly the C_{T2}-profiles calculated by our theory are larger in the middle of the channel and smaller near the boundary layers because C_{T2} depends on the mean grid width by $h^{=4/3}$ and on the dissipation only by $\varepsilon^{-1/3}$. Following these predicted C_{T2}-profiles one has to conclude that subgrid scale models should be used for the grids of cases 5 and 7, and that for cases 10 and 11 direct simulations without subgrid scale models should be possible.

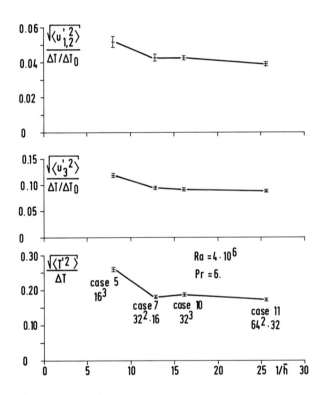

Fig. 7: Influence of the mean grid width $\overline{h} = (\Delta x_1 \, \overline{\Delta x_2} \, \overline{\Delta x_3})^{1/3}$ on the maxima of turbulent fluctuation values. $\overline{\Delta x_3} = D/N_3$, $\Delta T = \langle T_{max} - T_w \rangle$.

This predicted resolution limit can be tested by the experience gained from the direct simulations on these four grids. The results which have been found to be most sensitive to insufficient resolution are the root-mean-square values for the temperature and velocity fluctuations. In figure 7 the maxima of the vertical profiles of these data are given as a function of the reciprocal mean grid width. (The complete profiles and numerous other statistical data obtained from the simulation on the finest grid are documented in /5, 16/). Both, the velocity fluctuations in the horizontal directions (i= 1,2) and in the vertical direction (i = 3), as well as the temperature fluctuations are more or less constant except for the coarsest grid. For this one the subgrid scale models should be used. Thus, from this figure it follows that the total resolution of all relevant scales is reached with grids 7, 10, and 11. This is in sufficient agreement with the predictions of the theory to calculate C_{T2}, see Fig. 6.

4. CONCLUSIONS

A method is presented to calculate the coefficients of subgrid scale models. As in an earlier theory this method includes all details of the finite difference scheme as it is coded including all interpolations required, and also all averaging operators. This helps to compensate for most non-physical, numerical deficiencies resulting from the discretization on a staggered grid. On the other hand the strong dependence of the coefficients on the finite difference approximations indicates that the values of coefficients for subgrid scale models may only be used in another code if the same model is discretized with the same finite difference scheme.

An important extension of the method given here is that the subgrid scale coefficients depend on local flow variables. In all simulations in the transition range between direct numerical simulations and large eddy simulations with constant coefficients one needs an approximation for the dissipation profile to predict the dominant subgrid scale coefficients. This extension makes the subgrid scale model fully self-adaptive. It automatically switches the subgrid scale models gradually or totally off.

This theory is implemented in the TURBIT-3 code. It is used to calculate a new set of coefficients for each simulation. The correction factors remaining in the method have been fixed in 1976. Since that time very different flows have been simulated without any need to refit the correction factors. For example, large eddy simulations have been performed for Reynolds numbers between 2×10^4 and 3×10^5 for flows in plane channels and annuli with different ratios of radii, for smooth and partly roughened walls, flows without and with considerable buoyancy contributions, and on very coarse and on finer grids.

The most crucial test for the method has been given here with the application to low Prandtl number flows. According to the large thermal conductivity of the fluid the theory switches the subgrid heat flux model gradually or totally off in those cases in which the total resolution of all scales in the temperature field is reached. For the transition range between large eddy and direct simulations this results in radially strongly non-constant coefficients. Nevertheless, none of the data evaluated from these simulations seem to be non-physical. In contrast, this positive result could not be obtained applying radially constant coefficients.

A second verfication of the method has been performed by applying it as a tool for spatial resolution checks of given grids which shall be used for direct simulations. Comparing the results of direct numerical simulations on different grids for an internally heated horizontal fluid layer shows that the theoretically predicted spatial resolution limit is also gained by the simulation results. Thus this method to calculate the coefficients may also be used for reliable checks of the resolution capabilities of grids in advance of direct numerical simulations. The applicability of the method is not restricted to the types of flows considered in this paper. It can be applied to any type of flow by choosing appropriate dissipation profiles.

Finally, one might expect that this method can also help to circumvent the problems found in large eddy simulations resolving the viscous sublayer directly. The exceedingly strong damping of turbulence near the wall found otherwise in applying constant model coefficients can be reduced by applying the same idea as realized here for the coefficient of the subgrid scale heat flux model also to the coefficient of the subgrid scale shear stress model.

REFERENCES

/1/ P.R. Voke, M.W. Collins:
Large eddy simulation: Retrospect and prospect
Physico Chemical Hydrodynamics 4, 1983, p. 119 - 161.

/2/ R.S. Rogallo, P. Moin:
Numerical simulation of turbulent flows
Ann. Rev. Fluid Mech. 16, 1984, p. 99 - 137.

/3/ P. Moin, J. Kim:
Numerical investigation of turbulent channel flow
J. Fluid Mech. 118, 1982, p. 341 - 377.

/4/ U. Schumann, G. Grötzbach, L. Kleiser:
Direct numerical simulation of turbulence
In: Prediction Methods for Turbulent Flows, Ed.: W. Kollmann, Hemisphere Publ. Corp. 1980, p. 123 - 258.

/5/ G. Grötzbach:
Direct numerical and large eddy simulation of turbulent channel flows
in: Encyclopedia of Fluid Mechanics, Ed.N.P. Cheremisinoff, Gulf Publishing, Vol. 6 (in press), preprint as
KfK report 01.02.06.P49A, 1984.

/6/ U. Schumann:
Subgrid scale model for finite difference simulations of turbulent flows in plane channels and annuli
J. Comp. Physics 18, 1975, p. 376 - 404.

/7/ G. Grötzbach, U. Schumann:
Direct numerical simulation of turbulent velocity-, pressure-, and temperature-fields in channel flows
Turbulent Shear Flows I, Ed.F. Durst et al., Springer 1979, p. 370-385.

/8/ D.K. Lilly:
The representation of small-scale turbulence in numerical simulation experiments
Proc. IBM Scientific Comp. Symp. on Environmental Sciences, IBM Form No. 320-1951, 1967, p. 195-210.

/9/ U. Schumann:
Ein Verfahren zur direkten numerischen Simulation turbulenter Strömungen in Platten- und Ringspaltkanälen und über seine Anwendung zur Untersuchung von Turbulenzmodellen
Dissertation, Universität Karlsruhe, 1973, KfK 1854,
English translation in NASA-TT-F-15391.

/10/ G. Grötzbach:
Direkte numerische Simulation turbulenter Geschwindigkeits-, Druck- und Temperaturfelder bei Kanalströmungen
Dissertation, Universität Karlsruhe, 1977, KfK 2426,
English translation in DOE-tr-61.

/11/ M.S. Uberoi, L.S.G. Kovasznay:
On mapping and measurement of random fields
Quart.Appl.Math. 10, 1952, p. 375-393.

/12/ G. Grötzbach:
Spatial resolution requirements for direct numerical simulation of the Rayleigh-Bénard convection
J. Comp. Physics 49, 1983, p. 241-264.

/13/ G. Grötzbach:
Numerical simulation of turbulent temperature fluctuations in liquid metals
Int.J. Heat Mass Transfer 24, 1981, p. 475-490.

/14/ C.J. Lawn:
Turbulent temperature fluctuations in liquid metals
Int.J. Heat Mass Transfer 20, 1977, p. 1035-1044.

/15/ M. Jahn:
Holographische Untersuchung der freien Konvektion in einer Kernschmelze
Dissertation, Techn. Universität Hannover, F.R.G., 1975.

/16/ G. Grötzbach:
Direct numerical simulation of the turbulent momentum and heat transfer in an internally heated fluid layer
Heat Transfer 1982, Ed. U. Grigull, E. Hahne, K. Stephan, J. Straub, Hemisphere 1982, Vol. 2, p. 141-146.

Direct Simulation of High-Reynolds-Number Flows by Finite-Difference Methods

Kunio KUWAHARA
The Institute of Space and Astronautical Science
Komaba, Meguro-ku, Tokyo, Japan
Susumu SHIRAYAMA
Department of Aeronautics, University of Tokyo
Hongo, Bunkyo-ku, Tokyo, Japan

Introduction

At high Reynolds numbers, it is very difficult to solve the Navier-Stokes equations because of its numerical instability. This difficulty is due to the very small viscous diffusion. The eddy viscosity model introduces a rather large diffusion into the system, which stabilizes the computation. It is natural to ask whether the high-Reynolds-number flow fields can be obtained without introducing a turbulence model or sub-grid modeling.

Another way to overcome the numerical instability in high-Reynolds-number flow computation is to use an upwind scheme. The stability of the first-order upwind scheme is quite good, but it has a strong diffusive effect similar to the effect of molecular viscosity. Thus, it is not suitable for our purpose. The second-order upwind scheme is better in this sense, but it is more unstable and causes undesirable propagation of errors.

Recently, a third-order upwind scheme was developed. Using this scheme, the flow around a circular cylinder in the critical regime was successfully simulated without using any turbulence model[1]. This computation suggests that direct simulation without any turbulence model may be possible if the computational scheme is stable and if its numerical diffusion does not conceal the effect of viscosity. Although at high Reynolds numbers, the grid spacing needed to resolve the smallest eddies in the boundary layer is too fine for the computers available at the present time, the above computation using a relatively coarse grid has captured a global qualitative structure of the turbulent flow. Also, computations by LES have shown that the grid spacing need not be so fine as the theoretical assumption requires[2]. This means that the small scale structure may not have much influence on the large scale structure.

These results suggest that if we are interested only in the large scale structure of turbulence, its direct numerical simulation may be possible. Since the flow characteristics of engineering interest are usually determined by the large scale structures, direct simulation of high-Reynolds-number flow may give interesting results.

In the present paper, various high-Reynolds-number flows are simulated without using any turbulence model. The present method, does not require any special assumptions as do other schemes, appears to capture the Reynolds numbers dependence, and permits us to follow the transition to turbulence. Numerical results agree reasonably well with experiments.

Finite-Difference Schemes

The unsteady incompressible full Navier-Stokes equations written in generalized coordinates are solved directly without any turbulence model. All spatial derivatives except those of nonlinear terms are approximated by central differences. The nonlinear terms are approximated by a third-order upwind scheme[1]:

$$(u \frac{\partial u}{\partial x})_i = u_i (u_{i+2} - 2u_{i+1} + 9u_i - 10u_{i-1} + 2u_{i-2})/6h$$

$$\text{for } u_i > 0,$$

$$u_i (-2u_{i+2} + 10u_{i+1} - 9u_i + 2u_{i-1} - u_{i-2})/6h$$

$$\text{for } u_i < 0.$$

$$= u_i (-u_{i+2} + 8(u_{i+1} - u_{i-1}) + u_{i-2})/12h$$
$$+ |u_i| (u_{i+2} - 4u_{i+1} + 6u_i - 4u_{i-1} + u_{i-2})/4h .$$

From the last expression it is clear that this scheme has a numerical diffusion approximately expressed by fourth-order derivative.

The computational schemes for compressible flow are based on the implicit factorization method with improved accuracy[3]. An artificial diffusion term of fourth-order difference is added to the explicit term, as it is usually done, to remove the aliasing errors and to stabilize the computation. Three types of different schemes are used for the inversion of the matrix for the implicit part. These are a bidiagonal scheme[4] and a block tridiagonal scheme and a block pentadiagonal scheme[5] according to the accuracy of the implicit terms. The bidiagonal scheme is most efficient while the block pentadiagonal one is most accurate. The block tridiagonal scheme is essentially the Beam-Warming-Steger's[3]. The explicit part is the same in all schemes, thus all of them should give the same steady state solution.

For the convective terms on the right hand side, fourth-order differencing is used in order to improve the accuracy in the inviscid region. The Jacobian and the metric terms are also formed by using fourth-order differencing.

Flow around a Circular Cylinder

Figure 1 shows the streamlines at Reynolds numbers Re = 2000 and 40000, illustrating the flow patterns before and after the drag crisis. The grid is 80x80; the roughness is distributed only near the separation points. Figure 2 shows the corresponding lift and drag coelficients; the dependence of the drag on the Reynolds number is shown in Fig.3. Figure 4 shows the same flow case but the roughness is equally distributed on the surface with 200x100 grid points. The typical flow pattern after the drag crisis is clearly shown for Re = 60000. Figure 5 shows the flow around a circular cylinder without any roughness for Re = 670000. The number of grid points is 500x100. Even without surface roughness, the drag crisis is well captured.

At very high Reynolds numbers, the flow speeds become high and sometimes the compressibility may have some effect on the flow structure. Therefore, the same flow was simulated by solving the compressible full Navier-Stokes equations. The Mach number is 0.3 and the Reynolds number ranges from 1×10^5 to 7.83×10^6. The scheme is essentially what is used in the Beam-Warming-Steger method, but the accuracy is improved by using fourth-order differencing in the nonlinear terms. The second-order artificial diffusion terms are not employed at the sacrifice of the stability. The drag coefficients are shown in Fig.6 together with data of some experiments. The agreement between the calculations and the experiments is excellent up to the Reynolds number 10^6. Figures 7 and 8 show the comparison of the flows at four different Reynolds numbers[7]. The drag crisis of a smooth circular cylinder is clearly obtained. The transition to the transcritical regime is also obtained. No essential difference was found between incompressible computation and the compressible one at Mach number 0.3.

Flow around an Automobile

The method is applied to compute an incompressible flow around a two-dimensional automobile. Figure 9 shows the streamlines at $Re=10^5$. It was found that, if the number of grid points exceeds a certain limit, the reattachment of the boundary layer is well simulated and that the computed results agree well with experiment.

Fully-Developed Turbulent Flow in a Duct

A turbulent flow in a duct was simulated by integrating the incompressible Navier-Stokes equations without any subgrid modeling[8]. The initial and boundary conditions are essentially the same as those of Deardorff[2], Moin & Kim[9] and Horiuti & Kuwahara[10]. In most cases the number of mesh points is 30x20x30. The Reynolds number, based on the distance between the two walls and mean velocity at the center of the duct, is 20000. The average velocity, the Reynolds stress u'v' and the turbulence intensities are shown in Fig.10. All profiles agree reasonably well with experimental results and with the computational results obtained by Moin & Kim.

Transition to Turbulence

The simulation of the transition to turbulence is very difficult by LES, because initially the flow is laminar and the eddy viscosity model cannot be used.

Figure 11 shows space-averaged velocity profiles from impulsive start to turbulence[6]. Up to the non-dimensional time of t=560, the flow is almost laminar and the velocity distribution is becoming parabolic, but near t=620 an asymmetry indicating the beginning of the transition develops. The transition to turbulence takes place during a rather short time period as shown in Fig.11(b). Figure 12(a) shows the time- averaged velocity profile after the transition. Good agreement with Laufer's experiment was obtained. Fig.12(b) displays the averaged velocity near the wall.

Fig. 13 and 14 show the transition at different Reynolds numbers. The initial condition is the flow at t=575 of the previous case; it is just the beginning of the transition. At Reynolds number 5000 (Fig.13), a similar transition takes place, but for Re = 2000 (Fig.14) the transition is very slow and we can not observe any symptoms of transition from the initial velocity profile. The turbulence intensity, however, suggests the development of the disturbance. These findings correspond well to experimental observations.

Figure 15 shows the contour lines obtained by the computation using a larger number of grid points (60x40x50); the initial condition was determined by using the data observed by Laufer with a disturbance twice as large as that observed by Laufer. Figure 16 shows the transition for the same grid system. Essentially the same results were obtained with the above relatively coarse grid system.

The computation time needed to compute 46000 time steps in this simulation of transition to turbulence (using 24x14x20 mesh points) is only ten minutes on a HITACHI S810 (vector) computer, and that when using a refined grid (60x40x50) is two hours. For an effective use of the S810, three-dimensional arrays are rewritten to two-dimensional ones, thus resulting in very long vectors and in computational speeds of about 300 MFLOPS.

Turbulent Flow in a Circular Pipe

Figure 17 shows the turbulent flow in a circular pipe; the computational conditions are essentially the same as the previous ones. Typical turbulent flow patterns were obtained.

Flow around Three-Dimensional Bluff Bodies

Figures 18,19,20,21 show examples of three-dimensional flows around bluff bodies using the same method in generalized coordinates[11]. Figure 18 shows the equi-pressure surfaces around a cone on a plane. The Reynolds number is 10^3, the number of grid points is 40x32x22. Figure 19 shows the equi-pressure surfaces around a sphere at non-dimensional time t = 52.27 after impulsive start. The number of grid points is 30x28x24, the Reynolds number is 10^4. Figure 20 shows also equi-pressure surfaces behind a circular cylinder with a slanted base. The number of grid points is 30x32x30 and the Reynolds number is 10^4. A pair of vortex tubes separated from the base is clearly seen. Figure 21 shows the equi-pressure surfaces around a rocket at the angle of attack 90°. The number of grid points is 40x33x30. The Reynolds number is 10^4.

From these computations, it becomes clear that large structures of high-Reynolds-number flows can be captured at relatively high Reynolds numbers without any turbulence model.

Flow around an Airfoil at Low Angle of Attack

Figure 22 shows the solutions for the flow around a NACA 0012 airfoil at an angle of attack of 2 degrees[12]. The range of the Reynolds numbers is from 10^5 up to 6.7×10^6, the Mach number is 0.75. At the leading edge the minimum grid spacing normal to the surface was set to 0.1/Re. The minimum spacing at the trailing edge is two times larger. The CFL numbers were suppressed to about 40 in order to obtain better accuracy for the unsteady computation. The CPU time for 321x80 grid points was about one hour per thousand steps on a FUJITSU scalar computer M380. It took a total of 20 hours for the unsteady computation in the case of Re = 6.7×10^6.

The unsteady vortex separation was observed in every case. The coefficient of pressure obtained numerically for the case of R = 6.7×10^6 is compared with the experimental data[13] (Fig.23a). In this case the grid still needs refinement in order to capture turbulent phenomena. Figure 23b shows the numerical solutions obtained using the Baldwin-Lomax turbulence model[14]. The lift coefficient was 0.3, while that obtained by experiment was 0.29. In this case, the numerical results are in quite good agreement with experiment, although only a steady state can be predicted.

Flow around an Airfoil at High Angle of Attack

Since the flow is highly unsteady in this case, the most accurate pentadiagonal scheme is employed with a 320x80 grid [5]. The Reynolds number is 10^6 and the Mach number is 0.4. The CPU time is 2 hours on the S810. The time dependence of the density contours and the pressure coefficient curves for the case of an angle of attack of 15 degrees are presented in Fig.24. Small separation bubbles stick to the leading edge and this leads to the strong nearly steady suction peak there which agrees with experimental observation. The computation with a coarser grid of 160x40 could not capture this phenomenon well.

Conclusions

From these computations, it has become clear that the large-scale structure of turbulent flows can be simulated by directly integrating the Navier-Stokes equations. The numerical diffusion plays a very important role in getting reasonable results. Numerical diffusion of second-order-derivative type which appears in the first-order upwind scheme or as an implicit diffusion in the Beam-Warming-Steges method is similar to the molecular diffusion and conceals the dependence of the flow on the Reynolds number and is not suitable for high-Reynolds-number flow computation. On the other hand, the numerical diffusion of fourth-order-derivative type is of short range and does not conceal the effect of molecular diffusion and stabilizes the computation very well. At the present stage, this may be the best way to overcome the numerical instability in high-Reynolds-number flow computation.

The above results indicate that turbulent flows can be directly simulated, and that the number of grid points need not to be so large. Also, the transition to turbulence can be simulated, which is very difficult by LES.

The present methods are based on ordinary finite-difference methods and the codes are written in generalized coordinates. Therefore the boundary conditions can be very easily imposed. Moreover the code can be very easily and effectively vectorized. On the other hand, computations by LES are usually based on pseudospectral techniques, and it is not easy to impose boundary conditions, especially in generalized coordinate systems.

Even direct simulation of unsteady transonic flows at high Reynolds numbers would become feasible soon, with the development of supercomputers. The numerical simulation of turbulent flow may be much simpler than it was previously believed.

Acknowledgements

This work has been done in cooperation with Dr. Tetuya Kawamura, Dr. Katsuya Ishii, Mr. Shigeru Obayashi, Mr. Yoshifumi Shida and Mr. Keisuke Kamo of the University of Tokyo and Dr. Satoru Ogawa of the National Aerospace Laboratory and Mr. Ryutaro Himeno of Nissan Motor Co. Ltd. and Dr. Wei J. Chyu of NASA Ames Research Center.

References

1) T. Kawamura and K. Kuwahara; 1984 Computation of High Reynolds Number Flow around a Circular Cylinder with Surface Roughness, AIAA paper 84-0340.
2) J. W. Deardorff; 1970 A numerical study of three-dimensional turbulent channel flow at large Reynolds numbers. J. Fluid Mech. vol.41, pp.453-480.
3) J. L. Steger; 1979 Implicit Finite-Difference Simulation of Flow about Arbitrary Two-Dimensional Geometries, AIAA Journal, Vol.16, No.7, pp.679-686.
4) S. Obayashi and K. Kuwahara; 1984 LU Factorization of an Implicit Scheme for the Compressible Navier-Stokes Equations, AIAA Paper 84-1670.

5) Y. Shida and K. Kuwahara; 1985 Computational Study of Unsteady Compressible Flow around an Airfoil by a Block Pentadiagonal Matrix Scheme, AIAA paper 85-1692.
6) K. Ishii and K. Kuwahara; 1984 Computation of Compressible Flow aruond a Circular Cylinder, AIAA paper 84-1631.
7) K. Ishii, K. Kuwahara, S. Ogawa, W. J. Chyu and T. Kawamura; 1985 Computation of Flow aound a circular Cylinder in a Supercritical Regime, AIAA paper 85-1660.
8) T. Kawamura and K. Kuwahara; 1985 Direct Simulation of a Turbulent Inner Flow by Finite-Difference Method, AIAA paper 85-0376.
9) P. Moin and J. Kim; 1982 Numerical investigation of turbulent channel flow. J. Fluid Mech. vol.118, pp.341-377.
10) K, Horiuti and K. Kuwahara; 1982 Study of Incompressible Turblulent Channel Flow by Large Eddy Simulation, Proc. 8th ICNMFD, Springer-Verlag.
11) R. Himeno, S. Shirayama, K. Kamo and K. Kuwahara; 1985 Computational Study of Three-Dimensional Wake Structure, AIAA paper 85-1617.
12) S. Obayashi, H. Kubota and K. Kuwahara; 1985 Computation of Unsteady Shock-Induced Vortex Separation, AIAA paper 85-0183.
13) K. Takashima; to appear in Technical Memorandum of National Aerospace Laboratory, Japan.
14) B. S. Baldwin and H, Lomax; 1978 Thin layer Approximation and Algebraic Model for Separated Turbulent Flows, AIAA paper 78-257.

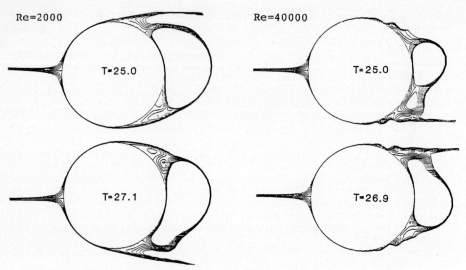

Fig.1 Detailed flow pattern near a circular cylinder.

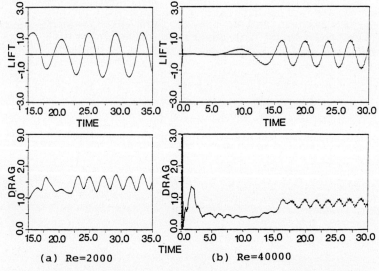

Fig.2 Time histories of the drag and lift coefficients.

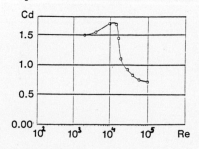

Fig.3 Variation of the drag coefficient with Reynolds number.

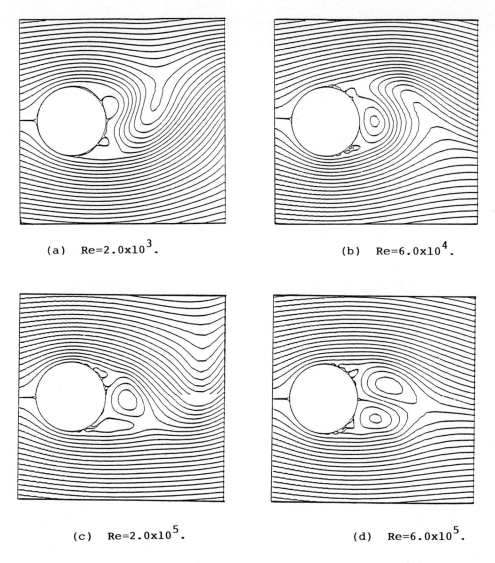

(a) Re=2.0x10^3. (b) Re=6.0x10^4.

(c) Re=2.0x10^5. (d) Re=6.0x10^5.

Fig.4 Stream lines around a circular cylinder with uniformly distributed surface roughness.

Fig.5 Stream lines at Re=6.7x10^5.

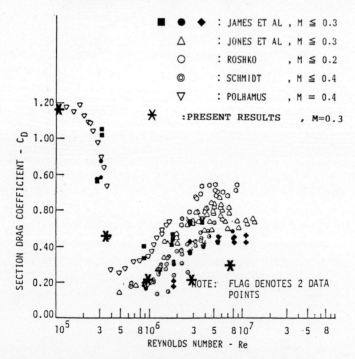

Fig.6 The drag coefficients of the present computation with experimental data.

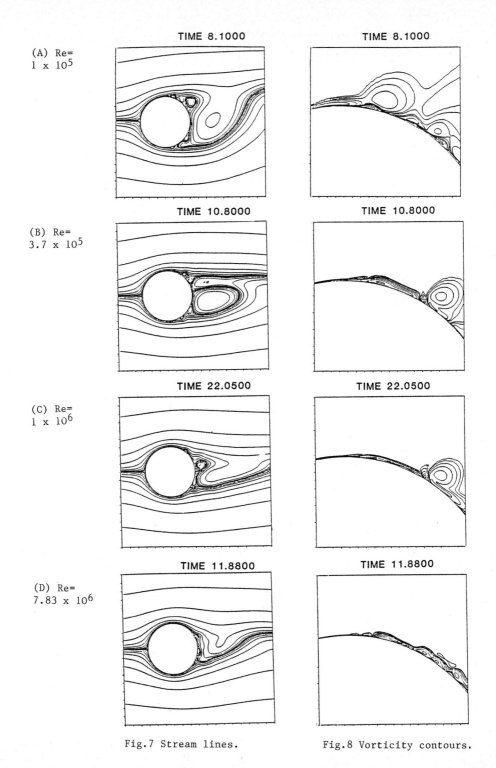

Fig.7 Stream lines.　　　Fig.8 Vorticity contours.

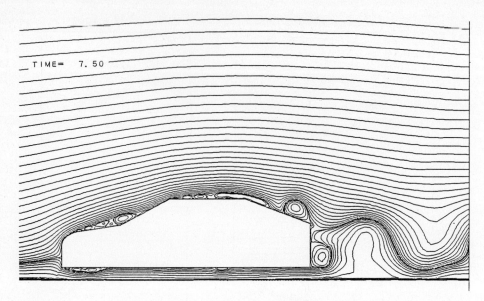

Fig.9 Stream lines at Re=10^5.

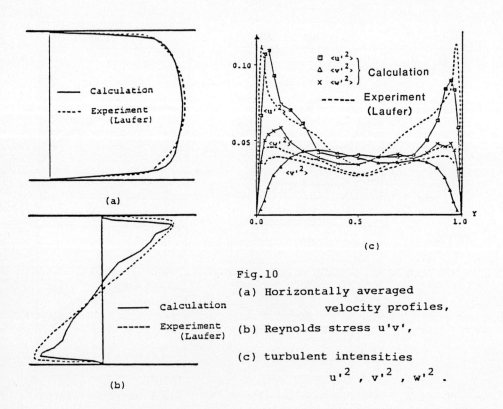

Fig.10
(a) Horizontally averaged velocity profiles,
(b) Reynolds stress u'v',
(c) turbulent intensities u'^2, v'^2, w'^2.

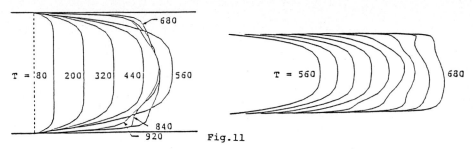

(a) Horizontally averaged velocity profiles.
(b) Velocity profiles showing the transition from laminar to turbulent flow.

Fig.11

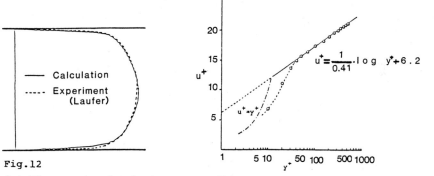

Fig.12
(a) Time averaged velocity profile after the transition.
(b) Mean velocity profiles near the wall.

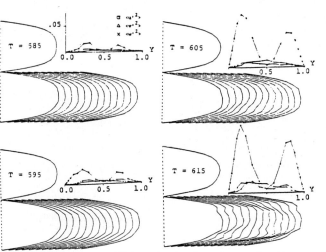

Fig.13 Mean velocity profiles, instantaneous velocity profiles and turbulent intensities at Re=5000.

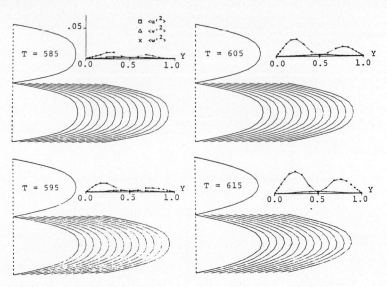

Fig.14 Mean velocity profiles, instantaneous velocity profiles and turbulent intensities at Re=2000.

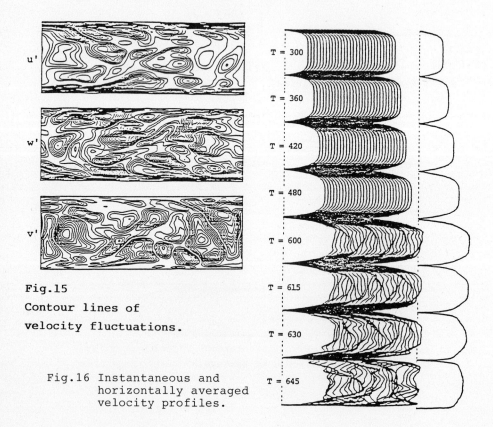

Fig.15 Contour lines of velocity fluctuations.

Fig.16 Instantaneous and horizontally averaged velocity profiles.

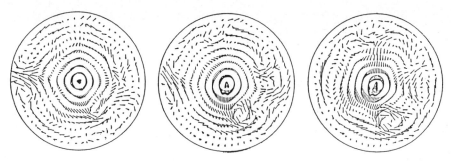

Fig.17(a) Instanteneous velocity vectors at some grid planes.

Fig.17(b) Averaged velocity vectors.

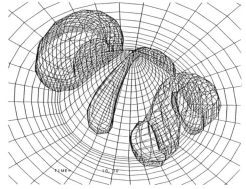

Fig.18 Equi-pressure surfaces.

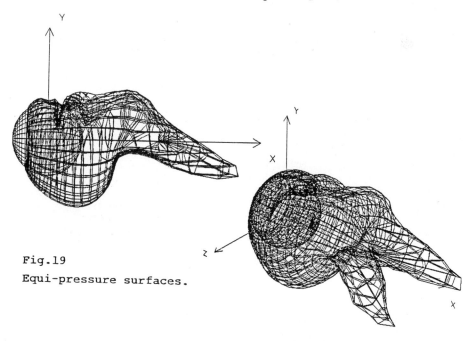

Fig.19 Equi-pressure surfaces.

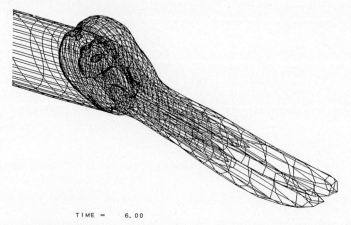

TIME = 6.00

Fig.20 Equi-pressure surfaces.

Fig.21 Equi-pressure surfaces around the rocket.

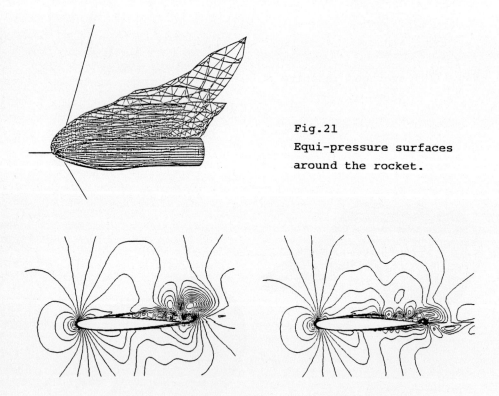

(a) Re=1.0x10^5 (b) Re=5.0x10^5

Fig.22 Density contours around NACA0012.

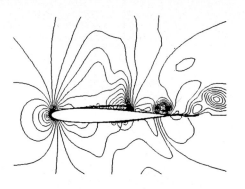

(c) Re=1.0x10^6

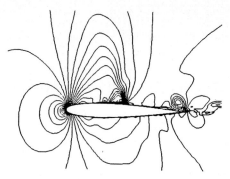

(d) Re=3.0x10^6

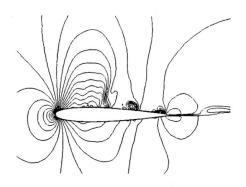

(e) Re=6.7x10^6

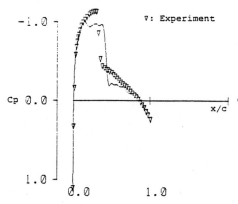

Fig.23(a)
Upper surface pressure coefficient distributions averaged in time: T=18-24 Re=6.7x10^6.

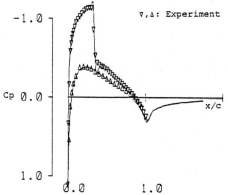

Fig.23(b)
Pressure coefficient distributions with turbulent model.

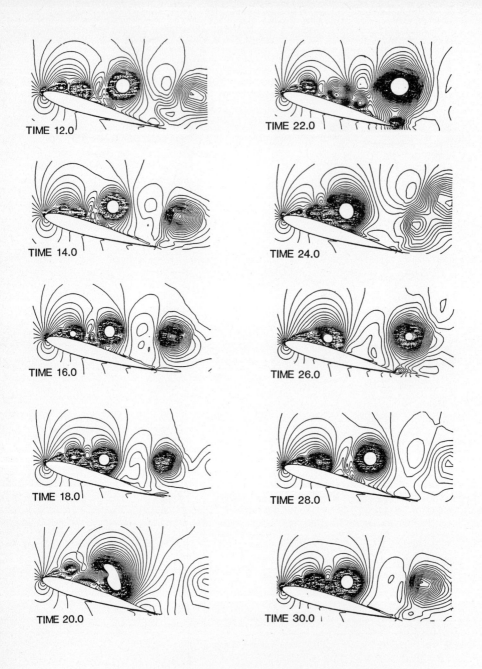

Fig.24 Time development of the density contours
No turbulence model. Reynolds number is 1.0×10^6.

DIRECT SIMULATION OF STABLY STRATIFIED TURBULENT HOMOGENEOUS SHEAR
FLOWS

U. Schumann, S.E. Elghobashi[*] and T. Gerz
DFVLR, Institute of Atmospheric Physics,
D-8031 Oberpfaffenhofen, Federal Republic of Germany
[*]University of California at Irvine, California 92717

SUMMARY

The paper presents results of a numerical solution of the exact time-dependent three-dimensional Navier-Stokes and temperature equations with moderate Reynolds numbers for a turbulent flow possessing uniform gradients of the mean velocity and temperature in a direction parallel to that of the gravitational vector. Comparison with earlier results for isotropic turbulence demonstrates the accuracy of the method. Shear and buoyancy effects on the transport of heat and momentum are examined and compared with available experimental data and those obtained via turbulence closure models.

INTRODUCTION

The objective of this study is to investigate the effects of stable stratification and shear on homogeneous turbulence. For this purpose we perform direct numerical simulations by solving the exact time-dependent three-dimensional Navier-Stokes equations in a cubical domain and in time. We assume constant viscosity ν and density ρ except for buoyancy due to temperature fluctuations (Boussinesq-approximation). The study is restricted to moderately large Reynolds numbers. The paper describes a new numerical method which allows the simulation of homogeneous turbulence under shear in an Eulerian framework and uses a combination of second-order finite difference and pseudo-spectral approximations.

The study is motivated by the need to understand and model the structure of atmospheric boundary layers. For the purpose of this study we disregard effects of spatial inhomogeneity. The interaction between shear and stratification is a typical feature of such layers. Shear forces and buoyancy due to stratification contribute to the generation or reduction of turbulence kinetic energy. In unstably stratified convective layers, heated from below, both forces amplify turbulence. In stably stratified layers (decreasing density with height) buoyant forces convert kinetic energy into potential energy and thus diminish turbulence while shear always tends to amplify the turbulent motion. Both forces anisotropically induce energy at scales where the

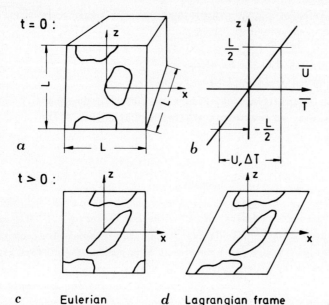

Figure 1. (a) A sketch of the cubical flow domain, (b) mean velocity and temperature profiles, (c) sketch of a flow-field in Eulerian frame after some time, (d) same flow field in Lagrangian frame. Here we use the Eulerian frame.

fluxes are large. Thus they produce anisotropic and scale-dependent structures. In strong stably-stratified layers turbulence can be converted to wave-motion with reduced turbulent diffusion and dissipation. Existing statistical turbulence models like the second-order turbulence closure-models proposed by Launder [6,7] have been shown to be well suited for shear flows and convective layers but inherently unable to account for the effects of coherent structures or the onset of waves. Thus our ultimate goal is to determine the structure of turbulent stratified shear flow and to provide data for calibration of second-order turbulence closure-models.

Direct simulations have been applied successfully in the past to study simple homogeneous turbulent flows, first without shear or stratification. Such flows can be represented in a cubical computational domain with periodic boundary conditions in all three space-directions. Since the invention of efficient Fourier-transform algorithms, Fourier-spectral approximations were the natural choice for such simulations. Examples are studies of isotropic [4, 8, 15] and axisymmetric turbulence [14, 16]. Homogeneous stratified turbulence has been numerically simulated by Riley, Metcalfe & Weissman [9]. Uniform stratification can be included easily in a spectral code by splitting the stratified scalar field into mean and fluctuating parts. The mean-part enters the equations by a constant gradient term only while the fluctuating part is still consistent with periodicity.

The inclusion of uniform shear in the simulations is non-trivial because of the boundary conditions. To be specific, we consider a mean velocity field $\mathbf{u}(x,y,z)$ with $\mathbf{u}=(\bar{U}(z),0,0)$, $d\bar{U}/dz=$const.

Under such shear an initially periodic flow field does not remain periodic in the shear direction because two fluid parcels with equal properties located initially above each other on the z-axis at a distance L initially are increasingly separated in the direction x with time t by the mean shear flow, see Figure 1. Rogallo [10] and Shirani et al. [17] used a time-dependent coordinate transformation to account for this behaviour. The transformation corresponds to a Lagrangian reference frame which distorts with the mean shear, see Figure 1. The Lagrangian frame allows the use of standard periodic boundary conditions and Fourier-spectral methods accordingly but requires remapping of the flow-field at a frequency of order $d\bar{U}/dz$. This remapping causes interpolation errors of the aliasing type [17].

Here we use "shear-periodic" boundary conditions with respect to the original Eulerian reference-frame [13] as has been proposed by Baron [1]. Baron applied his proposal for a shear flow without buoyancy. Shear-periodicity means that

$$f(x,y,z+L,t) = f(x-Ut,y,z,t) + (\partial \bar{f}/\partial z)L \tag{1}$$

together with standard periodicity in x- and y-direction. Here $\bar{f}$ is the ensemble average, L is the length of periodicity, and $U/L \equiv d\bar{U}/dz$. The term (x-Ut) is taken as modulo L.

In contrast to Baron, we split all fields into mean and fluctuating parts. The advection by the mean field is approximated pseudo-spectrally and only the derivatives of the fluctuating parts are discretized by finite differences of second order. Baron's implementation of the shear-periodic boundary condition was simplified by restricting the time-step Δt and the grid-spacing Δx to satisfy $\Delta t L(d\bar{U}/dz) = \Delta x$. This eliminates the need for any interpolation between grid points but may conflict with stability constraints on Δt. Therefore we generalize this approach by using a Fourier interpolation to determine $f(x-Ut)$ if the argument does not coincide with discrete grid points.

The shear-periodic boundary condition also influences the solution algorithm of the Poisson-equation for the pressure field. Baron used a time-consuming iterative scheme which we replace with a more efficient direct solver [11,13]. Finally we have set-up a rather general algorithm to define anisotropic initial values [12].

The present paper describes the integration scheme. Further we report on recent tests which show that the finite difference method gives results of comparable accuracy to those of the spectral method at least for low order statistics like the kinetic energy. For higher order statistics like the velocity-gradients skewness-coefficient, the finite difference solution requires double the resolution for the same accuracy at least for isotropic turbulence. Finally we present some results of the analysis of Komori's experiment [5] which extend those of Elghobashi, Gerz & Schumann [3].

MATHEMATICAL DESCRIPTION

Computational Domain

We consider a finite cubical domain inside the homogeneous shear flow with side-length L. The coordinates are x_1, x_2, x_3 or x, y, z and the velocity components u_1, u_2, u_3 or u, v, w, respectively. Both the mean temperature $\bar{T}$ and mean velocity in the x-direction $\bar{U}$ possess a uniform and constant gradient in the z-direction as shown in Figure 1.

Governing Equations

The turbulent flow under study is governed by the instantaneous Navier-Stokes equations in three dimensions. With the Boussinesq approximation the equation set reads:

$$u_{i,t} + (u_j u_i)_{,j} = -(p/\rho_o)_{,i} + \nu u_{i,jj} + \alpha(T-T_{ref})g\delta_{i3} \qquad (2)$$

$$T_{,t} + (u_j T)_{,j} = \gamma T_{,jj} \qquad (3)$$

$$u_{i,i} = 0, \qquad (4)$$

where p denotes the dynamic pressure, ρ_o the constant fluid density, ν the kinematic viscosity, γ the thermal diffusivity, and g the gravitational acceleration. The reference temperature T_{ref} at any horizontal plane, z=const., is set equal to the mean temperature $\bar{T}(z)$ so that buoyancy does not induce a mean gradient of the pressure. The isobaric volumetric expansion coefficient α is defined as

$$\alpha = -(1/\rho_o) \ (\partial\rho/\partial T)_{p=const.} \qquad (5)$$

The boundary conditions are of the shear-periodic type explained above with period L. Initial conditions prescribe a divergence-free velocity-field and the temperature-field at time t=0.

The Non-Dimensional Equations

The set (1-4) is normalized with respect to reference scales L, U, $\rho_o U^2$, ΔT for length, velocity, pressure, and temperature, respectively. Here $U \equiv Ld\bar{U}/dz$, $\Delta T \equiv Ld\bar{T}/dz$. The non-dimensional fields (u_1, u_2, u_3, T) are split into the ensemble mean values $(Sx_3, 0, 0, sx_3)$ and their deviations (u'_1, u'_2, u'_3, T'). Here

$$S = (L/U)d\bar{U}/dz = 1, \text{ or } 0; \qquad s = (L/\Delta T)d\bar{T}/dz = -1, 0, \text{ or } 1 \qquad (6)$$

for sheared and unsheared flow, and for unstable, neutral, and stable stratification, respectively. In the non-dimensional coordinates x_i, the computational domain covers a unit cube which we define for convenience as ($-0.5 \leq x_i < 0.5$, $i=1,2,3$). Subsequently, the fields denote non-dimensional fluctuating quantities and the primes are omitted. The non-dimensional equations read:

$$u_{i,t} + (u_j u_i)_{,j} + Sx_3 u_{i,1} + Su_3 \delta_{i1} = -p_{,i} + Re^{-1} u_{i,jj} + |Ri| T \delta_{i3} \qquad (7)$$

$$T_{,t} + (u_j T)_{,j} + Sx_3 T_{,1} + su_3 = (RePr)^{-1} T_{,jj} \qquad (8)$$

$$u_{i,i} = 0. \qquad (9)$$

The characteristic non-dimensional quantities are the Reynolds-number $Re = UL/\nu$, gradient Richardson-number $Ri = s\alpha g \Delta T L/U^2$ (positive for stable, negative for unstable stratification), and Prandtl-number $Pr = \nu/\gamma$.

For any non-dimensional field f, the boundary conditions require for arbitrary integers i,j,k:

$$f(x_1+i, x_2+j, x_3+k, t) = f(x_1-Skt, x_2, x_3, t). \qquad (10)$$

The Numerical Solution Method

The solution method uses a combination of second-order energy-conserving finite-difference approximations in time and space on a staggered grid and a pseudo-spectral treatment of mean advection. The number of grid cells N is 32 or 64 for the present computations, and $\Delta x = 1/N$. Typically, the time interval varies between $\Delta x/3$ and Δx. Diffusion, buoyancy and advection due to fluctuating velocities are treated explicitly by means of the Adams-Bashforth method. The Adams-Bashforth scheme is stable if the time step satisfies the Courant criterion. (In the code we check up the magnitude of the amplification eigenvalue λ_{max}; for small diffusion the method is weakly unstable but this poses no problem in practice since $(\lambda_{max}-1) \ll 1/N_t$ where

N_t is the number of time steps.) Inclusion of the mean advection in the finite difference formulae resulted in rather small time steps and in moderately large errors which show up most strongly in the velocity-derivative skewness results. Therefore, mean advection is treated by Fourier interpolation. Pressure is determined implicitly. The discrete explicit accelerations are

$$r_i = Re^{-1}\delta_j\delta_j u_i + |Ri|\,\overline{T}^3 \delta_{i3} - \delta_j(\overline{u_j}^i \overline{u_i}^j) - S\overline{u_3}^{-1}\delta_{i1}. \tag{11}$$

Here the usual discrete difference operators δ_j and averaging operators $\overline{}^j$ are used. From these accelerations the integration of velocity from time-step n to n+1 proceeds in three steps:

$$u^*_i = u_i^n + \Delta t[g_1 r_i^n - g_2 r_i^{n-1}], \tag{12}$$

with $g_1=1$, $g_2=0$ for $n=0$; $g_1=3/2$, $g_2=1/2$ for $n>0$,

$$\widetilde{u}_i(x_1) = u^*_i(x_1 - \Delta t S x_3) \tag{13}$$

by Fourier interpolation (see Appendix) and

$$u_i^{n+1} = \widetilde{u}_i - \Delta t \delta_i p^{n+1}. \tag{14}$$

The temperature is integrated likewise with $T^{n+1} \equiv \widetilde{T}$. The new pressure p^{n+1} is the solution of the Poisson equation in finite difference form

$$\delta_i \delta_i p^{n+1} = \delta_i \widetilde{u}_i / \Delta t. \tag{15}$$

Special attention is required for the highest Fourier mode with respect to the x-direction on a discrete grid (see Appendix) because only the cosine part of this mode is defined uniquely. Therefore, for this mode standard periodicity is applied in the z-direction. For the same reason, this mode is not affected by the mean advection.

Schumann [13] has developed a direct and efficient solution algorithm for the Poisson equation with shear-periodic boundary conditions. Originally it was described for $\Delta t = \Delta x$ but it can easily be generalized for any ratio $\Delta t/\Delta x$ by taking $\mu \equiv N\Delta t/\Delta x$, see [13], as a non-integer value. (In [11], eq. 45, N has to be taken as this non-integer value μ.)

Computations have been performed on a CRAY-1S. The CPU-time per time-step amounts to 3.1 seconds for N=64. The available storage for data (680000 words) does not suffice (for N=64) to keep all data required per time step. Therefore, the fields are segmented into slices and stored either in main memory or on disks. The input/output time is of the same order as the CPU time for N=32 but up to 14 times larger for N=64.

The code has been extensively tested by formal tests (e.g. mass and momentum conservation and comparison with analytical solutions). Other tests concerned comparison with the results of Shirani et al. [17] for a pure shear case and sensitivity studies of the results with respect to resolution (32^3 or 64^3 grid points). The results support the validity of our simulations.

RESULTS AND DISCUSSION

I. Comparison with spectral method for isotropic turbulence

The code has been first applied to reproduce the results of [15], case I1, for *isotropic* turbulence at a Reynolds number $Re_{\lambda o} \equiv \lambda_o v_o/\nu = 36.1$ based on the initial values of the Taylor-microscale λ_o and root-mean-square (r.m.s.) velocity v_o. The initial velocity field is generated from Gaussian random numbers with an energy spectrum of the form

$$E(k) = 16(2/\pi)^{1/2} v_o^2 k_p^{-5} k^4 \exp[-2(k/k_p)^2]. \qquad (16)$$

In the units of the present paper (which differ from those of [15] where the box size is $L=\pi$) we have $k_p = 3\pi$. The spectral simulations had been performed with a resolution of 32^3 points in real space. The present code has been applied for this case with N=32 and N=64.

Figure 2 shows that both the spectral method (SM) and the present finite difference method (FD) give about the same results even for N=32 with respect to the decay of energy. Better agreement is hardly to be expected because the random initial values though having the same spectra are not identical.

Higher order statistics are more sensitive than second order quantities are with respect to approximation errors. Figure 3 shows the time variation of the velocity-derivative skewness S_u defined here as:

$$S_u = -<\Sigma(\partial u_i/\partial x_i)^3/3> <\Sigma(\partial u_i/\partial x_i)^2/3>^{-3/2}. \qquad (17)$$

Here <> denotes a ensemble mean over all discrete points in space. Comparison of the first three curves listed in the figure legend shows that the finite-difference results for N=32 differ notably from the results of the spectral method but are in very good agreement for N=64. Thus the finite-difference code requires about double the resolution for the same accuracy as the spectral method - a result which was to be expected. But the loss of accuracy for equal resolution is small and outweighted by the code's simplicity and generality with respect to the boundary conditions.

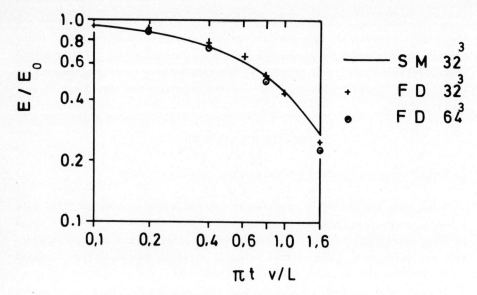

Figure 2. Mean kinetic energy $E(t)$ normalized with initial value $E(0)$ versus normalized time. Results for isotropic turbulence, case I1, as reported for the spectral method (SM with 32^3 grid points in real space) from Schumann & Patterson [15]. With FD the results of the present finite difference method are denoted for $N=32$ and $N=64$.

Figure 3 also shows results for the same physical situation but with a constant advection velocity $U = 2.36$ v superimposed (Galilean transformation). All other parameters including the time step are kept constant ($U\Delta t/\Delta x = 0.25$). The mean advection strongly reduces the accuracy and decreases the skewness. However if the mean velocity is treated with the Fourier-interpolation method corresponding to eq. (13), then the results are fairly insensitive to the additional advection. For simulation of turbulence with homogeneous shear it is crucial to use an integration scheme which is rather insensitive to mean advection. The results show that our code with Fourier-treatment of mean advection satisfies this condition.

II. Stratified turbulent shear-flows results

Here we compare the results of our direct simulation with the experimental data of Komori et al. [5] for a stably stratified turbulent shear flow. The stratification in that experiment was achieved by imposing a positive vertical temperature gradient in the outer layer of an open channel flow. The authors indicate that their stratified outer layer was free of wall

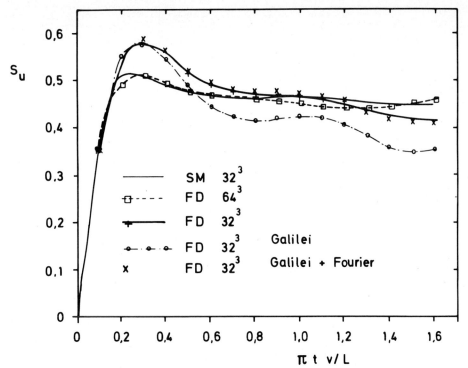

Figure 3. Velocity derivative skewness S_u versus time for isotropic turbulence, case I1. The first three curves in the legend denote the results without mean advection for the spectral and finite difference method with different resolution. The last two curves show the results with superimposed mean advection without and with Fourier-treatment in the present code.

effects and can be presumed close to a nominally homogeneous free shear flow. Because of the difference in the shapes of the vertical profiles of the temperature (T) and the horizontal velocity (u) it was possible to obtain turbulence measurements for a wide range of values of the local gradient Richardson number (Ri = 0. to 1.). The Reynolds number $4 R U_{av}/\nu$ varied from 9100 to 17000, where R is the hydraulic radius of the channel; U_{av} is the cross-sectional mean velocity. We performed computations for several cases whose initial conditions were identical except for the value of Ri (0. to 1.).

Because of the lack of information reported about the experimental values of the initial (t=0) turbulence correlations or energy spectra we had to assume initially isotropic turbulence with the following form of the three-dimensional energy spectrum E(k,t):

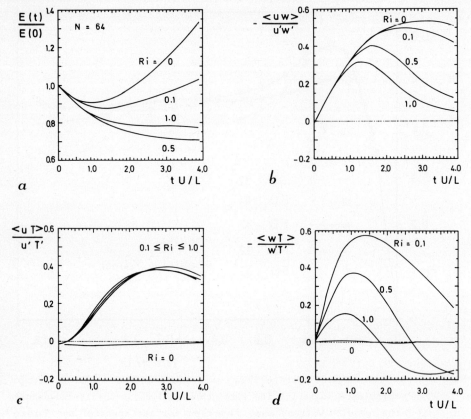

Figure 4. Time variation of the turbulent kinetic energy E(t) normalized by its inital value E(0), and the mean one-point correlations $-\langle uw\rangle/u'w'$, $\langle uT\rangle/u'T'$, and $-\langle wT\rangle/w'T'$ normalized by the actual root-mean-square values for different Richardson numbers Ri and for N=64.

$$E(k,0) = v_o^2 (k/k_p^2) \exp(-k/k_p), \qquad (18)$$

where k is the wave-number, k_p is the wave-number of maximum energy, and v_o is the dimensionless initial r.m.s. velocity fluctuation. This spectrum is appropriate for the initial period of decay [2, p.155]. The initial temperature field is generated with the same spectrum and a r.m.s. value $T'=v_0$. The following table contains the initial values employed in all the runs:

v_o	k_p	$Re_{\lambda o}$	Re	Pr
0.016	6	24.7	58050.	5.

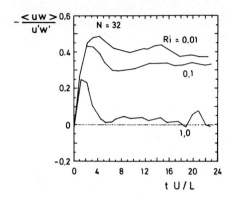 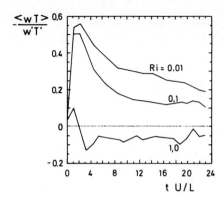

Figure 5. Time variation of mean one-point correlations $-\langle uw\rangle/u'w'$, and $-\langle wT\rangle/w'T'$ normalized by the actual root-mean-square values for different Richardson numbers Ri and for N=32.

The initial dimensionless microscale is λ_o=0.026 and the integral length scale is ℓ_o=0.049. The resolution is N=64, $\Delta t = \Delta x$, and the integration extended over 250 time steps.

Figure 4 and Figure 5 show the dynamic behaviour of the solutions in terms of second-order one-point correlations. The evolution of kinetic energy, Figure 4a, shows that the simulations do not come to a steady state. Although we cannot expect ever to get a truly steady state because the integral length scale grows indefinitely for homogeneous shear flow, we would consider a situation "steady" when the changes in kinetic energy are small during a time period of order L/U. For Ri=0 (neutral stratification) steady state in this sense would require a balance between shear production $-\langle uw\rangle d\overline{U}/dz \sim c_p v^2 d\overline{U}/dz$ and dissipation rate $\varepsilon \sim c_\varepsilon v^3/\ell$ where c_p and c_ε are approximately constants of order unity. Thus stationarity for Ri=0 requires that the ratio (c_p/c_ε)Sh of both rates becomes unity where

$$\text{Sh} \equiv (d\overline{U}/dz)\ell/v \qquad (19)$$

is the shear number. In our simulation, Sh=3.01 initially and even increases slowly with time. Obviously the shear number is too large for steady state.

Although we do not reach steady state in absolute values, we do find much earlier a quasi-steady state in which the correlations normalized with corresponding r.m.s.-values become virtually constant. This is seen from Figure 4b-d and more obviously from Figure 5 which has been obtained with N=32 for much longer times.

Figure 4a also shows the effect of buoyancy. For increasing Ri-number, weakly stable stratification converts kinetic energy into potential energy. For strong stability (Ri=1), however, this trend inverts. In order to understand this change in trend we have to examine the dynamic behaviour of momentum flux <uw> and heat fluxes <uT> and <wT> as shown in Figure 4b-d. The time evolution of these fluxes can be explained by the dynamic equations given in Table 1, even if we consider the linear parts only.

Table 1. Non-dimensional equations for the Reynolds-stresses, heat fluxes and the temperature variance

$$Ri = \alpha g L \Delta T / U^2$$

$$\partial <u^2>/\partial t = -2S<uw> + \Phi_{11} - \varepsilon_{11}$$
$$\partial <v^2>/\partial t = \Phi_{22} - \varepsilon_{22}$$
$$\partial <w^2>/\partial t = 2|Ri|<wT> + \Phi_{33} - \varepsilon_{33}$$
$$\partial <uw>/\partial t = -S<w^2> + |Ri|<uT> + \Phi_{13} - \varepsilon_{13}$$
$$\partial <uT>/\partial t = -s<uw> - S<wT> + \Phi_{1T} - \varepsilon_{1T}$$
$$\partial <wT>/\partial t = -s<w^2> + |Ri|<T^2> + \Phi_{3T} - \varepsilon_{3T}$$
$$\partial <T^2>/\partial t = -2s<wT> - \varepsilon_{TT}$$

Notation:
Φ_{ij} = redistribution of $<u_i u_j>$ by pressure fluctuations
Φ_{jT} = redistribution of $<u_j T>$ by pressure fluctuations
ε_{ij} = dissipation rate of $<u_i u_j>$
ε_{jT} = dissipation rate of $<u_j T>$
ε_{TT} = dissipation rate of $<T^2>$
$S = (L/U) d\bar{U}/dz$ (non-dimensional shear)
$s = (L/\Delta T) d\bar{T}/dz$ (non-dimensional stratification)

For example, <uw> is excited negatively by the interaction of shear with <w²> but positively under stable stratification for <uT>Ri > 0. In fact, positive values of <uT> are to be expected from the linear part of the equations in Table 1 and found in the simulations, see Figure 4c, for Ri>0. Thus for large Ri-values the negative <uw>-values return to zero. The heat-flux <wT> tends to be negative initially due to -<w²>s but gets also a positive contribution from Ri<T²>.

The temperature variance <T²> is excited positively independently of Ri if the heat-flux follows the gradient-assumption so that this quantity grows faster than the correlations with velocity components (in particular for large Prandtl-number where the dissipation of temperature variance ε_{TT} is rather small). Thus it is reasonable that for large Ri (and not too small

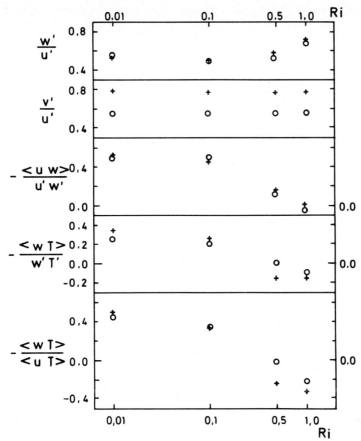

Figure 6. Turbulence quantities versus local gradient Richardson number, comparison of simulated results (+) for N=64, tU/L=3.516, with measurements (o) of Komori et al. [5].

Prandtl number) the heat-flux after some time changes sign and a counter-gradient heat-flux arises. This is what is shown in Figure 4d and Figure 5b: a counter-gradient heat-flux appears for strongly stable stratification. The sign-change comes earlier for Ri=1 than for Ri=0.5 as to be expected. Figure 5b shows that this sign change is of non-oscillatory nature. Thus at large Ri-numbers, the heat-flux no longer damps the turbulent motion rather than supplies additional kinetic energy from the reservoir of potential energy.

Next we compare our results with the experimental results of Komori et al. [5]. Because of the uncertainty with respect to the exact initial conditions in the experiment we cannot specify the required time of integration for getting results which are directly comparable with the measurements at the single station in the channel. The only information reported is that this station was located 75δ from the heating station, where δ is the flow depth; and that there the stably-stratified flow was fully developed. However, our experiences show that these uncertainties do not matter much if we compare normalized quantities after having reached quasi-steady state. Figure 6 shows this comparison between experimental and computed results based on the the simulations with N=64 at t=3.516. In Elghobashi et al. [3] we have shown a similar comparison from the results with N=32 and t=11.46. The difference between the two simulations is surprisingly small and the agreement with the experimental data is generally very good. The largest differences appear for $v' \equiv <v^2>^{1/2}$, the cross-stream fluctuations. Although speculative, it might be that this difference is due to contributions from scales too large to be covered in our finite computational domain. However, the comparison shows that these omitted fluctuations do not contribute largely to the fluxes. Both measurements and simulations show a strong decay of momentum flux with Ri-number and a weak counter-gradient heat-flux for Ri=1. Also the increase in w'/u' discussed in [3] for Ri≥0.5 is found in both cases. Thus the comparison shows that the direct simulation gives results which represent observable reality.

Table 2: Comparison between measured [5] and predicted values (from simulations with N=32 and N=64) of the anisotropy tensor components b_{ij}

Ri	Rf		b_{11}	b_{33}	b_{13}
0.01	0.0166	exp.:	0.192	-0.131	-0.163
for	N=32,	pred.:	0.187	-0.138	-0.108
for	N=64,	pred.:	0.189	-0.181	-0.147
0.1	0.124	exp.:	0.213	-0.162	-0.153
for	N=32,	pred.:	0.205	-0.162	-0.093
for	N=64,	pred.:	0.207	-0.199	-0.122
1.0	0.434	exp.:	0.158	-0.079	-0.017
for	N=32,	pred.:	0.147	-0.091	-0.021
for	N=64,	pred.:	0.139	-0.095	-0.023

The dependence of the components of the Reynolds stress anisotropy tensor b_{ij},

$$b_{ij} = \langle u_i u_j \rangle / \langle u_k u_k \rangle - \delta_{ij}/3 \qquad (20)$$

on the local flux Richardson number R_f is of particular importance in turbulence modelling, see Table 2. The flux Richardson number and the turbulent Prandtl number σ_t are defined by:

$$\sigma_t \equiv Ri/R_f \equiv \langle uw \rangle (dT/dz) / [\langle wT \rangle dU/dz]. \qquad (21)$$

It is seen from Table 2 that in a weakly stable flow the magnitudes of b_{11} and b_{33} increase with R_f where $b_{22} = -b_{11} - b_{33}$ is rather insensitive. Also the magnitude of b_{11} is always greater than b_{33}. For the case of $R_f = 0.434$ (Ri=1) we see the reverse trend for both b_{11} and b_{33}. This is significant since it disagrees with Launder's model [6,7] which predicts monotonic increase of magnitudes of b_{11} and b_{33} with R_f. (Although Launder probably did not intend to propose usage of his model for such strong stratifications and he had no experimental data to compare with for such situations.) Table 2 shows a comparison between our results for b_{ij} and those deduced from the measurements [5]. Our preliminary investigation of the causes of the monotonic increase of b_{11} and $-b_{33}$ by Launder's model indicates that although it predicts the correct trend of pressure-strain and pressure-temperature-gradient correlations (Φ_{ij}, Φ_{iT}) with increasing Ri it does overestimate their magnitudes significantly especially for Ri=1 where our predictions show nearly vanishing magnitudes except for Φ_{3T}.

The turbulent Prandtl number can easily be computed from Eq.(21) and Table 2. The results compare very well with Webster's [18] data. They do also agree well with Launder's [6,7] predictions even for Ri=1, however this agreement is incidental because the large deviations in the fluxes just cancel in the ratio.

CONCLUDING REMARKS

The paper has presented the results of a finite-difference method for direct simulation of turbulent homogeneous free shear flows including buoyancy effects for stable stratification.

The novel feature of this work is the application of "shear-periodic" boundary conditions to the computational domain thus avoiding the use of a coordinate transformation and the necessary remapping during time-integration. The method allows for adjustment of the time step Δt according to stability and accuracy requirements without the restriction $U\Delta t/\Delta x=1$ as in Baron [1]. Treatment of mean advection by Fourier interpolation has been found to be crucial for accuracy. Otherwise a second-order finite difference scheme gives sufficient accuracy.

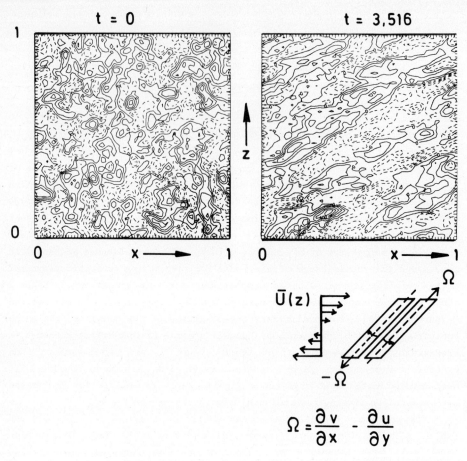

Figure 7. Creation of coherent structures under shear for Ri=0.1. The contour-line plots show the v-velocity in a z-x-plane at times t=0 and tU/L=3.516 for N=64. The initially isotropic field shows a dominant elongation and coherent structure in the direction of the diagonal. The result is consistent with the conceptional picture showing vortices which are bent by the shear.

Direct simulations are necessarily restricted to low Reynolds-numbers (here Re_λ=25 for N=64). Larger Reynolds-numbers can be achieved by using a subgrid-scale model. However even more important seems to be the restriction in the ratio of largest to smallest scales as measured by the integral length scale ℓ and the Kolmogorov-scale η. In our simulations for N=64, we have ℓ/η = 19 at t=0. Resolution of a wider scale spectrum can be achieved only by increasing N. The general agreement between our results for N=32 and N=64 however does show that the relative influences of shear and buoyancy can be investigated already with the present resolution.

In particular, we have shown that a "quasi-steady" state is reached quickly where the structure of the flow as measured in terms of normalized second-order correlations is represented by the simulations in good agreement with the experimental observations [5].

The simulations and the experiments show the development of a counter-gradient heat flux for $Ri \geq 0.5$. This phenomenon can be explained qualitatively already with linear models. Further, it has been shown that the magnitudes of the components b_{11} and b_{33} of the anisotropy tensor increase with Ri in the weak stability region and then decrease for higher values of R_f. Our results suggest that this is due to strong reduction in pressure-strain terms for large Ri. This finding shows limitations in Launder's [6,7] closure model which predicts a monotonic increase of b_{11} and $-b_{33}$ with R_f.

Finally it should be noted that much more information can be deduced from our direct simulations. As an example, we present in Figure 7 a contour-line plot of the velocity-component v at the initial and at a later time. It is obvious that the shear produces a structure at 45° inclination. The structure is consistent with excited vortices which have been bent by the shear in this direction. Analysis and interpretation of such structures are the subject of on-going work.

Acknowledgement. This work is supported by the Deutsche Forschungsgemeinschaft.

APPENDIX: DISCRETE FOURIER INTERPOLATION

For the Fourier-advection part, see eq. (13), and for implementation of the shear-periodic boundary condition for arbitrary Δt, see eq. (10), we need an algorithm to evaluate

$b(x) = a(x-d)$ for arbitrary d and real periodic data $a(x)$, $b(x)$.

In the discrete form we have the periodic sequences $a_j \equiv a(x_j)$ and $b_j \equiv b(x_j)$ with $j = 0, 1, 2, \ldots, N-1$, $x_j = j\Delta x$, $\Delta x = 1/N$. The discrete Fourier representation of a_j is

$$a_j = \Sigma \hat{a}_k e^{i2\pi jk/N} \quad \text{(sum from 0 to N-1)}$$

which can be interpreted as

$$a(x_j) = \Sigma \hat{a}_k e^{i2\pi(x_j)k}$$

so that

$$a(x_j-d) = \Sigma \, \hat{a}_k \, e^{i2\pi(x_j-d)k}$$

$$= \Sigma \, \hat{a}_k \, e^{-i2\pi(d/\Delta x)k/N} \, e^{i2\pi jk/N}$$

Thus the Fourier-modes of $b_j = \Sigma \, \hat{b}_k \, e^{i2\pi jk/N}$ are

(22) $\hat{b}_k = \hat{a}_k \, e^{-i2\pi(d/\Delta x)k/N}$, $k = 0,1,2,\ldots,N-1$.

We require a_j and b_j to be real. This means $\hat{a}_k$ forms a conjugate complex series ($\hat{a}_{N-k} = \hat{a}_k^*$) which implies that $\hat{a}_{N/2}$ has to be real, and correspondingly for $\hat{b}_k$. However, eq. (22) gives a real relationship only if $d/\Delta x$ is an integer value. Thus in the code, we use eq.(22) for integer values of $d/\Delta x$. Otherwise we use eq.(22) for k=0,1,2,...,(N/2-1), but set $\hat{b}_{N/2} = \hat{a}_{N/2}$, and set $\hat{b}_{N-k} = \hat{a}_k^*$ for k=(N/2+1),...,N-1. (The last relationship is used implicitely in the FFT algorithm.) This method is energy conserving and gives a good approximation if the magnitude of the highest independent Fourier-mode (k=N/2) is sufficiently small in comparison to the other modes.

REFERENCES

[1] BARON, F.: "Macro-simulation tridimensionelle d'ecoulements turbulents cisailles." These de Docteur-Ingenieur, Universite Pierre et Marie Curie, Paris 6 (1982).

[2] BATCHELOR, G.K.: "The Theory of Homogeneous Turbulence", Cambridge Univ. press (1953).

[3] ELGHOBASHI, S.E., GERZ, T. & SCHUMANN, U.: "Direct simulation of turbulent homogeneous shear flow with buoyancy." Proc. 5th Symp. on Turbulent Shear Flows, Aug. 7-9, 1985, Cornell Univ., Ithaca, New York, pp. 22.7-22.12 (1985).

[4] KERR, R.M.: "Higher-order derivative correlations and the alignment of small-scale structures in isotropic numerical turbulence." J. Fluid Mech. 153, 31-58 (1985).

[5] KOMORI, S., UEDA, H., OGINO, F. & MIZUSHINA, T.: "Turbulence structure in stably stratified open-channel flow." J. Fluid Mech. 130, 13-26 (1983).

[6] LAUNDER, B.E.: "On the effects of a gravitational field on the turbulent transport of heat and momentum." J. Fluid Mech., 67, 569-581 (1975).

[7] LAUNDER, B.E.: "Heat and mass transport." In "Topics in Applied Physics", Ed. Bradshaw, Vol. 12 (1976).

[8] ORSZAG, S.A. & G.S. PATTERSON, Jr.: "Numerical simulation of turbulence." In: M.Rosenblatt & C.Van Atta (eds.): "Statistical Models and Turbulence", Lecture Notes in Physics, Springer-V., Berlin, pp. 127-147 (1972).

[9] RILEY, J.J., METCALFE, R.W. & WEISSMAN, M.A.: "Direct numerical simulation of homogeneous turbulence in density-stratified fluids." In: B.J.West (ed.): "Nonlinear Properties of Internal Waves", AIP Conf. Proc. No. 76, American Institute of Physics, New York, pp. 79-112 (1981).

[10] ROGALLO, R.S.: "An ILLIAC Program for the Numerical simulation of homogeneous incompressible turbulence." NASA TM-73, 203 (1977).

[11] SCHMIDT, H., SCHUMANN, U., VOLKERT, H. & ULRICH, W.: "Three-dimensional, direct and vectorized elliptic solvers for various boundary conditions." DFVLR-Mitt. 84-15 (1984).

[12] SCHUMANN, U.: "Generation of random periodic velocity and temperature fields with prescribed correlation spectra." DFVLR Report No. IB 553/6/84 (1984).

[13] SCHUMANN, U.: "Algorithms for direct numerical simulation of shear-periodic turbulence." Proc. Ninth Intern. Conf. on Numerical Meth. in Fluid Dyn. (Soubbaramayer & J.P.Boujot, eds.), Lect. Notes in Phys. 218, Springer-V., Berlin, pp. 492-496 (1985).

[14] SCHUMANN, U. & HERRING, J.R.: "Axisymmetric homogeneous turbulence: a comparison of direct spectral simulations with the direct-interaction approximation." J. Fluid Mech. 76, 755-782 (1976).

[15] SCHUMANN, U. & G.S. PATTERSON, Jr.: "Numerical study of pressure and velocity fluctuations in nearly isotropic turbulence." J. Fluid Mech. 88, 685-709 (1978).

[16] SCHUMANN, U. & G.S. PATTERSON, Jr.: "Numerical study of the return of axisymmetric turbulence to isotropy." J. Fluid Mech. 88, 711-735 (1978).

[17] SHIRANI, E., FERZIGER, J.H. & REYNOLDS, W.C.: "Mixing of a passive scalar in Isotropic and sheared homogeneous turbulence." Rep. No. TF. 15, Mech. Eng. Dept., Stanford University (1981).

[18] WEBSTER, C.A.G.: "An experimental study of turbulence in a density stratified shear flow." J. Fluid Mech. 19, 221 (1964).

THE EFFECT OF COHERENT MODES ON THE EVOLUTION OF A TURBULENT MIXING LAYER

Ralph W. Metcalfe, Suresh Menon, and James J. Riley*
Flow Research Company
21414-68th Avenue South
Kent, Washington 98032 USA

We have performed direct numerical simulations of a temporally growing mixing layer, focusing on the interaction between the various unstable modes during transition to turbulence. Even at very low levels of initial excitation, the two-dimensional modes play a critical role in the subsequent evolution of the flow. There is a very strong secondary instability present in the flow that is characterized by the growth of counterrotating, streamwise vortices in the braids between the large spanwise vortices. The growth rate of the three-dimensional, secondary instabilities is a function of the amplitude of the spanwise, two-dimensional modes. It appears that the degree and nature of mixing in a reacting mixing layer can be significantly modified by appropriate low-amplitude forcing.

We have computed solutions to the Navier-Stokes equations in three spatial dimensions plus time. Using pseudospectral numerical methods, we have simulated the evolution of a temporally growing mixing layer. Periodic boundary conditions have been employed in the streamwise and spanwise directions and free-slip conditions transverse to the flow direction. A hyperbolic tangent mean velocity profile consistent with experimental data is used to initialize the flow field. A perturbation velocity field derived from the most unstable mode of the linear Orr-Sommerfeld equation and its subharmonics is used. To this is added a low-amplitude, pseudorandom velocity field with a broad-band spectrum. This initialization has been described in more detail elsewhere [1, 2].

Some of the most important features of the modal interaction in the early stages of evolution of this flow are present in the following two simulations, which are representative of other realizations we have performed. In the first simulation, Run A, the most unstable two-dimensional mode and its subharmonic were present initially, as was a spanwise perturbation with the same streamwise wavelength as the most unstable mode and a spanwise wavelength with this same value. There was no random component in this run. The energy histories of the Fourier components $E_{i,j}$ corresponding to these modes are shown in Figure 1. Here, i denotes the streamwise wavenumber and j the spanwise wavenumber, with i = 1 corresponding to the wavenumber of the most unstable two-dimensional mode. Thus, quantities of the form $E_{i,0}$ represent the energy in purely two-dimensional modes with no spanwise-varying component. The E_{3D} term is the sum of the energy in all modes with $j \neq 0$. The initial Reynolds number, based on the mean vorticity thickness and mean velocity difference, is 400. The mean vorticity thickness is defined as the maximum mean vorticity divided by the mean velocity difference. The early stages of this flow are dominated by the most unstable two-dimensional mode, $E_{1,0}$. The nonlinear attenuation of its growth rate (saturation) occurs by about t = 10, where t is nondimensionalized by the mean velocity divided by the mean vorticity thickness. The subharmonic, which starts at a much lower amplitude initially, does not roll up until about t = 40.

---

*Department of Mechanical Engineering, University of Washington, Seattle, Washington 98195 USA

One of the most significant aspects of the interaction between the primary two-dimensional modes ($E_{1,0}$; $E_{1/2,0}$) and the secondary three-dimensional instabilities (E_{3D}) is the stabilization of the three-dimensional modes by the rollup and pairing of the two-dimensional modes. This can be seen in Figure 1 in the attenuation of E_{3D} between $t = 10$ and 20 during the saturation of the fundamental and between $t = 35$ and 50 during the saturation of the subharmonic. The nature of the three-dimensional instability is shown in Figure 2, which is a three-dimensional perspective plot of surfaces having a value of 50% of the peak of the sum of the absolute values of all three vorticity components. The mean flow direction is in the positive x direction for large z and in the negative x direction for small z. The large vortical structure is the subharmonic two-dimensional mode ($E_{1/2,0}$), which has strong spanwise (y) coherence. The secondary, three-dimensional instability (E_{3D}) appears mainly in the form of streamwise vortical structures on the braid between the two-dimensional vortices. A more detailed analysis shows that these are counterrotating vortex pairs. Structures very similar to these have been seen in laboratory experiments [1, 3].

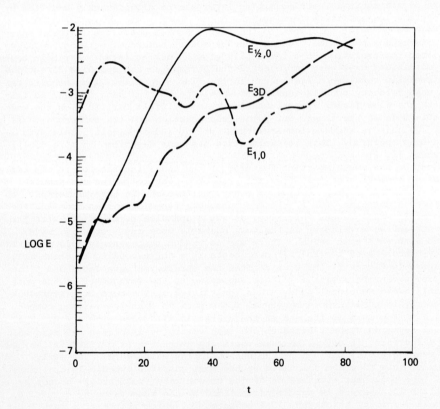

Fig. 1. A plot of the temporal evolution of the modal energies for Run A. $E_{1,0}$ corresponds to the most unstable two-dimensional mode, $E_{1/2,0}$ to its subharmonic, and E_{3D} to the sum of the energies in all modes with spanwise wavenumber greater than zero.

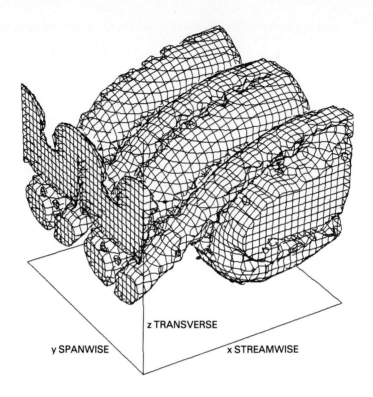

Fig. 2. A three-dimensional perspective plot of surfaces having a value equal to 50% of the peak of the sum of the absolute values of all three vorticity components for Run A at t = 48.

In the absence of the primary two-dimensional unstable modes, the flow can evolve much differently. In the second simulation, Run B, the initial perturbation field was an uncorrelated, random-phase velocity field with a broad-band three-dimensional energy spectrum. No coherent two- or three-dimensional modes were explicitly added to the velocity field. This field was convolved with a physical space function so that the relative turbulence intensity levels were consistent with those of experimental mixing layer data. However, the initial peak intensity level was about 3 orders of magnitude below the experimental values. Thus, the initial disturbance growth was in the linear regime.

Figure 3 shows the evolution of the modal energy for this initial condition. The $E_{1,0}$ term, which in this case does not explicitly contain the most unstable two-dimensional mode, initially grows at a slower rate than E_{3D} and does not appear to play a dynamically significant role in the flow evolution. However, in the absence of the two-dimensional vortex rollup process, the growth rate of E_{3D} is sharply reduced once the three dimensional modes become nonlinear, by about t = 30. This attenuation in the growth of E_{3D} permits the subharmonic $E_{1/2,0}$ to overtake these modes and become the dominant mode in the flow as it saturates by t = 80. The qualitative difference in the mixing layer evolution with and without the primary two-dimensional modes can

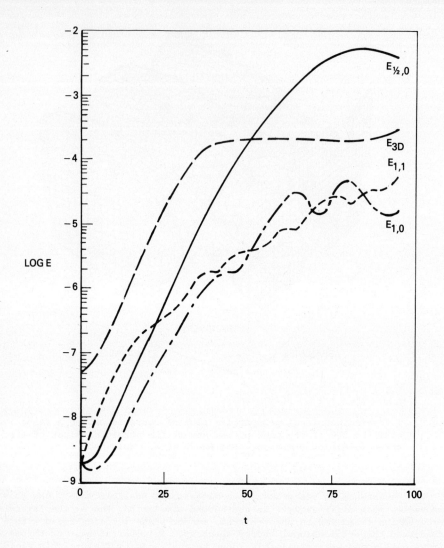

Fig. 3. A plot of the temporal evolution of the modal energies for Run B. $E_{1,0}$ corresponds to the most unstable two-dimensional mode, $E_{1/2,0}$ to its subharmonic, and E_{3D} to the sum of the energies in all modes with spanwise wavenumber greater than zero.

be seen by comparing the vorticity plots at t = 48 (Figure 4) with those at t = 96 (Figure 5). As the subharmonic saturates, the character of the flow changes dramatically, from a highly chaotic random state to one dominated by the presence of large-scale spanwise and streamwise vortical structures.

A more detailed analysis of these results [1] shows that, when the amplitudes of the three-dimensional modes are very small, mainly the high-wavenumber spanwise modes are damped by the vortex rollup and the lowest spanwise modes continue to grow. On the other hand, the two-dimensional vortex rollup and pairing appears to be the key mechanism for maintaining

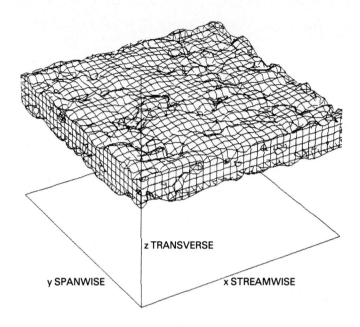

Fig. 4. A three-dimensional perspective plot of surfaces having a value equal to 50% of the peak of the sum of the absolute values of all three vorticity components for Run B at t = 48.

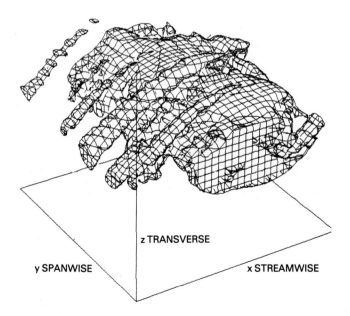

Fig. 5. A three-dimensional perspective plot of surfaces having a value equal to 50% of the peak of the sum of the absolute values of all three vorticity components for Run B at t = 96.

the observed rapid growth rates of mixing layer thickness. When this process is disrupted by the absence of two-dimensional modes, the rapid thickening of the mixing layer cannot be maintained. Thus, the growth of E_{3D} can be attenuated at low amplitudes by the rollup of the two-dimensional vortices (Figure 1) but also at high amplitudes by the absence of two-dimensional vortices (Figure 3).

The suppression of the low-amplitude three-dimensional secondary instabilities by primary vortex pairing may contribute significantly to the strong spanwise coherence that has been observed in laboratory experiments for flow near the splitter plate. At larger amplitudes, the low spanwise wavenumber modes grow into counterrotating streamwise vortices, while the higher spanwise modes decay. In a chemically reacting mixing layer [4] it is possible that the reaction rate could be increased by forcing the mixing layer to saturation (i.e., suppress pairing) immediately downstream of the splitter plate. This would enhance the growth of the three-dimensional modes (cf. Figure 3 for t < 30) and lead to more vigorous mixing of the reactants.

REFERENCES

[1] METCALFE, R. W., ORSZAG, S. A., BRACHET, M. E., MENON, S., and RILEY, J. J.: "Secondary instability of a temporally growing mixing layer," To be published in J. Fluid Mech. (1986).

[2] RILEY, J. J., and METCALFE, R. W.: "Direct numerical simulation of a perturbed, turbulent mixing layer," AIAA paper 80-0274 (1980).

[3] BERNAL, L. P.: "The coherent structure of turbulent mixing layers," Ph.D. thesis, Calif. Inst. Technology (1981).

[4] RILEY, J. J., METCALFE, R. W., and ORSZAG, S. A.: "Direct numerical simulations of chemically reacting turbulent mixing layers," Phys. Fluids, 29(2) (1986) pp. 406-422.

TURBULENT FLOW NEAR DENSITY INVERSION LAYERS

by D.J. Carruthers[+], J.C.R. Hunt and C.J. Turfus

Department of Applied Mathematics and Theoretical Physics,
University of Cambridge, Silver Street, Cambridge CB3 9EW.

[+]Department of Atmospheric Physics,
Clarendon Laboratory, Parks Road, Oxford OX1 3PU.

ABSTRACT

When a region of well-mixed turbulence adjoins stably stratified layers, the turbulence is distorted by the eddies impacting on the stable layer and the stable layer is perturbed by waves induced by the turbulence. These processes affect entrainment, the wave energy flux from the inversion and diffusion of scalars near the interface. A new theory for this interaction is outlined briefly here and is compared with the results of large eddy simulations of Deardorff & Moeng, and field measurements. The theory shows that L.E.S. cannot generally resolve some of the significant wave motions that occur near the interface. Nevertheless the agreement is reasonably satisfactory. Particle trajectories near interfaces have been computed by means of a two-dimensional simulation using random Fourier modes. It is shown that previous estimates of Lagrangian time scales for vertical velocity fluctuations are generally too small, which has implications for transfer processes near interfaces.

I. INTRODUCTION

There are many flows both natural and man-made in which there is an interface between regions of turbulence and layers of stably-stratified fluid. It is our aim in this paper to study the interaction between these regions by using large-eddy simulations, laboratory and field observations, and a theoretical model. This provides a nice illustration of how the large-eddy simulations can be used effectively in conjunction with other forms of investigation.

By way of introduction we show the most common naturally-occurring interfaces in Figure 1. This shows a typical ocean-atmosphere system. The lowest interface divides the turbulent Benthic (deep-ocean) boundary layer from the large body of the ocean which is stably stratified and is essentially non-turbulent (Richards 1985). This is capped in turn by an interfacial region, the thermocline (Turner 1973), above

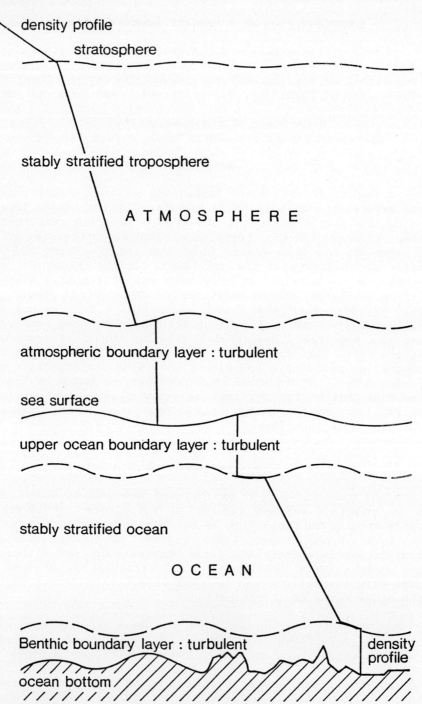

Fig.1. Interfaces between turbulent and stably stratified layers in the ocean and atmosphere.

which is the upper ocean boundary layer where turbulence is produced by the surface shear stress and/or the surface buoyancy flux. The atmospheric boundary layer above is bounded by an interface (atmospheric inversion layer) which is bounded by the mainly stable non-turbulent troposphere. Not shown in the figure are the regions of intermittent turbulence which may occur in the body of the ocean, in the troposphere and in the stratosphere. Over land the atmospheric inversion is also a very common and persistent phenomenon. Man-made flows of similar structure occur for example in the dispersion of dense gases (Hunt et al. 1984) and in the dispersion of warm, cooling water from power stations (Rodgers & Pickles 1985).

The interaction between the turbulent layer and the stably-stratified layer is of general importance because it can and often does control the movement of the interfaces (rate of entrainment) and the fluxes of momentum and heat and matter across the interface. The rate of entrainment determines the depth of the turbulent boundary layer and cloud layers (Deardorff 1980) which affects the dynamics of their turbulence, since the scale of energy-containing eddies are often determined by the depth of the turbulent layer. In the atmosphere the interface dynamics may control the fluxes of hydrogen peroxide into the boundary layer and therefore the oxidation of sulphur dioxide which may be present (Clark et al. 1984). Again, in the atmosphere, the humidity of entrained air is important in determining the microstructure of clouds which may be present in the boundary layer (Caughey et al.1982; Baker et al.1982). The interaction is important for two further reasons, since it determines: (i) the characteristics of the turbulence near the interface, and this determines the diffusion of particles in that region; (ii) the flux of wave energy in the stratified layers, which may be an important sink of energy from both the atmospheric and oceanic mixed layers.

In the next section (§2), we shall present some recent field observations and large-eddy simulations which show some of the main features observed near density interfaces. Two sorts of experiments show many features which we will attempt to explain using a theoretical model developed in §3 and described more fully in Carruthers & Hunt [1986a,b]. In §4 we shall make some comparisons between the theory and simulations, and in the light of these we suggest further developments for the large-eddy models.

II. SIMULATIONS AND OBSERVATIONS

The observations from the atmosphere and the ocean (Caughey &

Palmer 1979; Caughey et al. 1982; Brost et al. 1982; Scott & Kerry 1985; Shay & Gregg 1984) suggest that the interface has one of the following three forms with the turbulent structure dependent on the form (see Carruthers & Hunt 1986a,b): (i) the stable layer has a uniform density gradient and there is no mean shear; (ii) there is an intensely stratified layer marking the edge of the turbulent layer with fluid of lower stratification beyond. There is no mean shear; (iii) either of the above cases with mean shear. This is frequently most intense when it is associated with a narrow stably-stratified layer.

In this paper we concentrate on cases (i) and (ii) and here briefly outline simulations and observations of these cases. Figure 2 (from the numerical simulation of Deardorff [1980]) shows profiles of vertical and horizontal velocity variances for a number of different case studies. Note that $\overline{w^2}$ sometimes has a second maximum near the top of the boundary layer ($z=z_i$) and that this is found to occur only for cases (ii) and (iii) above where there is a strong inversion layer. When the stratification bounding the turbulence is uniform, $\overline{w^2}$ decreases monotonically from its maximum value near the centre of the turbulent layer. Figure 3. shows data from a large-eddy simulation performed by Dr. Moeng (see Moeng [1984] and qv) which is also further analysed in Carruthers & Moeng [1986]. The temperature structure shows a three-layer structure. In the contours of the instantaneous values of w there appears to be a dominant scale in the inversion layer which is smaller than the energy-containing scales of the turbulence. Using the theoretical model described in §3, we shall show that the motions of this scale are due largely to trapped waves.

Observations have thus far shown a less clear-cut distinction between cases (i) and (ii). The observations of Scott & Kerry [1985] (Figure 4) show trapped waves and an increase in $\overline{w^2}$ as evident from the short waves; the radar measurements of Gossard et al.[1982] also show increases in $\overline{w^2}$ near strong inversion layers, and such variations are commonly experienced while flying through the boundary layer. However, the observations of Caughey et al.[1982] have shown a steady decrease in $\overline{w^2}$ when a strong inversion existed.

III. THEORETICAL MODEL

A full presentation of the theory and results is given in the two papers by Carruthers & Hunt [1986a,b], so that in this

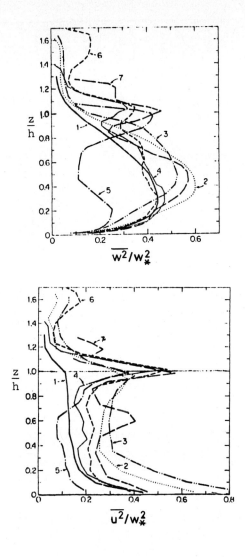

Fig.2. Vertical profile of $\overline{w^2}$ (a) and $\overline{u^2}$ (b) for different case studies from the numerical simulations of Deardorff [1980]. For details of cases 1-7, see Table 1.

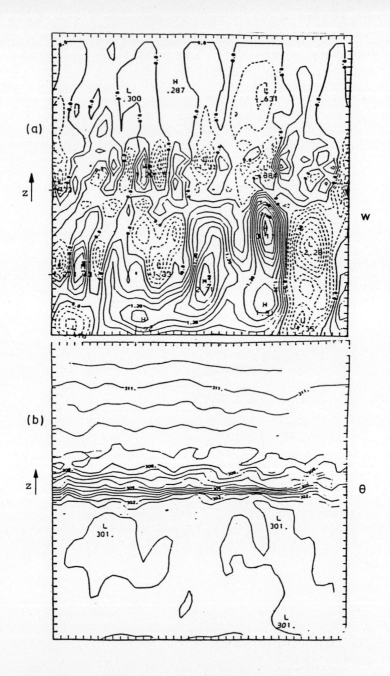

Fig.3. Interfaces between turbulent and stably-stratified layers in the ocean and atmosphere. Contours of instantaneous values of w(a), and θ (b) from LES & Moeng [1984].

discussion we shall emphasise the most important assumptions and conclusions, but omit the details of the theory. The structures we are considering are shown in Figure 5 (cases (i) and ((i) above). In case (i) the deep stratified layer has buoyancy frequency N; in case (ii) the strongly stratified layer of thickness h has buoyancy frequence N_1, while the deep stratified layer has buoyancy frequency N_1.

(i) <u>Turbulent region</u>

In the interior of the turbulent region ($z/L \ll -1$), the turbulence is assumed to be homogeneous and isotropic with wavenumber spectrum $\Phi_{ij}^{(H)}(\kappa)$ where

$$\Phi_{ij}^{(H)}(\kappa) = \int\int\int_{-\infty}^{\infty} \overline{u_i(x+r)u_j(x)} \, e^{-i\kappa \cdot r} \, dr \qquad (3.1)$$

which is assumed to be known. Following the observations of Kaimal et al.[1976] in the convective boundary layer, we assume that the energy dissipation rate, ϵ, is approximately constant with height, i.e.

$$\frac{\partial \epsilon}{\partial z} \approx 0 \, , \qquad (3.2)$$

from which it follows that

$$\frac{\partial \overline{\omega^2}}{\partial z} = 0 \, , \qquad (3.2)$$

where $\overline{\omega^2}$ is the mean square vorticity, so that the distortion of the turbulence near the interface is irrotational. Thus the simplest velocity field satisfying these constraints is

$$\mathbf{u} = \mathbf{u}^{(H)} + \mathbf{u}^{(S)} \, , \qquad (3.4)$$

where $\mathbf{u}^{(H)}$ is a homogeneous velocity field which is known and described by the wavenumber spectra and $\mathbf{u}^{(S)}$ is an irrotational velocity field. Assuming incompressibility we can then write

$$\mathbf{u} = \mathbf{u}^{(H)} - \nabla \phi \qquad (3.5)$$

where $\nabla^2 \phi = 0$ and ϕ is the velocity potential. Note that this solution is appropriate only when the shear in the turbulent region is small (buoyancy-produced turbulence dominates shear-produced turbulence).

(ii) <u>Stratified layers</u>

In the stratified layers the linearised momentum equation for the vertical motion is

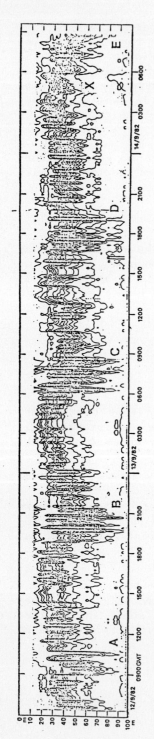

Fig.4. Thermistor chain isotherm data for the period 0600 on 12/9/82 until 0900 on 14/9/82. The temperature range shown is 12.0° to 16.8° in 0.4° steps. From Scott & Kerry [1985].

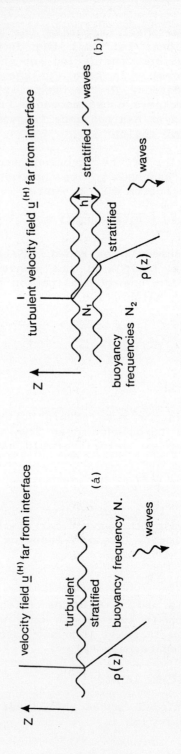

Fig.5. Flow fields described by the model.
(a) Two-layer model; single stratified layer;
(b) Three-layer model; two stratified layers.

$$\frac{\partial^2}{\partial t^2}(\nabla^2 w) - N_i \left[\frac{\partial^2}{\partial x^2} + \frac{\partial^2}{\partial y^2}\right] w = 0 \quad , \tag{3.6}$$

where N_i is N_1, N_2 or N depending on the layer we are considering. The horizontal components are easily obtained from the momentum equations

$$\rho \frac{\partial u}{\partial t} + \frac{\partial p}{\partial x} = 0 \tag{3.7}$$

$$\rho \frac{\partial v}{\partial t} + \frac{\partial p}{\partial y} = 0 \quad , \tag{3.8}$$

and $\rho \dfrac{\partial^2 w}{\partial z \partial t} = \left[\dfrac{\partial^2}{\partial x^2} + \dfrac{\partial^2}{\partial y^2}\right] p \quad .$ (3.9)

(iii) **Boundary conditions**

In the upper region as $z \to \infty$ the radiation condition is applied for $\omega < N_2$ or N and the condition that $w \to 0$ for $\omega > N_2$ or N. At the interface the vertical velocity (w) is continuous, and also the pressure. This implies that $\frac{\partial w}{\partial z}$ is continuous.

The system of equations is closed by defining a form for the joint wavenumber and frequency spectrum in homogeneous turbulence; $\chi_{ij}^{(H)}(\kappa,\omega)$. In homogeneous turbulence there have not even been measurements of the Eulerian frequency spectrum let alone $\chi_{ij}^{(H)}(\kappa,\omega)$, so we propose an approximate form in terms of the specified wave spectrum $\phi_{ij}^{(H)}(\kappa)$.

There are two main causes for the time variation in the velocity at a point moving with the mean flow: the change of velocity of a fluid element (which occurs in the Lagrangian timescale τ_L) and the random velocity of advection of fluid elements by the most energetic eddies (on a time scale ℓ/u_H where ℓ is the length scale of the advected eddy and u_H the velocity scale of the turbulence) (Tennekes 1975) (see Figure 6).

In physical terms the result of the use of the Lagrangian form would correspond to a model of waves being produced by eddies rising and falling, whereas the result using the Eulerian spectrum corresponds to a model of waves being produced by eddies moving randomly and horizontally along the interface. In the analysis presented here, we make use of the Eulerian spectrum since it is more energetic at moderate-to-high frequencies ($\omega > u_H/L_H$, where L_H is the longitudinal integral scale of the homogeneous turbulence) because the energy at frequency ω is largely being induced by eddies of wavenumber $|\kappa| \approx \omega/u_H$. Making use of the Eulerian form, the joint wavenumber and frequency spectrum has the approximate form

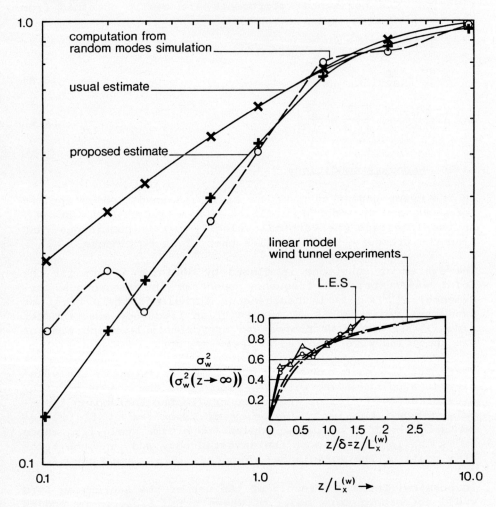

Fig.6. Ratios of time scales near a rigid boundary in shear-free turbulence to the time scales far from the boundary. The simulated values are obtained from the random mode simulation adapted from Kraichnan (1970).

$$\chi_{ij}^{(H)}(\kappa,\omega) = \Phi_{ij}^{(H)}(\kappa)\,\delta(\omega - u_H \kappa) \quad (3.10)$$

where $\kappa = |\kappa|$. We shall make use of the modified form

$$\chi_{ij}^{(H)}(\kappa,\omega) = \Phi_{ij}^{(H)}(\kappa)\,\delta(\omega - u_H \kappa_{12}), \quad (3.11)$$

where $\kappa_{12} = (\kappa_1^2 + \kappa_2^2)^{1/2}$. This allows the integrals to be calculated more easily, but it is shown in Carruthers & Hunt [1986a] that this has little effect on the solution.

(iv) <u>Three-dimensional simulation</u>

A complementary study has beeb the three-dimensional simulation of shear-free turbulence near boundaries to study particle motions and turbulent diffusion by Turfus [1985]. The velocity field u is generated as a set of random Fourier modes with a timescale of each mode linearly dependent on its spatial scale (or the Eulerian spectrum assumed above). Because no dynamical equations are solved, there is no interaction between the modes. But for studying diffusion, this interaction may not be critical (Kraichnan 1970).

For particles released at various source heights near a boundary in this simulated shear-free turbulence, we have computed their trajectories, and the statistics of their Lagrangian velocities. We have confirmed some analytical results for mean trajectories and dispersion (Hunt 1982). The generally assumed form for the Lagrangian integral timescale $\tau_L^{(w)}$ in inhomogeneous turbulence, in terms of the Eulerian integral scale $L_x^{(w)}$ of the vertical component, is

$$\tau_L^{(w)} \approx L_x^{(w)}/\sigma_w \quad (3.12)$$

(where σ_w is the rms value of w), even near an interface where $\sigma_w \to 0$. For a rigid shear-free interface, $L_x^{(w)} \propto z$, and $\sigma_w \propto z^{1/3}$, which implies $T_L^{(w)} \propto z^{4/3}$ as $z \to 0$. However, the results from the simulation show that a better approximation as $z/L_x^{(w)} \to 0$ is

$$\tau_L^{(w)} \approx L_x^{(w)}/\sigma_u, \quad (3.13)$$

where σ_u is the rms value of the <u>horizontal</u> velocity fluctuations. This is a smaller value (see figure 6), and shows that the vigorous horizontal velocity fluctuations advecting the eddies around near the interface determines the timescale even of <u>vertical</u> velocity fluctuations. This then suggests that the diffusivity is proportional to $\epsilon^{1/3} z^{5/3}/L_H^{1/3}$ rather than $\epsilon^{1/3} z^{4/3}$. We believe that other Lagrangian aspects of turbulence can probably be best studied by this technique of random modes. What the effects of interactions between the modes are, we do not know.

(v) Solutions and results

The equations and assumptions can be used (Carruthers & Hunt 1986a) to obtain expressions for the one-dimensional spectra, velocity variances and integral length scales in all the layers, whilst in the stratified layers we can also calculate the wave energy flux, the variances of the streamline displacements $\overline{\zeta^2}$ and its derivative $\overline{(\partial\zeta/\partial z)^2}$, and the variances of density $\overline{\rho^2}$ or temperature $\overline{\theta^2}$. $\overline{(\frac{\partial\zeta}{\partial z})^2}$ is important in determining whether there is any overturning in the stably-stratified layer.

In the case of the three-layer model, the eddies interact with the stratified layer in three different ways dependent on whether the frequency of the turbulent eddies is given by (i) $\omega < N_1, N_2$ (ii) $N_2 < \omega < N_1$ or (iii) $N_1, N_2 < \omega$. When $\omega < N_1, N_2$ waves excited by the turbulence can propagate in both stratified layers and energy is lost to the deep stratified layer. For the case $N_2 < \omega < N_1$, waves can propagate in the strongly stratified layer but are evanescent in the upper layer; thus there is the possibility of trapping of energy and resonance in the inversion layer. When $\omega > N_1, N_2$, waves are evanescent in both layers and there is no propagation of energy. In the two-layer model, there exist cases (i) and (iii) and there is no trapping of energy.

In the three-layer model, trapping and wave energy occur when

$$\cot(-\lambda_2/\mu_1) = (\mu_1^2+\lambda_2^2)/(\mu_1(\kappa_{12}+\lambda_2)) \tag{3.14}$$

where

$$\mu_1 = \left[\frac{N_1^2}{\omega^2} - 1\right]\kappa_{12} \text{ and } \lambda_2 = \left[1 - N_2^2/\omega^2\right]^{1/2} \cdot \kappa_{12}.$$

When $N_1 h/u_H < n\pi$ where n is integer, there exist, at most, n solutions the number decreasing with incrasing $N_2 (< N_1)$.

In order to obtain a finite-stationary solution when trapped waves exist, it is necessary to introduce damping. This was incorporated as Rayleigh friction, its magnitude being determined by a criterion developed in Carruthers & Hunt [1986b]. In physical terms we assumed that the trapped waves would grow until their magnitude was sufficient for convective instabilities to grow and develop, these instabilities limiting the further increase in amplitude of the waves.

As an example of a solution, we show the one-dimensional

spectrum of vertical fluctuations defined as

$$\theta_{33}(\kappa_1) = \int_{-\infty}^{\infty} \int_{-\infty}^{\infty} \Phi_{ij}(\mathbf{k}) d\kappa_2 d\kappa_3 \quad , \qquad (3.15)$$

for different z (Figure 7). In this case $N_1 L_H/u_H = 171$, $N_2 L_H/u_H = 10.5$ and $h/L_H = 0.01$, giving a resonant wave with frequency $\omega_* = 100 \, u_H/L_H$ (these values are typical of the atmosphere). The curve at $z = -\infty$ shows $\theta_{33}^{(H)}(\kappa_1)$ the assumed form for the homogeneous turbulence. At $z = 0, z = h$, the resonant waves are clearly seen on the spectra; note that at frequencies $\omega > N_1$, there is more energy at $z = 0$ than at $z = h$, since waves at these frequencies cannot propagate even into the strongly-stratified layer and thus have decayed at $z = h$ (a distance h from the turbulence). At $z = 0.4 L_H$ all frequencies $\omega > N_1$ have decayed and the spectrum is one of propagating internal gravity waves.

(iv) <u>Comparisons and discussion</u>

In Figure 8 we show the wave energy flux in the deep stratified layer for both the two- and three-layer model. The dashed curve represents the solution for the two-layer model, the solid lines the variation of the flux in the three-layer model with $N_2 L_H/u_H$ for different values of $N_1 L_H/u_H$. When $h/L_H \leq 0.01$, the solution is complicated by rapid changes in the flux for only small changes in the buoyancy frequency N_1, the maximum flux being greater than in the two-layer case.

Figures 9 (two-layer) and 10 (three-layer) show profiles of $\overline{w^2}$. In the two-layer case, a number of curves are shown corresponding to different values of the non-dimensional parameter NL_H/u_H where L_H and u_H are the longitudinal length scale and velocity scale of the turbulence. For the case $NL_H/u_H = 6$, comparisons are made with the atmospheric observations of Caughey & Palmer [1979] and the agreement is good. In the three-layer case, the curve corresponds to the atmospheric structure used to calculate the spectrum in Figure 7, and is typical of a number of structures we examined. Note the maximum in $\overline{w^2}$ due to trapped waves in the inversion layer. In Table 1 values of $\overline{w^2}$ calculated from the large-eddy simulation of Deardorff [1980] are compared with the predictions of the theory. In each case the agreement between the theory and simulations is quite good, and in particular both theory and simulation show a marked difference between the two-layer (cases 1-3) and three-layer (cases 4-7) structures; this emphasises the importance of the trapped waves. Further evidence that the increase in $\overline{w^2}$ is caused by trapped waves is

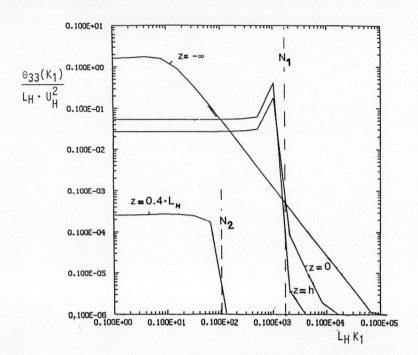

Fig. 7. One-dimensional spectra of fluctuations of vertical velocity. Three-layer model.

Table 1. Variance of vertical velocity from the simulation of Deardorff compared with the theory.
In cases 1-3 the atmosphere has a two-layer structure; in cases 4-7 the atmosphere has three-layer structure. $\Delta\Theta$ is the temperature step across the middle layer.

Cases	Z_i(m)	L_H(m)	h(m)	$\Delta\Theta$	u_H(ms)	h/L_H	$N_1 h/u_H$	$N_2 L_H/u_H$	$N_1 L_H/u_H$	$\overline{w^2}/u_H$ at $z=z_i$	
										simulation	theory
1	1300	650			1.26			8.0		0.2	0.23
2	1200	600			1.14			7.1		0.3	0.24
3	1220	610			0.90			8.6		0.32	0.22
4	1400	700	100	7	0.91	0.142	5.23	3.68	7.7	0.75	0.7
5	1795	898	200	2.4	0.95	0.22	4.7	17.9	9.45	1.0	0.5
6	1160	580	200	6	1.12	0.34	5.56	16.2	5.18	0.85	1.0
7	1550	775	100	3	0.62	0.13	5.02	38.9	12.5	1.1	0.8

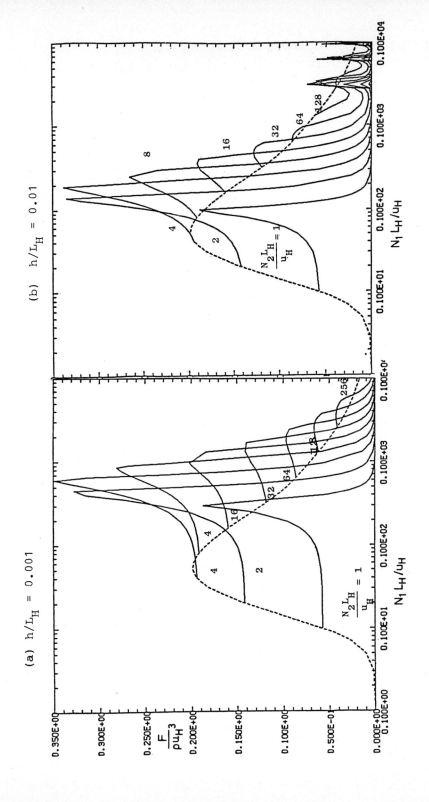

Fig.8. Calculated energy fluxes in the deep stratified layer: three layer model ———— two-layer model — — — —

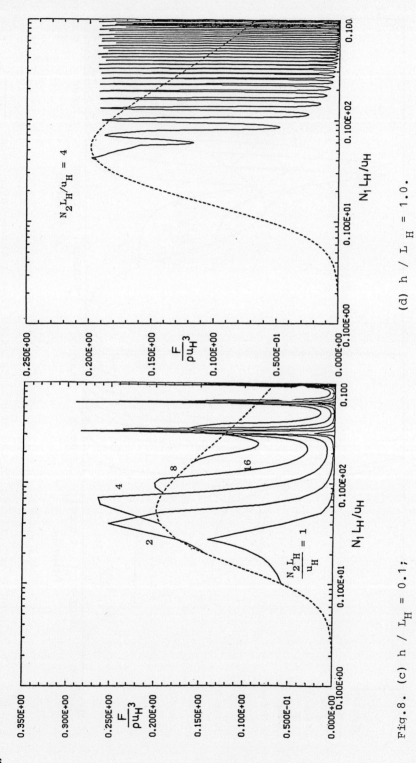

Fig.8. (c) $h/L_H = 0.1$; (d) $h/L_H = 1.0$.

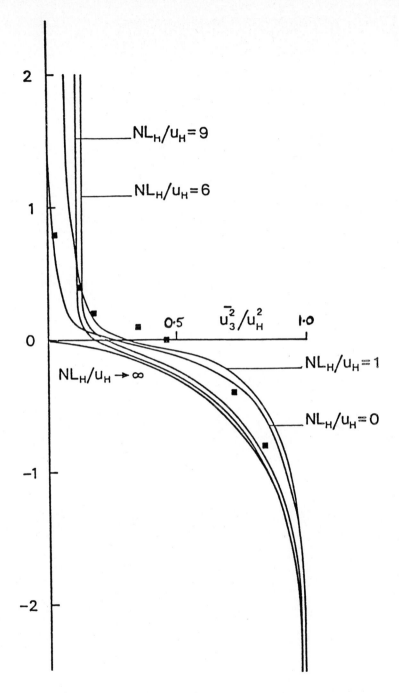

Fig.9. Two-layer model. Variance of vertical velocity $(\overline{w^2})$.
■ Observations of Caughey & Palmer (1979) $NL_H/u_H = 6$.

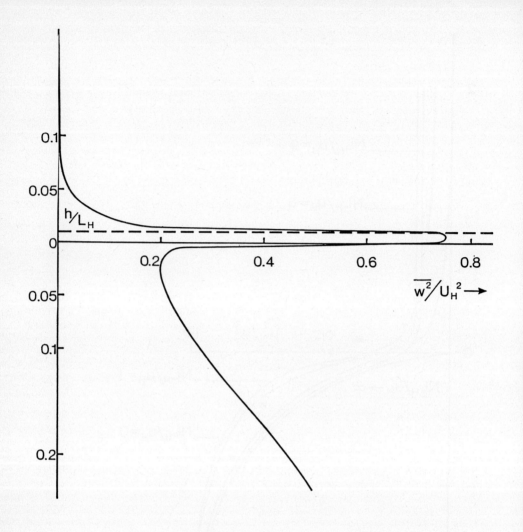

Fig.10. Three-layer model. Variances of vertical velocity (w^2).
$N_1 L_H / u_H = 161$; $N_2 L_H / u_H = 10.5$; $h/L_H = 0.01$.

given in Carruthers & Moeng [1986].

The theory shows that at interfaces very high gradients of velocity occur on scales smaller than those resolved by L.E.S. Furthermore, the subgrid scale motions are wave motions which are not adequately modelled by isotropic Gaussian small-scale motions. The other limitation of L.E.S. is its treatment of large-scale wave motions which are an important way in which inversion layers lose energy. Therefore entrainment which may be associated with breaking waves on or in the inversion layer, may not be adequately modelled by L.E.S.

Despite these possible limitations, the initial comparisons presented here and the results of the L.E.S. are encouraging. Further studies on turbulence near density inversion layers are needed, both theoretical and computational, on entrainment processes, wave energy flux and on diffusion of scalars near the interface.

Acknowledgement. We are grateful to Professors U. Schumann and R. Friedrich for organising a most successful Euromech and also to Dr. C-H Moeng for providing data from her large-eddy simulations. Dr. C. Turfus' work at Cambridge was supported by the British Gas Council. Dr. D. Carruthers' work was supported by the NERC at Cambridge and by NSF at NCAR during a visit there.

REFERENCES

Baker, M.B., Bluth, A.M., Carruthers, D.J., Choularton, T.W., Conway, B.J., Fullarton, G., Gay, M.J., Latham, J., Mills, C.S., Smith, M.H. & Stromberg, I.M. (1982) Field studies on the effect of entrainment on the structure of clouds, at Great Dun Fell. Q.J. Roy. Met. Soc. $\underline{108}$, 899.

Brost, R.A., Wyngaard, J.C. & Lenschow, A.H. (1985) Marine Stratocumulus layers. Part II. Turbulence budgets. J. Atmos. Sci. $\underline{39}$, 818.

Carruthers, D.J. & Hunt, J.C.R. (1986a) Density fluctuations between a turbulent region and a stably stratified layer. J. Fluid Mech. $\underline{165}$,

Carruthers, D.J. & Hunt, J.C.R. (1986b) Turbulence, waves and entrainment near density inversion layers. J. Fluid Mech. (to be submitted).

Carruthers, D.J. & Moeng, C-H. (1986) Waves in the overlying inversion of the convective boundary layer. J. Atmos. Sci. (to be submitted).

Caughey, S.J., Crease, B.A. & Roach, W.T. (1982) A field of

study of nocturnal stratocumulus. II. Q.J. Roy. Met. Soc. 108, 125.

Caughey, S.J. & Palmer, S.G. (1979) Some aspects of turbulence through the depth of the convective boundary layer. Q.J. Roy. Met. Soc. 105, 811.

Clark, P.A., Fletcher, I.S., Kallend, A.S., McElroy, W.J., Marsh, A.R.W. & Webb, A.H. (1984) Observations of cloud chemistry during long-range transport of power plant plumes. Atmos. Env. 18, 1849.

Deardorff, J.W. (1980) Stratocumulus capped mixed layers derived from a three-dimensional model. Boundary Layer Met. 18, 495.

Gossard, E.E., Chadwick, R.B., Neff, W.D. & Moran, K.P. (1982) The use of ground-based Doppler radar to measure gradients, fluxes and structure parameters in elevated layers. J. Appl. Met. 21, 211.

Hunt, J.C.R. (1982) In: Atmospheric Turbulence and Air Pollution Modelling. D. Reidel Publ.

Hunt, J.C.R., Rottman, J.W. & Britter, R.E. (1984) Some physical processes involved in the dispersion of dense gases. Proc. IUTAM Symposium 'Atmospheric dispersion of heavy gases and small particles'.

Kaimal, J.C., Wyngaard, J.C. Aaugen, D.A., Cote, O.R., Izumi, Y., Caughey, S.J. & Readings, C.J. (1976) Turbulence structure in the convective boundary layer. J. Atmos. Sci. 33, 2151.

Kraichnan, R. (1970) Diffusion by a random velocity field. Physics of Fluids 13, 22.

Moeng, C-H. (1984) A large eddy simulation model for the study of boundary layer turbulence. J. Atmos. Sci. 41, 2052.

Richards, K.J. (1985) Proc. IMA Conf. on 'Models of turbulence and diffusion in stably stratified regions of the natural environment'.

Rogers, I.R. & Pickles, J.H. (1985) Proc. IMA Conf. on 'Models of turbulence and diffusion in stably stratified regions of the natural environment'.

Shay, T.J. & Gregg, M.C. (1984) Turbulence in the oceanic convective mixed layer. Nature 310, 282.

Scott, J.C. & Kerry, N.J. (1985) Internal waves in Bay of Biscay. Internal report ARE Portland.

Tennekes, H. (1975) Eulerian and Lagrangian time microscales in isotropic turbulence. J. Fluid Mech. 67, 561.

Turfus, C.J. (1985) Ph.D. Thesis, University of Cambridge.

A LARGE EDDY SIMULATION MODEL
FOR THE STRATUS-TOPPED BOUNDARY LAYER

Chin-Hoh Moeng
National Center for Atmospheric Research
Boulder, Colorado, 80307, U.S.A.

SUMMARY

Because of the important impact of the stratus cloud regime on global climate, I have incorporated radiation and condensation processes into my large-eddy-simulation model to simulate the stratus-topped boundary layer. Such simulation helps to understand the turbulent structure within this type of planetary boundary layer (PBL), and can be used to test the statistical turbulence models which are used as PBL parameterizations in climate models.

A sample study for mixed-layer modeling shows that within the stratus-topped mixed layer about 60% of the total buoyant-generated kinetic energy is used for dissipation and the rest converts back to potential energy through thermally indirect motions.

I also study the pressure term in the scalar flux equation for second-order closure modeling. Decomposing the pressure fluctuations into different components according to their physical processes allows one to study each component separately. The result shows that in the scalar flux equation the buoyancy component of the pressure term is proportional to the direct buoyant production term; the proportionality constant is about 0.75 for the stratus-topped PBL. The nonlinear interaction component of the pressure term can be represented by Rotta's return-to-isotropy model, with a time scale profile proportional to $\overline{w'^2}/\epsilon$, where $\overline{w'^2}$ is the vertical velocity variance and ϵ is the kinetic energy dissipation rate.

1. INTRODUCTION

The increasing interest in the stratus cloud-topped boundary layer (CTBL), which is very important in climate dynamics, has stimulated intense research. Since the pioneer work by Lilly [1], many have applied mixed-layer modeling to study the CTBL. Others have approached this problem numerically with a higher-order closure technique.

The application of these statistical turbulence models to this problem, however, is still controversial, due to the complicated structure and dynamics of the CTBL and to a lack of observational data. Many have analyzed the few existing data sets, most of which are for the transient type of stratus cloud. Most of the persistent stratus cloud regimes exist over the ocean, where measurement problems can be quite severe.

Fortunately, with the increasing speed and memory of supercomputers, we are now able to carry out large-eddy simulations (LES) as "controlled field experiments" on the computer to generate a data base [2],[3],[4]. For this purpose I have developed an LES model that includes all of the turbulence, longwave radiation, and condensation processes. This paper briefly describes the model in Section 2, and gives the overall structure in Section 3. Using the model-generated data I then study some of the closure problems in the statistical turbulence models, e.g., the mixed-layer and second-order closure models; some preliminary results are given in Sections 4 and 5. The conclusions are set forth in Section 6.

2. MODEL DESCRIPTION

The basic dynamic framework of the LES model follows, generally, that of the Stanford group (e.g., [5],[6]) and is described in detail by Moeng [7]. A Gaussian filter is used to define the resolvable-scale variables, and hence the Leonard stress [8] is implicitly included in the horizontal directions. The model computes the horizontal derivatives by means of the pseudospectral method and the vertical derivatives by centered finite-differences. It solves the pressure field through the Poisson equation and parameterizes subgrid-scale effects through Deardorff's [4] turbulence energy model. The nonlinear advection terms conserve the integrals of the total kinetic energy and scalar variances. The model uses the Adams-Bashforth scheme for time advancement, and a staggered spacing in z so that the horizontal wind, scalar, and pressure fields are located at the levels between those for vertical velocity. To date, this LES model has been used to study the clear convective PBL [7],[9],[10],[11].

The longwave radiation parameterization follows that of Herman and Goody [12], and was described by Moeng and Arakawa [13]. The concept of "mixed emissivity" is used with Rodgers's [14] empirical formula for the water vapor emissivity, and the gray-body absorber assumption for the cloud-drop emissivity. The model applies the radiation parameterization at each vertical column for each time step.

In order to take into account the condensation process, I use two conservative quantities as the thermal and moisture prognostic variables: (1) the liquid water static energy $h_l = s_v - L\, q_l$; and (2) the total moisture mixing ratio $r = q_v + q_l$. L is the latent heat of condensation, s_v the virtual dry static energy, q_v the water vapor mixing ratio, and q_l the liquid water mixing ratio. If the total moisture mixing ratio in a grid box is greater than the saturation mixing ratio q_s, the model assumes that clouds fill that grid box; otherwise, it assumes the grid box to be cloud-free. The buoyancy flux is related to the fluxes of h_l and r.

In this simulation, I used a 1.5 s time interval with 40 x 40 x 40 grid points covering a domain 2.5 km x 2.5 km x 1 km. The sea surface temperature is 287 K, which gives a surface saturation mixing ratio of 9.8 g Kg^{-1}. The x-component geostrophic wind was 10 m s^{-1}, and the y-component was zero. The model has a prescribed large-scale subsidence motion corresponding to a constant large-scale divergence of 3 x 10^{-6} s^{-1}.

Initially, the mean environment was designed so that the mean PBL top was at 500 m, the inversion strength was 12 K, the inversion depth was 200 m, the moisture at the PBL top was slightly over saturation, the moisture above the inversion was 4 g Kg^{-1}, and the mean h_l and r were uniform within the PBL. At the initial time, I applied a small amount of random noise to this mean environment near the surface and made many test runs (e.g., dry clouds without radiation, dry clouds with prescribed radiation, and dry clouds with computed radiation), with the large-scale subsidence turned off, before simulating the wet-cloud case with fully interactive radiation. I made an 87.5-min simulation (3,500 time steps) for the wet-cloud case and recorded it every 100 time steps for the last 2,500 steps.

3. OVERALL STRUCTURE

The horizontally averaged cloud top height evolved steadily from 484 m to 534 m during the last 2,500 time steps. The CTBL growth rate is about 1.3 cm s^{-1} and hence, after subtracting the large-scale subsidence, the entrainment rate (w_e) is about 1.5 cm s^{-1}. The time average of the CTBL height (z_i) is about 511 m. The velocity scale w_*, defined by Deardorff [4] as the one-third power of the vertically averaged buoyancy flux, is about 0.9 m s^{-1}. Figure 1 shows an example of the instantaneous look of x-z cross sections of the vertical velocity, liquid water static energy, and liquid water mixing ratio fields. The internal gravity waves gradually build up in the layer above the cloud top, especially in the capping inversion, to a fairly large amplitude.

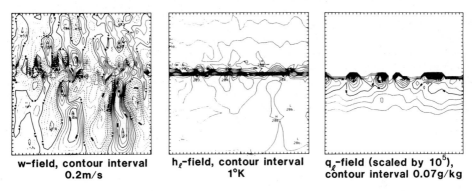

Fig. 1: Contour plots of the vertical velocity, liquid water static energy, liquid water mixing ratio in a vertical cross section on an x-z plane (y = 4Δy).

The statistics shown in the following are the averages over the x-y plane, and over the last 25 recorded time steps. The mean momentum, thermal, and moisture fields are given in Fig. 2. The conserved fields are almost well-mixed within the CTBL. However, note that the total moisture field decreases slightly with height, especially near the cloud top. Also note that the liquid water mixing ratio is only a few percent of the total moisture mixing ratio. If one computes the liquid water mixing ratio from the assumed well-mixed total moisture field, as it usually does in mixed-layer models, the computed liquid water mixing ratio can be totally wrong. The u-component wind has a local maximum above the CTBL which corresponds to a downward momentum flux there, as shown later. The liquid water mixing ratio increases linearly from the cloud base to about 0.9 z_i.

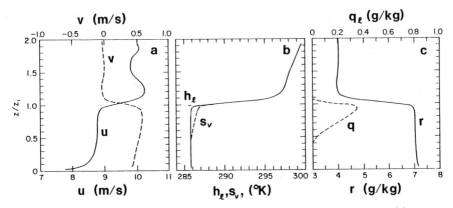

Fig. 2: Vertical profiles of the (a) mean winds, (b) thermal fields, and (c) moisture fields for the resolved scales.

Figure 3 presents the total kinetic energy for the resolvable-scale and subgrid-scale eddies, and three components of the resolvable-scale kinetic energy. The subgrid-scale eddies contain less than 15% of the total kinetic energy. The difference between x and y components is due to the mean wind. From Fig. 3b one sees the anisotropic feature of the CTBL turbulence. The large amount of the horizontal component of energy right at the top of the cloud layer is due to the fact that turbulent flow impinges on a very stably stratified layer and thus the vertical energy component yields to the horizontal component [15]. The local maximum on all three components at $z = 1.1\ z_i$ is due to the trapped gravity waves. Above the CTBL, where gravity waves dominate, the three energy components have approximately the same magnitude. Figure 4 shows the vertical fluxes of the momentum. The downward u-component momentum flux above the CTBL is probably due to internal gravity waves having a phase speed about the same as that of the mean wind, because there is not as much corresponding pressure

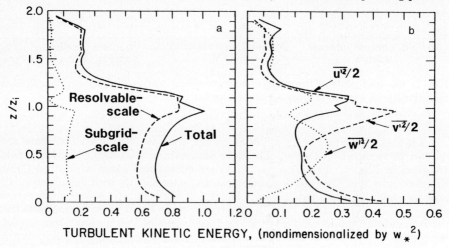

Fig. 3: Vertical profiles of the turbulent kinetic energy.

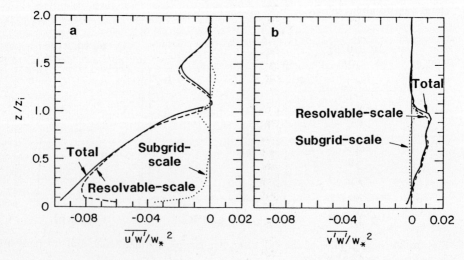

Fig. 4: Vertical profiles of the momentum fluxes.

flux there. Figure 5 gives the vertical total heat, buoyancy, and total moisture fluxes. The total heat flux $\overline{w'h'_l}$ has a strange profile; but when it combines with the radiative flux, the combined flux has a smooth, linear profile. The subgrid-scale contributions to the buoyancy and moisture fluxes are much smaller than those of the resolvable eddies except near the surface and near the top of the cloud layer. The local maximum of the heat flux right above the cloud top is probably due to gravity waves, even though it cannot be explained by the linear wave theory which gives zero heat flux; it may also be caused by truncation error. There is a corresponding local maximum of the subgrid-scale dissipation rate of the energy in this layer.

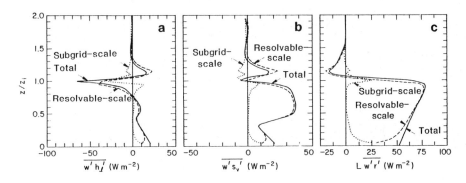

Fig. 5: Vertical profiles of the thermal and moisture fluxes.

4. AN IMPLICATION FOR MIXED-LAYER MODELING

Lilly [1] first applied the mixed-layer model approach to study the CTBL. This type of model assumes that the turbulent mixing is very effective so that the conserved mean fields are always well-mixed within the PBL, and hence the models can bypass the details of dynamics and physics involved in the CTBL. It takes only a single numerical layer and therefore is a proper PBL parameterization scheme to be used in larger-scale meteorological models (e.g., climate models). The only closure problem in mixed-layer modeling is to determine the entrainment rate.

The entrainment hypothesis, in most of the models, is based on the layer-averaged turbulent kinetic energy equation. Typically, the models assume a balance between the net buoyant generation B and viscous dissipation D in the energy equation, partition B into production part P (i.e., thermally direct motion part) and consumption part N (i.e., thermally indirect motion part), and relate D to P. In other words, the models assume that part of the thermally generated turbulent kinetic energy is used in dissipation and the rest returns to potential energy through thermally indirect motions. How to partition B and how to relate D to P are, unfortunately, difficult problems and are still under debate [16],[17] for modeling the CTBL.

From the definition, the thermal production of the kinetic energy is due to warm, rising and cold, sinking air parcels, while the thermal consumption is due to warm, sinking and cool, rising air parcels. Therefore, the exact method to partition B into P and N requires detailed information of the three-dimensional structure of the buoyancy and vertical velocity fields. There is no such detailed information in mixed-layer models, however, and hence mixed-layer models have to artificially implement the partitioning.

Using the LES data, one can calculate the almost-exact production and consumption rates which can then be used to investigate the partitionings implemented in mixed-layer models. I compute the large-eddy contribution to P by summing all positive products of w' and s'_v over all grid points at every level and N by summing all negative products. Figure 6 shows the vertical distribution of the production and consumption rates. For comparison, in the same figure I also show them for a clear convective PBL [7].

In the clear PBL, the production has a maximum near the surface due to surface heating, while the consumption has a maximum near the top of the mixed-layer due to entrainment. Note that these rates are from the resolvable-scale eddies only and therefore P is small near the surface.

In the CTBL, both production and consumption have their maximum near the top of the mixed layer, because the main energy source and sink are both there. These complicated distributions result from all turbulent parcels due to the radiative cooling, surface heating, condensation, and entrainment processes. Further examination of these profiles allows one to study the contribution of each physical process involved.

The time evolution of the layer-averaged production and consumption rates, as well as their ratio, is shown in Fig. 7. For the subgrid-scale contribution, P is computed as the area covered by the positive subgrid-scale flux (given in Fig. 5b) and N by the negative subgrid-scale flux, since there is no detailed structure of the subgrid-scale field available. The ratio is about 0.2 for the clear PBL and 0.4 for the CTBL.

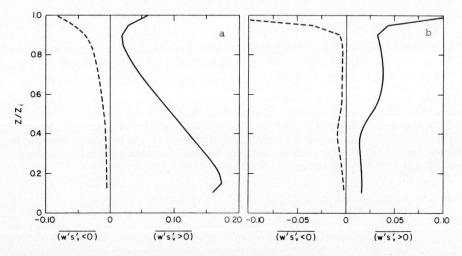

Fig. 6: Vertical distributions of the buoyant production and consumption for the (a) clear convective PBL and (b) stratus-topped PBL.

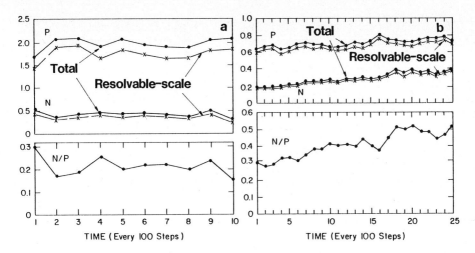

Fig. 7: Time evolutions of the layer-integrated buoyant production and consumption rates, and their ratio, for the (a) clear convective PBL and (b) stratus-topped PBL.

5. AN IMPLICATION FOR SECOND-ORDER CLOSURE MODELING

In second-order closure models, one uses predictive equations for Reynolds stress, heat flux, and other second-moment quantities. These equations contain moments of higher order which are parameterized in terms of the second- or first-order moments in order to close the whole set of equations. One of these parameterizations involves the pressure-related terms, such as the pressure-velocity terms in the Reynolds stress equations and the pressure-temperature terms in the heat flux equation. Unfortunately, it is very difficult to infer the behavior of these terms from laboratory or field observations.

The pressure fluctuation p in an incompressible fluid is governed by the Poisson equation, derived by taking the divergence of the equation for the fluctuating velocity field. The solution of the Poisson equation involves integration over the entire flow domain; this indicates that the pressure fluctuations at a certain point depend on the flow field not only at that point but also at other points over the whole flow. As a result, the pressure covariances are integrals of two-point correlations. The Poisson equation also indicates that there are four distinct sources of pressure fluctuations: turbulent-turbulent (nonlinear) interactions, mean shear, buoyancy, and the Coriolis effect. The first contribution is usually called the return-to-isotropy part since it is responsible for the gradual return to isotropy in the absence of other forcing. The sum of the last three contributions, which are linear in the fluctuating velocity and temperature fields, is called the rapid part.

Rotta [18] first proposed a return-to-isotropy parameterization for the pressure-velocity covariance in a shear flow. It is based on the assumption that the rate of energy transfer is proportional to the degree of anisotropy. His return-to-isotropy parameterization applied to the pressure-scalar covariance is $-\overline{w'c'}/\tau$, where τ is a time scale and c is any scalar. Here I call this parameterization for the pressure-scalar covariance the return-to-isotropy because it acts to reduce the vertical scalar transport and to relax to an isotropic situation.

To account for the buoyancy effect, Deardorff [19] added a correction $(-A\beta g\overline{s'_v c'}$, where A is a constant) to the return-to-isotropy parameterization. Under the assumption of isotropy one can show [2],[20] that $A = 1/3$. One can also derive this value from isotropic tensor modeling [21],[22].

In order to study the contribution to pressure from each physical process, I define five pressure components. The total fluctuating pressure is written as the sum of the turbulence-turbulence part p_T, the mean-shear part p_S, the buoyancy part p_B, the Coriolis part p_C, and the subgrid-scale part p_{SG}. Note that the last part is not precisely the subgrid-scale pressure field; rather, it is the contribution to the pressure field from our parameterized subgrid-scale term in the equation of motion, which could be different. Each component is solved by the Poisson equation with its only corresponding physical term on the right-hand side. For example, the buoyancy part is governed by

$$\nabla^2 p_B = \frac{\partial}{\partial z}\beta g s'_v. \qquad (1)$$

The boundary conditions are periodic in x and y, and zero vertical gradients of p at both the surface and the top of the numerical domain. The zero-gradient condition at both upper and lower boundaries implies that the horizontal fluctuations of all physical processes vanish at the surface and at the top of the numerical domain. Since we solve only for the fluctuating component of the pressure field, and the temperature and vertical velocity fluctuations are zero at both boundaries, this condition seems proper [23].

Figure 8 shows the correlations of the pressure gradient with the temperature and moisture fields. In order to show the structure inside the CTBL in more detail, I use a linear-logarithmic scale transformation of $y = x/\mid x \mid ln(1+ \mid x \mid /x_0)$ on the abscissa, where x_0 is a scaling parameter. The Coriolis component is by far the

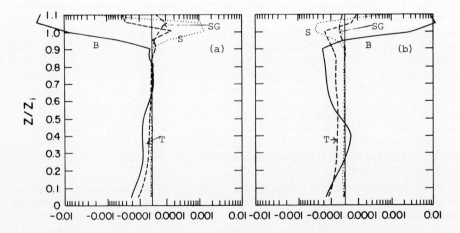

Fig. 8: Vertical distributions of the individual contributions of a) $-\overline{h'_l \partial p'/\partial z}$ and b) $-\overline{r' \partial p'/\partial z}$; T is the turbulence-turbulence component, B the buoyancy component, S the mean-shear component, and SG the subgrid-scale component.

smallest (and hence is not presented); by scale analysis one can show that this is the case simply because the turbulent Rossby number w_*/fz_i is large (~ 20) in the simulation. The subgrid-scale contribution to the resolvable pressure correlation is relatively small, except near the surface and at the top of the CTBL. The mean-shear and turbulence-turbulence contributions for the pressure-temperature covariance are of about the same order of magnitude as the subgrid-scale one and therefore are not considered to be reliable. For the pressure-moisture covariance, both the turbulence-turbulence and buoyancy parts are large within the CTBL. Near the top of the CTBL and in the inversion the buoyancy part dominates.

I first examine the buoyancy contribution. Isotropic tensor modeling predicts that the buoyancy contribution to the pressure-gradient/scalar covariance is proportional to the buoyant production of the vertical flux of a scalar [21]. The results (Fig. 9) indicate that the proportionality constant A is about 0.75. This implies that the pressure covariance reduces the direct buoyant production in the scalar flux equation by three-quarters throughout the CTBL. Moeng and Wyngaard [11] found that $A = 0.5$ for the clear convective PBL.

Above the cloud top the mechanisms causing the pressure covariance are mainly due to gravity waves, not turbulence, and therefore the result is not presented here. Above the PBL, the constant A depends mainly on the stability and the thickness of the inversion.

I next compare Rotta's return-to-isotropy parameterization with the turbulence-turbulence contribution, for the pressure-moisture covariance only. The profile for the dimensionless time scale $\tau*$, defined as the ratio of $-\overline{w'r'}(w_*/z_i)$ and $-\overline{r'\partial p'/\partial z}$, is given in Fig. 10, which can be approximated by $(\overline{w'^2}/2\epsilon)(w_*/z_i)$, where $\overline{w'^2}$ is the vertical velocity variance and ϵ is the kinetic energy dissipation rate.

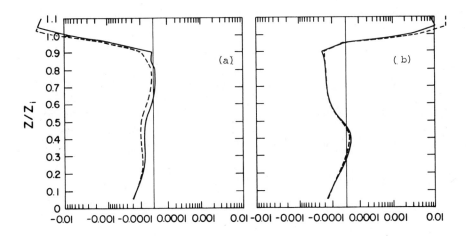

Fig. 9: Solid curves are the analyzed (a) $-\overline{h'_l \partial p'_B/\partial z}$ and (b) $-\overline{r'\partial p'_B/\partial z}$, and dashed curves are their parameterizations with $A = 0.75$.

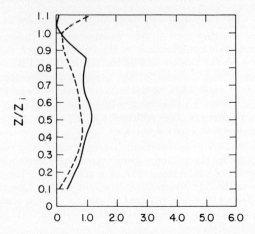

Fig. 10: Solid line is the computed dimensionless time scale, and dashed line is $(\overline{w'^2}/2\epsilon)(w_*/z_i)$.

As a result, a parameterization for the pressure-gradient/scalar covariance in the CTBL without significant mean wind shear is

$$-\overline{c'\frac{\partial p'}{\partial z}} = -A\beta g\overline{s'_v c'} - \frac{\overline{w'c'}}{\tau_*(z_i/w_*)} , \qquad (2)$$

where τ_* is given by $(\overline{w'^2}/2\epsilon)(w_*/z_i)$ and $A = 0.75$.

The analyzed total pressure-gradient/scalar covariance and this parameterization are compared in Fig. 11.

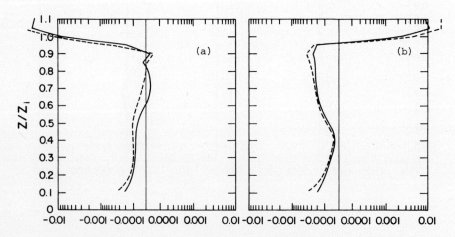

Fig. 11: Solid curves show the analyzed (a) $-\overline{h'_l \partial p'/\partial z}$ and (b) $-\overline{r'\partial p'/\partial z}$, and dashed curves show the proposed parameterization for these covariances.

6. CONCLUSIONS

This paper describes a large-eddy-simulation model which consists of the longwave radiation and condensation processes in addition to the turbulence process. This paper also reports the structure of the CTBL as seen through the large-eddy simulation, and presents some sample studies of using the computer-generated data to investigate the turbulence statistical models.

The results show that the resolvable-scale eddies contain at least 85% of the turbulent kinetic energy and support most of the buoyancy and moisture fluxes. Below $z = 0.9 z_i$, the mean conservative variables are rather well mixed; the momentum and moisture fluxes, and the sum of the total heat and longwave radiative fluxes are rather linear. Therefore, to a good approximation, using the mixed-layer modeling approach is proper for parameterizing this type of PBL.

However, the entrainment hypothesis in mixed-layer models for the CTBL is still controversial, especially on questions of how to partition the buoyancy effect into production and consumption of kinetic energy, and how much of the buoyant-generated kinetic energy is used in dissipation. Using the LES data, I exactly computed the production and consumption rates and found that about 60% of the buoyant-generated kinetic energy is used in dissipation and 40% converts back to potential energy. This information on the "exact" production and consumption rates will be used in further studies to examine the "approximated" partitionings implemented in mixed-layer models.

I also studied the correlations of the pressure-gradient with the temperature and moisture fields in second-order closure modeling. Splitting the pressure fluctuations into five components according to different physical processes (i.e., nonlinear interaction, buoyancy, rotation, mean shear, and subgrid-scale effect), I examined separately the buoyancy and nonlinear interaction contributions. The results show that the buoyancy contribution to the pressure-scalar covariance is proportional to the buoyancy production term in the scalar vertical flux equation, and the turbulence-turbulence nonlinear effect can be approximated by Rotta's return-to-isotropy assumption.

REFERENCES

[1] Lilly, D. K.: "Models of cloud-topped mixed layers under a strong inversion", *Q. J. R. Meteorol. Soc.*, **94** (1968) pp. 292-309.

[2] Deardorff, J. W.: "Numerical investigation of neutral and unstable planetary boundary layers", *J. Atmos. Sci.*, **29** (1972) pp. 91-115.

[3] Sommeria, G., and Deardorff, J. W.: "Subgrid-scale condensation in models of nonprecipitating clouds", *J. Atmos. Sci.*, **34** (1977) pp. 344-355.

[4] Deardorff, J. W.: "Stratocumulus-capped mixed layers derived from a three-dimensional model", *Boundary-Layer Meteorol.*, (1980) **41** pp. 2052-2062.

[5] Moin, P., Reynolds, W. C., and Ferziger, J. H.: "Large eddy simulation of incompressible turbulent channel flow", Report No. TF-12, Thermoscience Div., Stanford, Calif. (1978).

[6] Moin, P., and Kim, J.: "Numerical investigation of turbulent channel flow", *J. Fluid Mech.*, **118** (1982) pp. 341-377.

[7] Moeng, C.-H.: "A large-eddy-simulation model for the study of planetary boundary-layer turbulence", *J. Atmos. Sci.*, **41** (1984) pp. 2052-2062.

[8] Leonard, A.: "Energy cascade in large eddy simulations of turbulent fluid flows", Advances in Geophysics, vol. 18, Academic Press, New York (1974) pp. 237-248.

[9] Moeng, C.-H., and Wyngaard, J. C.: "Statistics of conservative scalars in the convective boundary layer", *J. Atmos. Sci.*, **41** (1984) pp. 3161-3169.

[10] Fiedler, B., and Moeng, C.-H.: "A practical integral closure model for mean vertical transport of a scalar in a convective boundary layer", *J. Atmos. Sci.*, **42** (1985) pp. 359-363.

[11] Moeng, C.-H., and Wyngaard, J. C.: "An analysis of closures for pressure-scalar covariances in the convective boundary layer", (Submitted to *J. Atmos. Sci.*, 1986).

[12] Herman, G. F., and Goody, R. M.: "Formation and persistence of summertime arctic stratus clouds", *J. Atmos. Sci.*, **33** (1976) pp. 1537-1553.

[13] Moeng, C.-H., and Arakawa, A.: "A numerical study of a marine subtropical stratus cloud layer and its stability", *J. Atmos. Sci.*, **37** (1980) pp. 2661-2675.

[14] Rodgers, C. D.: "The use of emissivity in atmospheric radiation calculation", *Q. J. R. Meteorol. Soc.*, **93** (1967) pp. 43-54.

[15] Carruthers, D. J., and Hunt, J. C. R.: "Velocity fluctuations near an interface between a turbulent region and a stably stratified layer". (to be published in *J. Fluid Mech.*, 1985).

[16] Stage, S. A., and Businger, J. A.: "A model for entrainment into a cloud-topped marine boundary layer. Part I: Model description and application to a cold-air outbreak episode", *J. Atmos. Sci.*, **38** (1981) pp. 2213-2229.

[17] Randall D. A.: "Buoyant production and consumption of turbulence kinetic energy in cloud-topped mixed layers", *J. Atmos. Sci.*, **41** (1984) pp. 402-413.

[18] Rotta, J. C.: "Statistische theorie nichthomogener turbulenz", *Z. Phys.*, **129** (1951) pp. 547-572.

[19] Deardorff, J. W.: "Three-dimensional numerical study of turbulence in an entraining mixed layer", *Boundary-Layer Meteorol.*, **7** (1974) pp. 199-226.

[20] Wyngaard, J. C.: "Boundary layer modeling. Atmospheric Turbulence and Air Pollution Modeling", Ed. by F. T. M. Nieuwstadt and H. van Dop. D. Reidel Publishing Company. Dordrecht, Holland/Boston, U.S.A./London, England 1982.

[21] Lumley, J. L.: Introduction. In "lecture series 76: Prediction Methods for Turbulent Flows.", Von Karman Institute, *Fluid Dyn.*, Rhode-St-Genese, Belgium 1975.

[22] Zeman, O., and Tennekes, H.: "A self-contained model for the pressure terms in the turbulent stress equations of the neutral atmospheric boundary layer", *J. Atmos. Sci.*, **32** (1975) pp. 11808-1813.

[23] Klemp, J. B., and Rotunno, R.: "A study of the tornadic region within a supercell thunderstorm", *J. Atmos. Sci.*, **40** (1983) pp. 359-377.

DECAY OF CONVECTIVE TURBULENCE,
A LARGE EDDY SIMULATION.

F.T.M. Nieuwstadt[*], R.A. Brost[+], T.L. van Stijn[*]

[*] Royal Netherlands Meteorological Institute
  De Bilt, The Netherlands.
[+] National Center for Atmospheric Research
  Boulder, Colo 80307, U.S.A.

SUMMARY

A large eddy model for the simulation of atmospheric turbulence is discussed. The model is applied to the case of the convective boundary layer in which the surface heat flux is suddenly turned off. The characteristics of the ensuing turbulence decay are studied.

INTRODUCTION

Large eddy simulations have played an important role in our understanding of the atmospheric turbulence. In particular the pioneering work of Deardorff [1],[2] on the convective boundary layer is noteworthy. Stimulated by the growing capability of present day supercomputers numerical simulation of atmospheric turbulence has recently enjoyed increasing attention and it has been applied successfully to new problem areas: for example, the dispersion of passive contaminants (see e.g. [3] and [4]).

One of the attractive features of large eddy simulations is that we are able to study variables or processes which are not as easily investigated in real-life experiments. A good example of such a variable is pressure. A reliable and accurate device to sense the fluctuating static pressure within a turbulent flow is yet to be designed. Nevertheless, so-called pressure terms play a dominant role in the dynamics of turbulence but much needs to be clarified about their parameterization. Another example is the topic of this paper: the decay of atmospheric turbulence. Decay processes are inherently non-stationary and therefore difficult to study directly in the atmosphere, where we usually rely on time-averages to describe a turbulent flow.

As already stated our topic is the decay of turbulence in the convective boundary layer. This type of boundary layer, which is characterized by intense turbulent activity, develops when the boundary layer is heated up at the surface. A typical example is the day-time boundary layer over land when the sun is not obstructed by clouds. Near sunset the solar heating of the surface, the source of convective turbulence, ceases and consequently turbulence starts to decay. In this paper we will discuss some of the characteristics of this decay process.

Decay of turbulence is usually studied within the context of homogeneous, isotropic and non-stratified turbulence. An extensive review may be found for instance in [5]. However, the decay of turbulence in the atmospheric boundary layer is considerably more complicated. First of all the effects of stratification cannot be neglected. A convective boundary layer is usually characterized by a slightly stable temperature gradient, in the order of 0.001 $^{\circ}$C m^{-1}, and this temperature gradient becomes larger during the decay process. The influence of even such a small gradient cannot be neglected. Such a conclusion follows also from the laboratory experiments discussed in [6], [7] and [8]. Due to this influence of stability turbulence in the atmospheric boundary layer can be considered neither isotropic nor homogeneous. Moreover, the confinement between a free surface at the top and a fixed surface at the bottom renders vertical homogeneity of the boundary layer impossible. In addition entrainment at the top of the boundary layer forms an extra complication. All of these factors turn out to be important for the decay of convective turbulence.

In the real atmosphere the transition around sunset is rather complex because the boundary layer is then governed by several time scales. A turbulence time scale needs to be considered but also the time scale which governs the decrease of the surface heat flux, our source of turbulence. To avoid dealing with the complete problem we will restrict ourselves to an idealized and more simple flow: the decay of turbulence in a convective boundary layer for which the surface heat flux is instantaneously set to zero. We believe, however, that this investigation is a useful first step towards a full understanding of the boundary layer during the evening transition.

In this paper we use the results of a large eddy model which has been developed by one of us (RAB) and which we will describe in some detail in section 2. The details and results of the decay are then discussed in section 3.

DESCRIPTION OF THE MODEL

Large eddy modeling of atmospheric turbulence was started by Deardorff ([1], [2], [9] and [10]). Most of the recent applications ([3], [4] and also the model described in this paper) are still based on his original work.

We depart from the Boussinesq equations for a stratified flow with velocity components u_i in a rotating frame of reference. They read

$$\frac{\partial u_i}{\partial x_i} = 0, \tag{1}$$

$$\frac{du_i}{dt} = -\frac{1}{\rho_o}\frac{\partial p}{\partial x_i} + \frac{g}{T}(\theta-\theta_o)\delta_{i3} - 2\varepsilon_{ijk}\Omega_j u_k + \nu\frac{\partial^2 u_i}{\partial x_j^2}, \tag{2}$$

$$\frac{d\theta}{dt} = \kappa\frac{\partial^2 \theta}{\partial x_j^2}, \tag{3}$$

where $d/dt = \partial/\partial t + u_j\partial/\partial x_j$ and all other notations have their usual meaning (see [11]). The first equation expresses continuity of an incompressible fluid; the second equation describes the rate of change of momentum due to the forces on the right-hand side which are pressure, buoyancy, Coriolis acceleration and viscous friction; the last equation is the energy equation expressed in terms of a potential temperature, θ.

In a large eddy model turbulence is explicitly calculated by numerically solving (1) - (3) on a three-dimensional grid with dimensions Δx, Δy and Δz. To remove all scales of motion smaller than the grid size a filter is applied to (1) - (3) as discussed in [12]. A filtered or resolvable variable is denoted by an overbar and the filtered equations read

$$\frac{\partial \bar{u}_i}{\partial x_i} = 0, \tag{4}$$

$$\frac{d\bar{u}_i}{dt} = -\frac{\partial \pi}{\partial x_i} + \frac{g}{T}(\bar{\theta}-\theta_o)\delta_{i3} - 2\varepsilon_{ijk}\Omega_j \bar{u}_k + \frac{\partial \tau_{ij}}{\partial x_j}, \tag{5}$$

$$\frac{d\bar{\theta}}{dt} = -\frac{\partial h_j}{\partial x_j}, \tag{6}$$

in which $d/dt = \partial/\partial t + \bar{u}_j \partial/\partial x_j$.

In comparison with (1) - (3) we note a number of changes. There is no viscous term in (5) because in our applications the Reynolds number based on the grid size can be generally considered as very large. In addition we have introduced a modified pressure, which is defined by

$$\pi = \frac{p-p_o}{\rho_o} + \frac{1}{3}(\overline{u_i^2} - \bar{u}_i^2), \tag{7}$$

where p_o is a reference pressure which is determined by requiring that the horizontally averaged vertical acceleration equals zero. Furthermore, two new terms appear in (5) and (6). These describe the influence of the subgrid scales on the resolved motion. Their definition reads

$$\tau_{ij} = -(\overline{u_i u_j} - \bar{u}_i \bar{u}_j) + \frac{1}{3}(\overline{u_k^2} - \bar{u}_k^2)\delta_{ij}, \tag{8}$$

$$h_j = \overline{u_j \theta} - \bar{u}_j \bar{\theta}. \tag{9}$$

To solve the set of equations (4) - (6) with (7) - (9) we need to specify closure relationships. We take

$$\tau_{ij} = K_m \left(\frac{\partial \bar{u}_i}{\partial x_j} + \frac{\partial \bar{u}_j}{\partial x_i}\right), \tag{10}$$

$$h_j = -K_h \frac{\partial \bar{\theta}}{\partial x_j}, \tag{11}$$

where K_m and K_h are exchange coefficients to be discussed below.

Our approach implies that the Leonard terms, which read

$$\tau_{ij}^L = \{\overline{\bar{u}_i \bar{u}_j} - \bar{u}_i \bar{u}_j\}, \tag{12}$$

are not taken into account explicitly. Our justification is that the influence of these terms is not yet clear. According to [13] they may be responsible for about 30% of the energy transfer from large to small scales. However, a study discussed in [14] showed that the Leonard terms do not contribute to the energy transfer and that the simulation improves by not including them.

Next we consider the transport coefficients K_m and K_h which were introduced in (10) - (11). They are expressed by

$$K_m = 0.12 \, \ell \, E^{\frac{1}{2}} \tag{13}$$

$$K_h = (1 + \frac{2\ell}{\Delta}) \, K_m, \tag{14}$$

where $E = \frac{1}{2}(\overline{u_i^2} - \overline{u}_i^2)$ is the subgrid energy and where Δ is given by $(\Delta x \, \Delta y \, \Delta z)^{1/3}$. The mixing length scale is chosen depending on the local temperature gradient as

$$\ell = \Delta, \quad \text{if} \quad \partial\overline{\theta}/\partial z \leq 0 \tag{15}$$

$$\ell = \min\,(\Delta, \; 0.5 \, E^{\frac{1}{2}}/\{\frac{g}{T}\frac{\partial\overline{\theta}}{\partial z}\}^{\frac{1}{2}}), \quad \text{if} \quad \partial\overline{\theta}/\partial z > 0. \tag{16}$$

The latter expression for ℓ implies that for stable conditions ℓ can become very small so that subgrid transport becomes negligible.

To calculate (13) - (14) we need an additional equation for the subgrid velocity scale $E^{\frac{1}{2}}$. Such an equation which can be derived directly from the subgrid energy equation, is here approximated as

$$\frac{dE^{\frac{1}{2}}}{dt} = \frac{1}{2E^{\frac{1}{2}}}\{\tau_{ij}\frac{\partial \overline{u}_i}{\partial x_j} + h_j\} + \frac{\partial}{\partial x_j}\{2K_m \frac{\partial E^{\frac{1}{2}}}{\partial x_j}\} - \frac{\varepsilon}{2E^{\frac{1}{2}}}. \tag{17}$$

The first two terms on the right hand side of (17) are related to the production terms in the energy equation: namely production due to local shear and to local buoyancy flux; the third term can be interpreted as a transport term, which is modeled according to the gradient transfer hypothesis; finally, the ε in the last term stands for the viscous dissipation and it is parameterized by

$$\varepsilon = f \, [0.19 + 0.51 \frac{\ell}{\Delta}] \, \frac{E^{2/3}}{\ell}. \tag{18}$$

Here f is an empirical correction factor given by

$$f = 1 + \frac{2}{(z/\Delta z + 1.5)^2 - 3.3}. \tag{19}$$

It is designed to model the influence of the lower wall.

The expression (18) for ε insures that in stable conditions subgrid turbulence vanishes when the Richardson number exceeds a certain value. This so-called critical value is obtained by requiring that the ratio of the production terms to the dissipation term be larger than one. Using (10) - (19) and substituting $\ell/\Delta \to 0$ for very stable conditions we find

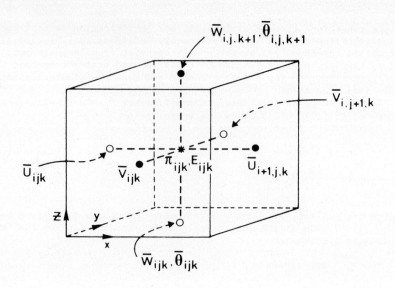

Fig. 1 The definition of the variables on a staggered grid.

$$Ri_{cr} \equiv \frac{\frac{g}{T}\frac{\partial \bar{\theta}}{\partial z}}{\frac{1}{2}(\frac{\partial \bar{u}_i}{\partial x_j} + \frac{\partial \bar{u}_j}{\partial x_i})^2} = 0.14, \quad (20)$$

where we have also assumed that $f = 1$.

The set of equations discussed above are solved in a rectangular domain with a horizontal dimension of 5 km and a vertical dimension of 2 km. In each coordinate direction we use 40 grid points so that the gridsize is 125 m in the horizontal and 50 m in the vertical direction.

The variables are defined on a staggered grid which is illustrated in Fig. 1. The resolved velocity, potential temperature and subgrid energy equations are solved in time by a leapfrog technique. The advection terms are discretized according to a finite difference scheme proposed in [15]. The diffusion terms are expressed in central differences. An Asselin filter [16] is used to prevent time splitting.

Continuity is satisfied by solving a Poisson equation for the pressure, which is readily derived from (4) and (5) as

$$\frac{\partial^2 \pi}{\partial x_i^2} = \frac{\partial R_i}{\partial x_i} - \frac{\partial}{\partial t}(\frac{\partial \bar{u}_i}{\partial x_i}), \quad (21)$$

where R_i stands for all the terms in (5) except for the pressure term and the local time derivative. The last term on the right hand side of (21), which should be zero for an exact solution, is carried along to insure that the divergence does not grow beyond round-off errors [9]. The solution of (21) is performed by a Fast-Fourier Transform technique in the horizontal directions and by a tridiagonal matrix decomposition in the vertical.

The horizontal boundary conditions are periodic for all variables exept for pressure, which is given a trend along the y-axis in order to impose a mean geostrophic wind of 5 m s^{-1} in the positive x-direction.

The upper boundary condition is a zero gradient for all velocity variables and a prescribed gradient for the temperature. The subgrid energy is set equal to a minimum value. To avoid reflection of gravity waves against the top of the model we include a so-called sponge layer which extends across the ten upper levels. It means that a linear relaxation term is added to the right hand side of (5) and (6), which forces the solution at every gridpoint toward the horizontal average and thus dampens all fluctuations. The time scale in the relaxation equation is taken as 50s at the upper level and it is increased by a factor 5 for each lower level.

At the lower boundary we set the vertical velocity and the gradient of the subgrid energy equal to zero. The boundary condition for w implies that the horizontally averaged vertical velocity will vanish at each level due to the continuity equation and the cyclic lateral boundary conditions. The surface-layer similarity profiles of [17] with a roughness length of 1 cm are employed at the lowest level to calculate the momentum and temperature flux from the velocity and temperature variables. However, profiles in the surface layer vary quickly especially during unstable conditions and cannot be resolved on a grid of 50 m. To remedy this shortcoming the temperature difference between $z = \Delta z$ and 0 is diminished by a correction factor in order to calculate the temperature flux at $\Delta z/2$.

Furthermore, it turns out that the shear production term in (17) is much larger than would follow from surface-layer similarity profiles. This is the reason for adopting the wall correction factor f in (18), otherwise the subgrid energy would grow too large here. On the whole it must be concluded that the formulation of the lower boundary condition is not completely satisfactory. However, a fortunate circumstance seems to be that the overall results of our model are not very sensitive to these lower boundary conditions.

DECAY EXPERIMENTS

Let us now consider some results of our large eddy model. We start with a fully developed convective boundary layer, which we obtain by running the model while increasing the surface temperature at a constant rate. The turbulence of this initial condition can be considered as statistically stationary. At time t = 0 the temperature gradient between

Table 1 Characteristic Boundary-layer Parameters

Experiment	1	2	6	10
w_* (m/s)	2.4	1.7	0.97	2.3
h (m)	1400	1300	1400	1350

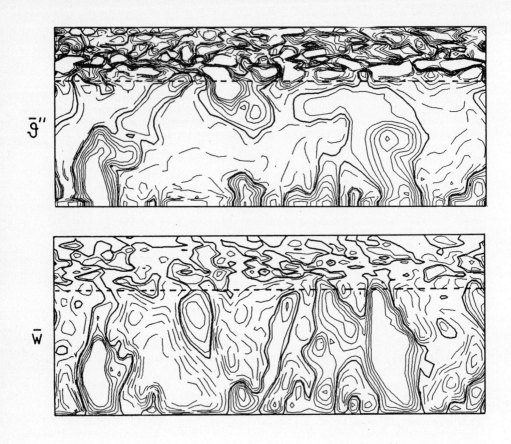

Fig. 2   Isolines of temperature and vertical velocity fluctuations in a vertical cross section of the calculation domain for experiment 1 at $tw_*/h=0.06$. Solid lines indicate positive fluctuations and broken lines negative fluctuations. The values of the isolines are -0.33, -0.27, -0.21, -0.15, -0.09, -0.03, 0.03, 0.09, 0.15, 0.21, 0.27, 0.33 °C for temperature fluctuations and -1.8, -1.4, -1.0, -0.6, -0.2, 0.2, 0.6, 1.0, 1.4, 1.8, m s^{-1} for the vertical velocity fluctuations. The thick solid line denotes the zero fluctuation. The dashed horizontal line is the mean boundary-layer height.

the surface and the grid point at $z = \Delta z$ is set equal to zero and kept zero afterwards. Consequently, the subgrid portion of the heatflux at $z = \Delta z/2$ vanishes and since the resolved part of the heatflux is generally small at $z = \Delta z/2$, we may consider the total surface heatflux as negligible after $t = 0$. In other words, we have removed the source of convective turbulence and as a result turbulence starts to decay.

At this stage it is convenient to introduce some notations on several averaging operators. In the previous section we have already encountered

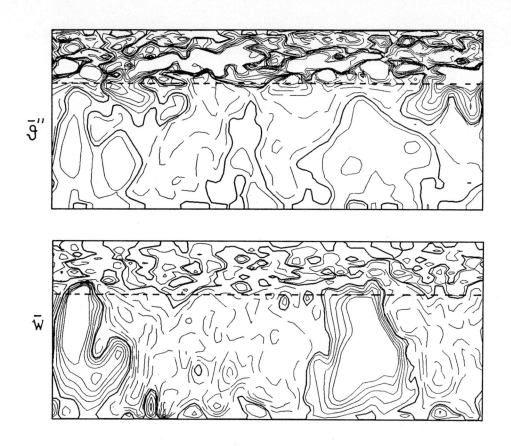

Fig. 3 Same as Fig. 2 for $tw_*/h=0.67$.

the overbar, which stands for the filtering of subgrid fluctuations. A horizontal average across the whole calculation domain is indicated by angular brackets: $\langle \ \rangle$ and deviations from this average are indicated by a double prime: e.g. $\overline{f}'' = \overline{f} - \langle \overline{f} \rangle$. Since we apply cyclic lateral boundary conditions a horizontal average is equivalent to an ensemble average. Finally, square brackets [ ] denote a vertical average across the convective boundary-layer depth.

We will discuss here the results of four decay experiments. Some characteristics of the initial conditions of each experiment are given in table 1. The velocity scale w_* is a measure for the magnitude of velocity fluctuations. It is defined by

$$w_* = (\frac{g}{T} H_\theta h)^{1/3}, \qquad (22)$$

where $H_\theta = \langle \overline{w}\ \overline{\theta}''\rangle_{z=0} - \langle K_h \partial\overline{\theta}/\partial z\rangle_{z=0}$ is the surface temperature flux and h

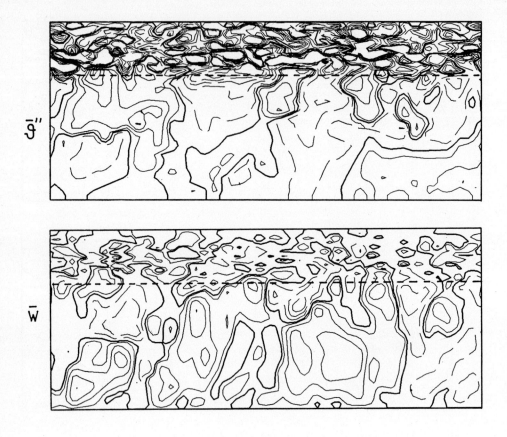

Fig. 4 Same as Fig. 2 for $tw_*/h=1.69$.

is the boundary-layer height.

Let us consider now the results of our experiments. We start with a qualitative approach by showing in Figs 2, 3 and 4 some vertical cross-sections of the instantaneous fluctuations $\overline{\theta}''$ and $\overline{w}''$ at several stages during decay experiment 1. Fig. 2 shows the initial condition, which is characterized by warm rising thermals that span the whole boundary-layer depth. In this figure one may also identify near the top of the boundary layer areas of so-called entrainment. At the time $tw_*/h = 0.67$ shown in Fig. 3 most of the temperature fluctuations have died out whereas in the vertical velocity fluctuations the structure of thermals is still recognizable. Also some evidence of entrainment remains visible here. Finally at $tw_*/h = 1.69$ shown in Fig. 4 most velocity and temperature fluctuations have disappeared, although some areas of relatively intense fluctuations remain present near the top of the boundary layer.

Next we consider the behavior of several variables as a function of time in somewhat more detail. We start with the resolved turbulent kinetic energy which is defined by

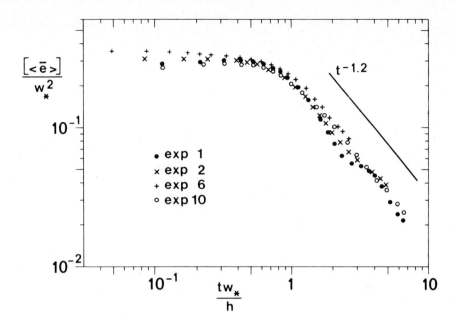

Fig. 5 The resolved turbulent kinetic energy averaged over the boundary layer as a function of tw_*/h.

$$\bar{e} = \tfrac{1}{2}(\overline{u''^2} + \overline{v''^2} + \overline{w^2}), \tag{23}$$

where we have used that $\langle \bar{w} \rangle = 0$ so that $\bar{w} = \overline{w''}$. The vertical average of $\bar{e}$ non-dimensionalized by w_*^2 is plotted in Fig. 5 as a function of tw_*/h. We find that all four experiments follow more or less one single curve. In other words the decay process is self-similar when scaled in terms of w_* and h. Physically it means that the initial state of turbulence governs the ensuing decay process.

The picture of Fig. 5 indicates that the turbulent kinetic energy stays approximately constant for one non-dimensional time unit and then drops according to a power law with an exponent close to -1.2. This exponent lies in the range usually observed for the decay of isotropic turbulence [5]. However, this fact should be considered as fortuitous because the turbulence in our case is far from isotropic.

This latter remark may be substantiated if we consider the behavior of the individual velocity components separately. In Figs. 6 and 7 we show the results for the horizontal velocity fluctuations along the y-axis, $\overline{v''}$, and for the vertical velocity fluctuations, respectively. Both components stay initially constant but at $w_* t/h \sim 1$ they start to decay at a different rate: $\bar{w}$-fluctuations faster than the $\overline{v''}$-fluctuations. The result is that $\bar{w}$-fluctuations, which are initially larger than the $\overline{v''}$-fluctuations, become smaller than the $\overline{v''}$-fluctuations near the end of our experiment. Turbulence is thus clearly non-isotropic. The explanation of this behavior is two-fold. First, the mean temperature gradient in the boundary layer is slightly stable. Second, as we have seen

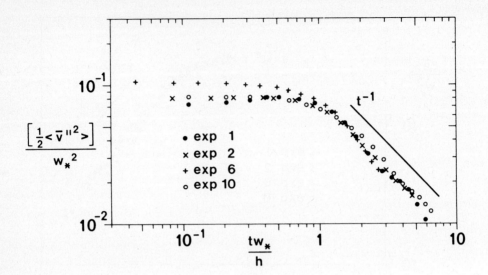

Fig. 6 The resolved horizontal velocity variance averaged over the boundary layer as a function of tw_*/h.

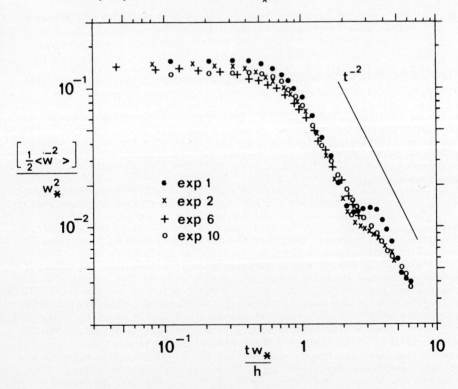

Fig. 7 The resolved vertical velocity variance averaged over the boundary layer as a function of tw_*/h.

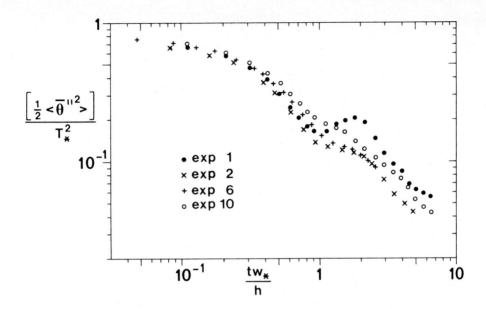

Fig. 8 The resolved temperature variance averaged over the boundary layer as a function of tw_*/h.

in Fig. 3 entrainment continues during the decay process. Both effects act preferentially to destruct $\bar{w}$-fluctuations rather than $\bar{v}''$-fluctuations.

Finally we show in Fig. 8 the behavior of the temperature fluctuations as a function of time. We find that the temperature fluctuations start to decay almost immediately, which is in agreement with the results shown in the isoline plots of Fig. 2-4. In addition we notice that some deviations from similarity occur. In particular around $tw_*/h \simeq 2$ the curves for the four experiments behave quite differently with e.g. an increase in temperature fluctuations in the case of experiment 1. A corresponding increase in the vertical velocity fluctuations may be also observed in Fig. 7. It turns out that this behavior is the result of demixing of entrained air from the boundary layer, and for a more detailed discussion we refer to [18].

CONCLUSIONS

We have discussed in some detail a large eddy model for the simulation of atmospheric turbulence. This model has been applied to a study of the decay of turbulence in a convective boundary layer. Our main conclusion is that as a first approximation the decay process can be scaled in terms of variables such as w_* and h, which characterize the initial condition. The decay hierarchy is that temperature fluctuations decay first, whereas velocity fluctuations initially persist. Next the vertical velocity variance decays and finally the horizontally velocity variance.

REFERENCES

[1] Deardorff, J.W., 1974a: Three-dimensional numerical study of the height and mean structure of a heated planetary boundary layer. Bound.-Layer Meteor., 7, 81-106.

[2] Deardorff, J.W., 1974b: Three-dimensional numerical study of the height and mean structure of a heated planetary boundary layer. Bound.-Layer Meteor., 7, 199-226.

[3] Wyngaard, J.C. and R.A. Brost, 1984: Top-down and bottom-up diffusion of a scalar in the convective boundary layer. J. Atmos. Sci., 41, 102-112.

[4] Moeng, C.H. and J.C. Wyngaard, 1984: Statistics of conservative scalars in the convective boundary layer. J. Atmos. Sci., 41, 3161-3169.

[5] Monin, A.S. and A.M. Yaglom, 1975: Statistical Fluid Mechanics: mechanic of turbulence, Vol. 2, J.L. Lumley (ed.), The M.I.T. press.

[6] Dickey, T.D. and G.L. Mellor, 1980: Decaying turbulence in neutral and stratified fluids. J. Fluid. Mech., 99, 13-31.

[7] Britter, R.E., J.C.R. Hunt, G.L. Marsh and W.H. Snyder, 1983: The effects of stable stratification on turbulent diffusion and the decay of grid turbulence. J. Fluid. Mech., 127, 27-44.

[8] Stillinger, D.C., K.N. Helland and C.W. van Atta, 1983: Experiments on the transition of homogeneous turbulence to internal waves in a stratified fluid. J. Fluid. Mech., 131, 91-122.

[9] Deardorff, J.W., 1973: Three-dimensional numerical modeling of the planetary boundary layer. Workshop of micrometeorology, Amer. Meteor. Soc., Boston.

[10] Deardorff, J.W., 1980: Stratocumulus-capped mixed layers derived from a three-dimensional model, Bound.-Layer Meteor., 18, 495-527.

[11] Businger, J.A., 1982: Equations and concepts. Atmospheric Turbulence and Air Pollution Modelling, F.T.M. Nieuwstadt and H. van Dop (eds.), D. Reidel, Dordrecht.

[12] Wyngaard, J.C., 1982: Boundary-layer modelling. Atmospheric Turbulence and Air Pollution Modelling. F.T.M. Nieuwstadt and H. van Dop (eds.), D. Reidel, Dordrecht.

[13] Leonard, A., 1974: Energy cascade in large eddy simulations of turbulent fluid flows. Advances in Geophysics, Vol. 18A, Academic Press, 237-248.

[14] Antonopoulous-Domis, M., 1981: Large-eddy simulation of a passive scalar in isotropic turbulence. J. Fluid. Mech., 104, 55-79.

[15] Piacsek, S.K and G.P. Williams, 1970: Conservation properties of convection difference schemes. J. Comput. Phys., 6, 392-405.

[16] Asselin, R., 1972: Frequency filtering for time integrations. Mon. Wea. Rev., 100, 487-490.

[17] Businger, J.A., J.C. Wyngaard, Y. Izumi and E.F. Bradley, 1971: Flux-profile relationships in the atmospheric surface layer. J. Atmos. Sci., 28, 181-189.

[18] Nieuwstadt, F.T.M. and R.A. Brost, 1986: The decay of convective turbulence. To appear in J. Atmos. Sci.

SUMMARIZING STATEMENTS ON RESULTS, TRENDS AND RECOMMENDATIONS

The Colloquium was ended with a concluding session. The chairmen had asked several leading scientists to express their personal view on "what have we learned, what has been recommended, what are the trends" with respect to:

1. Numerical methods, boundary and initial conditions
2. Filtering and subgrid scale models
3. Basic picture of turbulent flow structures
4. Engineering applications
5. Atmospheric flows
6. Impacts on statistical models
7. Transition to turbulence.

Here the write-ups are collected as provided by the speakers. Two statements have been provided with respect to point 3.

NUMERICAL METHODS, BOUNDARY CONDITIONS, AND INITIAL CONDITIONS
John Kim
NASA Ames Research Center, Moffett Field, California 94035

I was originally asked to summarize the papers topically related to numerical methods, boundary conditions, and initial conditions presented during the meeting. However, there was not much discussion on these subjects during the meeting. I will, therefore, attempt to present a personal point of view regarding these topics in this summary. It appears that the diminished research interest in these subjects has certain implication that enough confidence in numerical algorithms has been achieved in reproducing realistic turbulence. This is indicative of the maturity of this field - both direct and large-eddy simulation of turbulence - and it appears that turbulence simulation has now become a truly predictive and effective tool in turbulence research. However, this is only true for simple turbulent flows at relatively low Reynolds numbers. Much more work is needed to extend our capability into more complicated flow at high Reynolds numbers. In the years to come, the following subjects, at least in my opinion, will be the major pacesetting research areas in the numerical simulation of turbulence.

NUMERICAL METHODS

Spectral vs. Finite Difference methods

Spectral methods will continue to prevail in computing turbulent flows involved with simple geometries. A carefully written spectral code can provide both accuracy and efficiency with little extra cost in computing these flows. However, it is not clear at this point that the application of spectral methods to complex geometries will provide any advantages over the finite difference method. For large scale computations, the difficulty involved with complex geometries is intensified by more elaborate data management systems required to perform the simulation. Spectral methods require all data points (global) in contrast with finite difference methods which require only adjacent points (local). Furthermore, there is some indication that only twice as many computational nodes (as those required for spectral methods) are sufficient for a second-order finite difference method to achieve the same accuracy as in a spectral method for turbulence simulations. It seems apparent that the finite difference codes will be more readily used for consideration of more complicated turbulence problems.

Time-advancement

There has not been much progress in this subject. Most people use an explicit scheme for the non-linear terms and an implicit scheme for the viscous terms, and the time advancement of all these schemes is limited by the Courant-Friedrichs-Lewy (CFL) condition. Although the requirement of time accuracy in turbulence simulations does not allow an order of magnitude improvement in time step, there is still a room for improvement in this area by exploring the possibility of eliminating the CFL condition. Time advancement has always been a critical issue because several tens of CRAY CPU hours are typically required to obtain good statistics in direct simulations of low Reynolds number flows. The work by LAURENCE in handling the convective terms looks promising in this respect.

BOUNDARY CONDITIONS

Wall Boundary Condition

In a numerical simulation of a wall-bounded shear flow, one has to use a very fine mesh system near the wall to resolve the small scales existing in this region. This requirement of fine mesh near the wall becomes more severe as the Reynolds number increases. In most cases, however, we do not have to resolve the fluid motions in the vicinity of the wall, if we are only interested in the global characteristics of the flow. This is well demonstrated by the earlier works by DEARDORFF and by SCHUMANN in which they used a synthetic boundary condition rather than the natural no-slip boundary condition thereby avoiding explicit computation of the near wall region. By implementing well designed synthetic boundary conditions, we can dramatically reduce the fine-grid requirement for high Reynolds number flows. This will be the key to a successful application of numerical simulations to practical engineering problems. Not much progress has been made since the pioneering work of DEARDORFF and SCHUMANN. Now that much more is known about the large eddy structures near the wall owing to both experiments and numerical simulations, it should be relatively easy to devise a better synthetic boundary condition. A few possible candidates for this approach include: 1) inclined eddy structures near the wall; 2) the use of the bursting process; and 3) the near-wall splatting phenomena.

Inflow-Outflow

All the computational work presented during the meeting, except the work of SCHMITT, RICHTER & FRIEDRICH, has used periodic boundary conditions to avoid inflow-outflow boundary conditions. Periodic boundary conditions convert a spatially growing flow into a temporally growing flow. For an internal flow, such as the flow in a channel or in a pipe, this type of approximation does not seem to affect the flow characteristics. However, the flow characteristics of an open flow, such as a turbulent mixing layer or a turbulent jet, seems to be affected more by the change from a spatially growing flow to a temporally growing flow. For example, there is some indication that a temporally growing flow may be more dependent on initial conditions than a spatially growing counterpart. Needless to say, a whole lot of work is required to find the means of implementing a reasonable inflow boundary condition. A good inflow boundary condition obviously requires only a short computational domain; however, for an unreasonable inflow condition to develop into a realistic one, a longer computational domain will be necessary. Again, constructing an inflow condition based on large scale structures, such as horseshoe vortices for a wall-bounded flow, may produce a better flow field than that simply matching low order statistics. This has been attempted in the past without much sucess. The outflow boundary condition may not impose as serious a problem as the inflow condition since the upstream effect will be limited near downstream boundaries.

INITIAL CONDITIONS

It appears that initial conditions do not seem to affect the final results of the internal flows with the exception of the time required to reach the statistically steady state. Better initial conditions obviously, result in shorter transient time. For a temporally growing open flow, such as a mixing layer, this may not be the case. However, there is no hard evidence in either way.

In summary, some strong capabilities of numerical simulations have been demonstrated during the colloquium. The premise of large-eddy simulation has been validated, and the future of numerical simulation of turbulence looks promising. Computer generated data will be used more and more in studying turbulence. The next pacing items will be in developing better wall boundary conditions for high Reynolds number flows and inflow-outflow boundary conditions for spatially growing flows. Once we achieve these goals, the field of numerical simulation of turbulence will have an unmatched potential for predicting, understanding and manipulating turbulent flows having wide practical applications.

FILTERING AND SUBGRID SCALE MODELLING
B. Aupoix
ONERA / CERT / DERAT, Toulouse, France

FILTERING

When I was asked to make this review, I thought it will be only devoted to subgrid scale models. So I had the great pleasure to find during this meeting people still asking questions about filters.

The paper by LAURENCE has pointed out the filtering operation due to the numerical method. He proposed a way to determine this filter and to use only it in the computations. As the numerical filter is close to a Gaussian filter, this could renew studies on Gaussian filters instead of low-pass filters used by most of people developing subgrid scale models.

The paper by MASON (see J. Fluid Mech., Vol. 162, 1986, 439) brought into evidence the difficult balance between mesh, filter, subgrid scale model and numerical scheme. His study was aimed at channel flow simulations with a finite difference method. With the SMAGORINSKY model, he proposed to interpret the product $C_s \Delta$ where C_s is the model constant and Δ the mesh scale as a measure of the filter scale. To be sure to capture the energy producing eddies in the wall region, this length scale $C_s \Delta$ must be smaller than 5 % of the channel half-width; otherwise, the model does not describe the correct turbulent damping. Moreover, MASON proposed to interpret the required value of the constant C_s as a measure of the numerical resolution.

SUBGRID SCALE MODELLING

Figure 1 gives a very restricted view of the state of art in subgrid-scale models, limited to models used or referenced during this meeting. Horizontal axis refers to the degree of sophistication of the mathematics involved in subgrid scale modelling. Vertical axis refers to the complexity of the flow. Dotted lines show model evolutions while continuous lines bring into evidence some role played by the subgrid scales, found with one theory and introduced in another model.

The first class of model is mainly based on KOLMOGOROV's hypothesis and dimensional analysis so I called them "semi-empirical" models. The first and very popular model is due to SMAGORINSKY who developed it for inhomogeneous flow computations. The model constant has been calibrated with respect

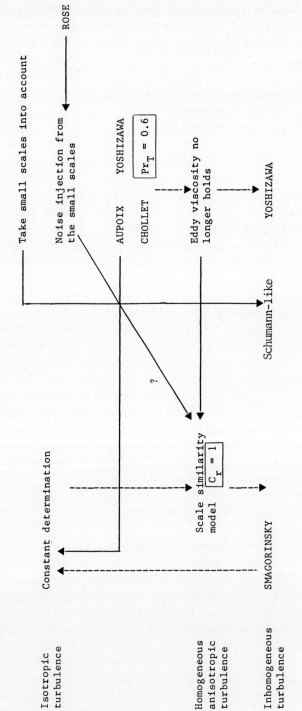

to isotropic turbulence decay and a refinement was proposed by BARDINA to improve the model by comparison with subgrid scale terms obtained in direct simulation of homogeneous turbulence. LAURENCE has proposed a modification of BARDINA's scale similarity model to model together both LEONARD and subgrid terms and to insure Galilean invariance. Independently, SPEZIALE found that the constant in BARDINA's model should be equal to one if the variation of the subgrid term under Galilean motion is to cancel that of the LEONARD term when this one is separately modelled or computed.

Another kind of semi-empirical model is what I labelled as "Schumann-like" as a lot of people reported using SCHUMANN's model or a variant of it. The main characteristics of these models are

- use of a filter which is defined as the mean value over one computational grid cell

- separation of stresses into those induced locally ("isotropic" part) and those due to mean shear ("inhomogeneous" part)

- using length scales and model coefficients which account for the grid geometry and are applicable also for non-cubic grid cells

- use of a transport equation for the subgrid-scale kinetic energy

As shown in the paper by GRÖTZBACH, this model was also derived for inhomogeneous turbulent scalar fields and constants have been settled by comparison with simple inhomogeneous experiments.

The second class of models is based on two-point closures or statistical models, following pioneering papers by KRAICHNAN or LESLIE and QUARINI. The models proposed have been derived in a step by step approach, starting from isotropic turbulence and increasing the flow complexity.

For isotropic turbulence, statistical models bring into evidence two major features of subgrid scale terms: first, they are linked to the small scales so that some knowledge of the small scale, at least in a statistical way, is needed; secondly they inject random noise or phase in the large scales. The first point leads to models taking the small scales into account in ways similar to SCHUMANN's model. The second point has also been brought into evidence by ROSE, using renormalization group theory. It has been usually discarded by most of the modellers but BERTOGLIO and LAURENCE proposed ways to inject stochastic noise from the small scales into the large scales.

For isotropic turbulence, YOSHIZAWA and myself have shown that simple models derived from statistical theories are equivalent to SMAGORINSKY's model under restrictive assumptions. At last, the paper by CHOLLET, dealing with passive scalar, showed that the turbulent Prandtl number can be assumed constant in subgrid scale modelling when the molecular Prandtl number is about one.

For anisotropic turbulence, BERTOGLIO has shown that eddy viscosity models no longer apply as energy transfer becomes highly anisotropic. My paper confirms these results. BERTOGLIO proposed sophisticated models to inject noise from the small scales and modelled the anisotropy of energy exchange. I have tried to bring into evidence the fact that this anisotropy is linked to the small scale anisotropy and to model it in a simple way.

Up to now, only YOSHIZAWA extended models for inhomogeneous turbulence derived from statistical models. The model he proposed is much more complex than standard models used in inhomogeneous flow LES, but seems to be able to mimic the anisotropy of energy exchange even in the wall region.

As renormalization group theory (RNG) papers were not presented, I shall not comment them.

I would like to point out finally two kinds of trends.

Some opposition exists between people developing models and people using them. Developers often use statistical theories, so the use of low-pass filter is the more natural choice for them. As they get deeper knowledge into the role played by the small scales, they want to develop very complex models for homogeneous turbulence. They are dealing with homogeneous turbulence, so they use Fourier methods and have little numerical problems. On the other hand, people using subgrid scale models are not always using low-pass filters but Gaussian filters or "numerical" filters as they are more suited to their numerical codes. Moreover, they hope to be able to capture the essential features of the flow in the resolvable scales and are looking for easy-to-implement and inexpensive subgrid scale models. At last, their numerical method often does not have the accuracy of Fourier methods to evaluate space derivative and their subgrid model has to balance these defficiencies. Papers by GRÖTZBACH and LAURENCE presented subgrid scale models developed to be well suited to the numerical codes.

For inhomogeneous flows, which are the final goal of LES, two kinds of models are mainly used, the SMAGORINSKY or scale-similarity models on one hand, the SCHUMANN-like models on the other hand. The scale similarity models take advantage of the properties of the Gaussian filter while SCHUMANN-like models are better adapted to the discrete solution scheme. The YOSHIZAWA model or simplified forms of it could be promising challengers.

THE CONTRIBUTION OF DIRECT AND LARGE EDDY SIMULATION TO OUR BASIC
UNDERSTANDING OF TURBULENCE 'STRUCTURES'

J. C. R. Hunt

University of Cambridge, Department of Applied Mathematics
and Theoretical Physics, Cambridge, United Kingdom

This successful colloquium has certainly reinforced my belief that computation of turbulent flows by direct (DS) and large eddy simulation (LES) can improve our understanding of all scales of turbulent motion. Some progress has been made, but there is a lot more that could be done, and I will make some suggestions.

This understanding should lead to a qualitative picture of the structure or mechanism of turbulent flow and help our search for methods of describing different parts of turbulent flows with simple formulae or simple computations. However, one needs to think carefully how best to achieve these two objectives. As a number of papers at this colloquium have shown, one can categorise the methods for applying DS and LES to gain understanding into two groups.

KINEMATICAL/STATISTICAL ANALYSIS

There are many different aspects of turbulent flows that can be usefully understood from a direct kinematical and statistical analysis of the results of computations. But the kind of kinematical analysis of the data depends very much on the aspect that needs to be understood or the particular application to be understood. An interesting feature of this colloquium has been the realisation that the traditional statistical methods of spectra moments and correlations are not enough.

(a) Orthogonal mode decomposition, which has been discussed in a number of papers, is a useful technique for describing and ordering in an unambiguous manner which motions contain most energy; we have learnt that especially for low Reynolds number flows, the number of modes may be less than 10. But it may not be the best way of describing the trajectories of fluid elements or of the convolution of fluid surfaces which are sensitive to relative motions.

(b) Lagrangian motions of particles and surfaces need to be better defined if we are to understand or model diffusion, mixing and reactions. (In our own paper by CARRUTHERS, HUNT & TURFUS we showed that a very simple kinematic simulation suggests that the conventional estimates for the Lagrangian time scale near a surface or interface in a shear-free flow is wrong, along with the conventional idea about the motion of fluid elements in such a region.) The rate at which pairs of fluid elements move apart is of course a measure of how the turbulence can stretch vortex lines. New flow visualisations, tracer and analytical techniques for measuring Lagrangian quantities, and new techniques of random-flight modelling of diffusion and mixing should provide a stimulus to this application of DS and LES

(c) Defining the location and nature of singular points is in general not very useful for turbulent flows, but where the turbulence has some regular spectral and temporal structure, the main features of the structures can be defined by the singular points of the streamlines. Then it becomes possible to compare different computations more precisely than by the too frequent process of looking for superficial similarity! (PERRY has shown that by giving an actual flow, e.g. a jet or wake, a small forced oscillation ('jitter'), these basic structures appear strongly out of the noise. It would be interesting to see if the same is found in computed flows using DS or LES).

Rotational (or vortical) and irrotational motions both play an essential role in turbulence; in homogeneous turbulence, KERR's DS have shown how irrotational straining motions surround regions of high vorticity, as one would expect, because the former provide the energy for the latter. We have seen in this colloquium nice examples of concentrated vortex motion in boundary layers, but it would also have been interesting to have learnt about the straining motions surrounding these hair-pin vortices. (It is as if we have seen Hamlet without the ghost.) In our research at Cambridge, we have been finding that the way turbulence usually adjusts to the presence of rigid surfaces or density interfaces, in shear-free flows, is by generating irrotational motions (as MOENG's LES largely confirms). This idea is an old one in turbulence (BRADSHAW & TOWNSEND refer to these as 'inactive motions'), but until DS and LES methods become available, it could not be directly tested.

COMPARING COMPUTATIONS WITH SIMPLE MODELS AND CONCEPTS

This is the other main way of using DS and LES to improve our understanding, assuming that we mostly want to know the simplest way of describing the essential features of turbulence relevant to any particular situation or application. So DS and LES should enable us to know where the simple models and concepts are or are not appropriate; for example it is straightforward to test mixing length models by comparing shear stress and velocity gradients, but more difficult to test the mixing length concept. Do fluid elements move across streamlines, interact with their new surrounding, and accelerate, as PRANDTL envisaged; or do large fluid volumes interact with the whole scale of the flow, as is believed by the 'coherent structure school' of turbulence. It appears that the latter phenomena are often occurring in flows whose statistics can be described by models based on local gradients and mixing length or transport equation methods. But why? I believe these are the dynamical questions that need answering if we are to understand the physics of turbulence better and also know when models based on a statistical physics concept of turbulence are appropriate.

Another different class of 'simple' models are those based on the rapid distortion theory (R.D.T.), in which the turbulence structure is calculated from the linear interaction between the forcing influence on the flow and some initial or base state of turbulence, e.g. stratification, boundaries, shear, etc. This approach is essentially one based on the idea that turbulent eddies have a long life and 'remember' their history - quite different to the statistical physics idea of turbulence. These models are formally valid when the distortion time scale is small compared with the turnover or non-linear time scales of the turbulent eddies. There have been several useful DS and LES comparisons with R.D.T., including those mentioned at this colloquium. But I would say that such comparisons have not been carried out with the detail that enables us to know what the level of error is in any given sitation in using R.D.T. That is something that I await with eager anticipation!

The interaction of motions at different scales is at the heart of the fascination of studying turbulence, and is crucial to many important practical problems. LES computation schemes are based on some concept or model of the interaction of small (sub-grid) and large scales. We have heard about several different approaches to modelling S.G.S. at this conference. The fact that they differ so much suggests we need to understand more about these interactions. This is where both DS can help and also analytical models based on different kinds of simplification or assumption, e.g. renormalisation, eddy-damped quasi-normal modes, and models for which idealised spectra (such as where the energy is confined to two different regions of wave number space; cf. DHANG's presentation.)

More generally, DS (and to a lesser extent) LES should also help with understanding how the small scales affect the large scales. It is obvious that they can act as an eddy viscosity, but can they transmit fluctuating effects upscale, helicity or scalar quantities? These are exciting questions for the future.

ON THE BASIC PICTURE OF TURBULENCE
D. K. Lilly
University of Oklahoma, School of Meteorology, Norman, Oklahoma
73071

Having come to this conference with relatively little recent contact with several of the fields represented, I may have been overly impressed by the more dramatic results presented. With that caveat, I would nevertheless characterize the sessions on inhomogeneous turbulence as containing near-breakthrough contributions to the understanding of turbulence generation and dissipation, both interestingly based on eddy geometry. Whether the "hairpin" instability structure is as universally relevant as suggested by MOIN and KIM, or is just easily identified from simulation, remains to be shown, but the results are sure to stimulate further work. I was also much impressed by the statistics presented by SHTILMAN et al. showing the strong inverse correlations between helicity and dissipation in simulated decaying turbulence. I have argued the importance of helical flows in long-lived buoyant connective storms in the atmosphere (LILLY, 1986), and this comparison shows the value to fluid dynamics of frequent information exchanges across the increasingly broad span of our field. A key remaining problem is to show how the instability induced eddies which generate turbulent energy are transformed or reduced to the helical eddies which characterize decaying turbulence.

I would like to comment also on the dramatic results shown by OGAWA et al. (documented in the paper by KUWAHARA & SHIRAYANA) in their simulation of classical and applied flows, in which subgrid-closure is eliminated and apparently successfully replaced by a third order upwind-difference scheme containing implicit fourth-order damping. Again I believe analogous results have been found in atmospheric simulations, though not so sharply defined. I interpret these results as indicating not so much that interaction of resolved and sub-grid scales are insignificant, but that accurate simulations of the interaction of resolvable scales are crucial. We may be forced to admit that these previously interactions are severely suppressed and distorted by most of the sub-grid closure schemes previously espoused. Potential users should, however, recognize that the upwind difference schemes are indefensible in at least one respect - their damping phase distort rates depend on flow velocity and are therefore not independent of a Galilean transform.

Lilly, D.K., 1986: On the structure, energetics and propagation of rotating convective storms. Part I and II. J. Atmos. Sci. 43 (1986) 113-140.

LARGE EDDY SIMULATIONS OF THE PLANETARY BOUNDARY LAYER
P.J. Mason
Meteorological Office, Meteorlogical Research Unit, R.A.F.
Cardington, Bedford, United Kingdom

Flow in the planetary layer is complex but in principle just an example of the many flows for which LES can be used to describe. It follows that the other statements summarising the proceedings have, apart from some points of special difficulty, dealt with all that is involved in LES of the planetary boundary layer. Difficulties in dealing with the planetary layer arise from the wide range of dynamics and length scales involved.

To date, simulations have considered cases dominated by either homogeneous fluid shear instability or buoyant convection. The former studies have much in common with studies of turbulent channel flow and although some success has been obtained there is uncertainty over the reliability of the technique and the results. The latter studies have been most extensive and most successful. It is clear that the large scale eddies which characterise buoyant convection are easy and natural to represent. In contrast to most one point closure methods the results show good agreement with observations.

The most difficult aspect of the simulation of the planetary boundary layer concerns flows with a stable static stability. Such flows may occur in the whole night-time boundary layer or may be a local, but important, process at an elevated inversion capping a neutral or unstable boundary layer. In both these different cases the scale of the turbulence is reduced and most transports may occur on a subgrid scale. As a result of this difficulty there are no successful simulations of the nocturnal boundary layer. The problems at an elevated inversion are hard to ascertain; simulations of convective boundary layers with a capping inversion give plausible values to the entrainment and suggest that the entrainment may largely be determined by the resolved scale motions. If this is the case, the problems with the small scale processes may not be of such practical importance. There is an urgent need to provide experimental verification of the behaviour of motions on various scales in the region of the capping inversion.

A related problem concerns the finite difference resolution of the capping inversion. Not only does the turbulence occur on a small scale but also the mean gradients are large. The capping inversion is thus often represented as a gridscale discontinuity with values of local Richardson number dependant on the mesh resolution. Urgent work is needed to ascertain whether the large consequent finite difference errors affect the results.

When these uncertainties over the representation of statically stable flows are resolved, the Large-eddy simulations of the planetary boundary layer will be much more certain to give valuable results.

ON ENGINEERING APPLICATIONS

Kunio Kuwahara

The Institute of Space and Astronautical Science, Tokyo 153, Japan

To treat turbulent flows in engineering applications, approaches somewhat different from a puristic scientific research are needed. The problems themselves are so complex in general that rigorous solution methods are inapplicable. Since very often the codes are used by non-fluid dynamicists they should be as simple as possible and robust.

If the flow boundaries are not necessarily composed of straight lines nor pure circular curves, then the code is expected to be written in a generalized coordinate system so that one coordinate line is always parallel to the wall.

In complicated flows, even if the Reynolds number is very high, the flow is not always nor everywhere turbulent, but it is sometimes or somewhere laminar. This means that transition from laminar to turbulent flows and from turbulent to laminar flow can occur. The code should therefore be capable to compute both laminar and turbulent flow.

Usually large scale structures are more important than small scale ones in engineering applications. At least, we have to capture the large scale structures.

The accuracy of finite difference solutions depends on the number of grid points chosen. Usually we have to concentrate them in the neighbourhood of rigid boundaries or within regions of strong shear, like e.g. wakes. Often it is practically impossible do deal with a sufficient number of grid points, so that the finite-difference solution does not give the proper answer. On the other hand, vortex methods are grid-free and suitable to treat the vortex-dominated high-Reynolds-number flow. Turbulent wakes can be more easily simulated by a vortex method. Three-dimensional vortex methods have been well developed recently and even the three-dimensional far-wake structure has been studied by a vortex method.

To see the complicated flow structure, a three-dimensional colour graphic system is inevitable. By using this, we can visualize the flow structure to an extent which cannot be done by experimental flow visualization.

Recent supercomputers of pipeline architecture are very suitable for fluid dynamical computations. If the code is well vectorized, a speed-up by a factor from 20 to 60 compared with the original scalar code can be easily achieved, and vectorization of the Navier-Stokes codes is not difficult. If better computational capability becomes available then we need less modelling and are able to compute more directly the flow from the basic equations without special assumptions and the results should become more reliable.

IMPACT OF DIRECT AND LARGE EDDY SIMULATIONS ON STATISTICAL MODELS
W. Rodi
University of Karlsruhe, Karlsruhe, F. R. Germany

In spite of the advances in direct and large-eddy simulation methods presented at the meeting, these methods are not yet ready to be used in most practical calculations of complex engineering problems, a situation which is unlikely to change in the near future. As has been stated in the literature (see e.g. RODI 1980, FERZIGER 1983), the greatest short-term impact of these methods in engineering calculations will be their use for improving statistical models of turbulence. Indeed, both direct and LES methods seem ideal for providing information which turbulence modellers always wanted to know but could not obtain from measurements. The exact equations for the Reynolds stresses and turbulent heat or mass fluxes, which form the basis of the more advanced statistical turbulence models, contain higher-order correlations about which model assumptions have to be introduced. Many of these correlations, especially those involving the fluctuating pressure, are difficult if not impossible to measure. Hence, information on these correlations extracted from direct or large-eddy simulations is extremely useful for constructing realistic and generally applicable model assumptions.

The question then arises to what extent direct and LES calculations have so far been used for this purpose and what we have learnt about this aspect at the meeting. The answer is that we have learnt dissappointingly little, as there were only three or four papers that touched on the subject. This provokes of course the question why so little use is made of the wealth of information generated by these methods. Are the direct and LES modellers just preoccupied with doing their own calculations? Is it the excitement that these calculations offer vis a vis the somewhat pedestrian processing that would be necessary? Or is it just too difficult to extract proper statistical quantities from the direct and LES calculations that are needed in statistical modelling? Further, is it perhaps not so certain whether, in LES calculations, the correlations calculated, like for the pressure-strain term, represent the whole spectrum (as they should in the statistical model context) or represent only a part of the spectrum? The answer is to be given by the direct and large-eddy modellers, but an appeal is made here to make more use of the information generated by direct and large-eddy simulations for statistical modelling.

What have we heard about the impact of direct and large-eddy simulations on statistical models from the few papers that touched the subject at this meeting? The paper by MOENG addressed directly the extraction of information on statistical correlations from LES calculations. In particular, LES calculations of a stratified planetary boundary layer were processed to examine the pressure-scalar covariances appearing in the transport equations for turbulent heat and mass fluxes. The various contribtions to this covariance were examined such as those due to turbulence-turbulence interaction, mean shear, buoyancy, and coriolis, and model proposals for these individual contributions were examined. This evaluation work is described in more detail in MOENG and WYNGAARD (1985). The conclusion of this study is that the turbulence-turbulence and buoyancy contributions are by far the largest ones and that the former part can be modelled quite well with ROTTA's return to isotropy model while the latter is one half of the direct buoyancy production of the covariance. It is interesting to note that this is precisely the model introduced for the buoyancy part by LAUNDER (1975).

The paper by REYNOLDS and LEE (1985), presented by MANSOUR, reported a direct simulation of homogeneous turbulence subjected to irrotational strain and the relaxation from that strain. They reported a number of very interesting findings which undermine some of the popular turbulence-modelling concepts. Their simulations reveal rather strong anisotropy of the small-scale structure which casts doubt on the assumption of local isotropy, i.e. isotropy of the small-scale turbulent motion, made in most statistical models but also in subgrid-scale models. Their observation is, of course, based on calculations at rather low Reynolds numbers, and the question is what happens at large Reynolds numbers at which local isotropy is assumed to prevail. The calculations also cast doubt on the often used return-to-isotropy model of ROTTA which assumes that, in the absence of mean strain, the pressure-strain term in the Reynolds stress equations is proportional to the anisotropy of the Reynolds stresses. The calculations show that this anisotropy is not well correlated with the pressure-strain term, rather a strong correlation was found with the anisotropy of the small-scale motion, offering the basis for a new model for the pressure-strain term. These findings based on the calculation for one flow should perhaps be extended to other flows and hopefully to higher Reynolds numbers before the fairly established models are thrown overboard.

SCHUMANN, ELGHOBASHI and GERZ presented results of a direct calculation for a stratified homogeneous shear layer which indicate that the second-order closure model of LAUNDER (1975) has difficulties when the stratification is very strong. The direct simulation results also point to possible reasons for these difficulties. However, as LAUNDER's model was not really intended nor used for very strong stratification, further studies are necessary.

In general, flows with buoyancy and rotation are situations where direct and LES methods could have the largest impact on statistical models. For example, DANG's calculations of turbulence subjected to rotation could be usefully exploited to obtain information on how to model this situation. This was done at Stanford by BARDINA, FERZIGER and ROGALLO (1985) who, based on direct simulations, modified the length-scale-determining ε-equation (ε =

dissipation rate). Experiments have indicated that the decay of isotropic turbulence is reduced by rotation. These experiments posed one of the test cases at the 1980-81 AFOSR/HTTM Stanford Conference on Complex Turbulent Flows (KLINE, CANTWELL and LILLEY, 1982), and none of the turbulence models used in the calculations of that case produced any effect of rotation. All the models employed the standard form of the ε-equation. BARDINA et al. (1985) showed that with the ε-equation extended by an additional rotation term, the experiments can be simulated fairly well. In a similar way, direct simulations were used at Stanford (WU, FERZIGER and CHAPMAN 1985) to develop a modified turbulence model that accounts correctly for the effect of compression on turbulence. Models employing the standard ε-equation respond incorrectly to compression as they produce an increase of the length-scale under compression.

An appeal is made here to use direct and LES methods more extensively to study stratified flows. The methods are ideal for investigating the general influence of stratification on turbulence, and specifically the collapse of turbulence under stable stratification. It is not clear at present from experiments whether this collapse leads to the generation of gravity waves, to two-dimensional turbulence or to both phenomena simultaneously. This could be studied usefully with direct and LES methods. Both situations with and without mean shear should be considered. In the former case this would be an extension of the non-buoyant mixing layer calculations of METCALFE, MENON and RILEY presented at the meeting. In the case without mean shear, the decay of a turbulent patch in stratified surroundings should be simulated.

MOIN's paper opened up an avenue to an entirely different impact of direct and LES methods on statistical turbulence models, namely to making use of our knowledge about the structure in turbulent flows in statistical prediction models. So far all the many experimental studies on coherent structures have not in fact led to any new theory or model development, and a dead end seems to be reached. With the wealth of information that can be made available on structures from direct or large-eddy simulations, we are in a much better position to make use of the information on structures in theoretical model development.

References:

Bardina, J., Ferziger, J.H. and Rogallo, R.S., 1985, Effect of rotation on isotropic turbulence: computation and modelling. J. Fluid Mech., Vol. 154, pp. 321-336.

Ferziger, J.H., 1983. Higher-level simulations of turbulent flow. In Computational Methods for Turbulent, Transonic and Viscous Flows, J.A. Essers, ed., Hemisphere.

Kline, S.J., Lilley, G.M. and Cantwell, B.J., 1982. Proceedings of the 1980-81 AFOSR-HTTM-Stanford Conference on Complex Turbulent Flows, Dept. of Mech. Eng., Stanford University.

Launder, B.E., 1975. On the effect of a gravitational field on the turbulent transport of heat and momentum, J. Fluid Mech., Vol. 67, pp. 569-581.

Lee, M.J., Reynolds, W.C., 1985. On the structure of homogeneous turbulence. Proc. 5th Symp. on Turbulent Shear Flows, Cornell University, Ithaca, N.Y., USA, 17.7-17.12.

Moeng, C.H. and Wyngaard, J.C., 1985. A study of the closure problem for pressure-scalar covariances, Proc. 5th Symp. on Turbulent Shear Flows, Cornell University, Ithaca, N.Y., USA, 12.31-12.36.

Rodi, W., 1980. Turbulence models and their application in hydrauliics. Book publcation of the International Association for Hydraulic Research, Delft, The Netherlands.

Wu, C.T., Ferziger, J.H. and Chapman, D.R., 1985. Simulation and modelling of homogeneous, compressed turbulence. Proc. 5th Symp. on Turbulent Shear Flows, Cornell University, Ithaca, N.Y., USA.

ON TRANSITION TO TURBULENCE
T. Herbert

Virginia Polytechnic Institute and State University,
Department of Engineering Science, Blacksburg, Virginia 24061, USA

This meeting summarized new results on transition to turbulence as well as on fully turbulent flow.

Concerning transition, it is notable that the past couple of years bridged the gap between the slow viscous instability mechanism and the fast convective (secondary) instability mechanisms in shear flows. At the same time, a multitude of competing modes has been discovered. One of the major goals in the near future will be the unraveling of the nonlinear interaction between these modes. Theoretical computational and experimental work will reveal details of the energy transfer, mode suppression, and mode selection for given initial conditions. Nonlinear analysis also holds the promise to uncover the mechanisms of selfsustained disturbance growth and may, therefore, contribute to the understanding of selfsustained turbulence.

Much of the past progress has been due to close cooperation between experiment, simulation, and theory. Experiments have shown phenomena and parameter ranges of interest and have provided a scarce data base helpful for verification of numerical results. Computer simulations are increasingly capable of describing these phenomena in detail, although understanding requires tedious analysis of the overwhelming amount of data. Theoretical modeling combined with mathematical and numerical methods has held its place in revealing the structure, mechanisms, and classification of the phenomena.

In view of the finite number of mechanisms active in shear flows, the occurrence and explanation of vortical mechanisms in wall bounded flows is a relevant fact. Numerous phenomena observed in the Couette-like viscous sublayer are reminiscent of those in the later stages of transition. The main problem is to bridge the gap and to provide the link between these virtually related phenomena. Most of the stability analysis has been done in low-disturbance environments, and results are for the free or natural response of the fluid flow. In turbulent flow, however, the sublayer is massaged by the outer flow. The forced response of the flow will be of prime concern.

Looking more at the turbulent side of our topic, it is surprising to find (after some cancellations) only a few contributions in the area of dynamical systems. The prediction of a small number of bifurcations leading from laminar to chaotic/turbulent flow has encouraged the step-by-step analysis of the cascade of instabilities. Many of the concepts like renormalization or fractal dimensions, however, do not seem appropriate to the deterministic (although sensitive to initial data) transition phenomena. Applications seem to be more successful in homogeneous turbulence but may suffer in the neighborhood of walls. Layered or hybrid methods - an extended concept of large eddy simulation - may be a challenging subject.

Finally, this meeting revealed another problem: the dissemination of results. We have seen various more or less successful attempts to display the essentials of unsteady three-dimensional flow fields. Archival journals in color or stereo are not in sight, while computer graphics rapidly develop. It may be worthwhile to think about another Euromech Colloquium on the "Graphical Representation of 3-D Flow Fields."

LIST OF PARTICIPANTS

B. Aupoix; ONERA/CERT/DERAT 2, Avenue E. Belin, 31400 Toulouse, France

F. Baetke; Institut für Strömungsmechanik, TU München, Arcisstr. 21, 8000 München 2, Federal Republic of Germany

F. Baron; Electricite de France, Labor. Nat. d'Hydr., 6, quai Watier, 78400 Chatou, France

J.-P. Bertoglio; Lab. de Mech. des Fluides, Ecole Centrale de Lyon 36, av. G. de Collonques, 69130 Ecully, France

E. F. Brown; ONRL, 223 Old Marylebone Road, London NW1 5th, United Kingdom

D..Carruthers, Dep. of Atmosph. Physics, Clarendon Laboratory, Parks Road, Oxford OX1 3PU, United Kingdom

J.-P. Chollet; Institut de Mechanique de Grenoble BP 68, 38402 St-Martin-d'Heres, Cedex, France

K. Dang; ONERA, BP 72, 32 Av. Div. Leclerc, 92322 Chatillon, Cedex, France

T. Eidson; Georgia Inst. of Techn., School of Mech. Engin. Atlanta, Georgia 30332, USA

D. Etling; Inst. f. Meteorologie u. Klimatologie, Univ. Hannover, Herrenhäuser Str. 2, 3000 Hannover 21, West Germany

J.-M. Favre, CEA Etablissement T, BP No. 7, 77181 Courtry, France

G. Fiorese; 35, rue du Reveillon, 91800 Brunoy, France

T. R. Fodemski; Mechanical Eng. Dept., The City University, Northampton Square, London EC1V OHB, United Kingdom

R. Friedrich; Inst. f. Strömungsmechanik, TU München, Arcisstr. 21, 8000 München 2, West Germany

M. A. Fry; Science Applications Int., 1710 Goodridge Dr., McLean, Va. 22102, USA

S. Gavrilakis; Queen Mary College, London EL 4NS, United Kingdom

T. Gerz; Institut f. Physik d. Atmosphäre, DFVLR Oberpfaffenhofen, 8031 Wessling, West Germany

N. Gilbert; DFVLR, Institut f. Theoretische Strömungsmechanik, Bunsenstr. 10, 3400 Göttingen, West Germany

G. Grötzbach; Kernforschungszentrum Karlsruhe, Institut f. Reaktorentwicklung, Postfach 3640, 7500 Karlsruhe, West Germany

T. Herbert; Dept. Engineering Science, Virginia Polytechnic Inst. and State University, Blacksburg, VA 24061, USA

E. H. Hirschel; MBB, UFE122, Postfach 801160, 8000 München 80, West Germany

K. Horiuti; Inst. of Ind. Science, University of Tokyo, 7-22-1, Roppongi, Minato-ku, Tokyo 106, Japan

J. Hunt; Dep. of Appl. Mathematics and Theor. Physics, Silver Street, Cambridge CB3 9EW, United Kingdom

J. Kim; NASA Ames Res. Center, Moffet Field, Ca. 94035, USA

H. Klein; Inst. f. Strömungsmechanik, TU München, Arcisstr. 21, 8000 München 2, West Germany

L. Kleiser; DFVLR, Inst. f. Theoretische Strömungsmechanik, Bunsenstr. 10, 3400 Göttingen, West Germany

T. Kobayashi; Inst. of Ind. Science, University of Tokyo, 7-22-1, Roppongi, Minato-ku, Tokyo 106, Japan

S. Komori; DAMTP, University of Cambridge, Silver Street, Cambridge CB3 9EW, United Kingdom

K. Kuwahara; The Institute of Space and Astronautical Science, 6-1 Komaba A-chome, Meguro-ku, Tokyo 153, Japan

D. Laurence; Electricite de France, Laboratoire Nat. d'Hydr., 6, quai Watier, 78400 Chatou, France

M. Lesieur; Institut de Mechanique de Grenoble, BP68, 38402 St-Martin-d'Heres, Cedex, France

E. Levich; The City College of the City Univ. of New York, Steinman Hall 202, New York, N.Y. 10031, USA

D. K. Lilly; University of Oklahoma, School of Meteorology, 200 Felgar St., Norman, Ok 73071, USA

N. Mansour; NASA Ames Res. Center, Moffett Field, Ca. 94035, USA.

P. J. Mason; Meteorological Office, Meteor. Research Unit., R.A.F. Cardington, Bedford, MK 42 OTH, United Kingdom

W. D. McComb; Department of Physics, University of Edinburgh, Mayfield Road, Edinburgh EH9 3JZ, United Kingdom

R. Metcalfe; Flow Industries Inc., 21414 68th Avenue South, Kent, WA 98032, USA

C.-H. Moeng; National Center for Atmospheric Research, P. O. Box 3000, Boulder, Colorado 80307, USA

P. Moin; Computational Fluid Dynamics Branch, NASA Ames Res. Center, Moffett Field, Ca. 94035, USA

U. Müller; Kernforschungszentrum Karlsruhe, Inst. f. Reaktorbauelemente, Postfach 3640, 7500 Karlsruhe 1, West Germany

F. Nieuwstadt; Delft University of Technology, Dept. of Mechanical Engrg., Lab. for Aero and Hydrodynamics, Rotterdamseweg 145, 2628 AL Delft, The Netherlands

R. Pelz; D-309 Eng. Quad., Princeton Univ., Princeton, NJ 08544, USA

Pucher; Inst. f. Verbrennungskraftmaschinen u. Thermodynamik, TU Graz, 8010 Graz, Austria

K. Richter; Inst. f. Strömungsmechanik, TU München, Arcisstr. 21, 8000 München 2, West Germany

W. Rodi; Inst. f. Hydromechanik, Univ. Karlsruhe, Kaiserstr. 12, 7500 Karlsruhe 1, West Germany

M. Schmatz; MBB, UFE122, Postfach 801160, 8000 München 80, West Germany

H. Schmidt; Institut f. Physik d. Atmosphäre, DFVLR Oberpfaffenhofen, 8031 Wessling

L. Schmitt; Inst. f. Strömungsmechanik, TU München, Arcisstr. 21, 8000 München 2, West Germany

U. Schumann; Institut f. Physik d. Atmosphäre, DFVLR Oberpfaffenhofen, 8031 Wessling

S. Shirayama; Kuwahara Lab., The Institute of Space and Astronautical Science, 6-1 Komaba A-chome, Meguro-ku, Tokyo 153, Japan

D. J. Thomson; Meteorological Office (MET O 14), London Road, Bracknell, Berkshire, RG12 2SZ, United Kingdom

H. M. Tsai; Dept. of Nuclear Eng., Queen Mary College, London EI4N5, United Kingdom

A. Tsinober; Inst. of Appl. Chem. Physics, The City Coll., CUNY, T-202 Conventave & Iyoth Street, New-York, N.Y. 10031, USA

E. Tzvetkov; The City College of the City Univ. of New York, Steinman Hall 202, New York, N.Y. 10031, USA

P. R. Voke; The Turbulence Unit, Queen Mary College, London EI, United Kingdom

H. Wengle; FB Luft- und Raumfahrt, Hochschule d. Bundeswehr, Werner-Heisenberg-Weg 39, 8060 Neubiberg, West Germany

H. Werner; FB Luft- und Raumfahrt, Hochschule der Bundeswehr, Werner-Heisenberg-Weg 39, 8060 Neubiberg, West Germany

LIST OF AUTHORS

The number identifies the first page of the contribution

B. Aupoix	37, 321		E. Levich	82
R. A. Brost	304		D. K. Lilly	328
D. J. Carruthers	271		P.J. Mason	329
T. M. Eidson	188		W. D. McComb	67
S. E. Elghobashi	245		S. Menon	265
R. Friedrich	161		R. W. Metcalfe	265
S. Gavrilakis	105		C.-H. Moeng	291
T. Gerz	245		P. Moin	181
N. Gilbert	1		F. T. M. Nieuwstadt	304
G. Grötzbach	210		K. Richter	161
T. Herbert	19, 334		J.J. Riley	265
K. Horiuti	119		W. Rodi	331
J. C. R. Hunt	271, 325		L. Schmitt	161
M. Y. Hussaini	188		U. Schumann	245
M. Kano	135		S. Shirayama	227
J. Kim	177, 318		T. L. van Stijn	304
L. Kleiser	1		H. M. Tsai	105
T. Kobayashi	135		C. J. Turfus	271
K. Kuwahara	227, 330		P. R. Voke	105
D. Laurence	147		A. Yoshizawa	119
D. C. Leslie	105		T. A. Zang	188